Communications in Computer and Information Science 2204

Series Editors

Gang Li⬤, *School of Information Technology, Deakin University, Burwood, VIC, Australia*
Joaquim Filipe⬤, *Polytechnic Institute of Setúbal, Setúbal, Portugal*
Zhiwei Xu, *Chinese Academy of Sciences, Beijing, China*

Rationale
The CCIS series is devoted to the publication of proceedings of computer science conferences. Its aim is to efficiently disseminate original research results in informatics in printed and electronic form. While the focus is on publication of peer-reviewed full papers presenting mature work, inclusion of reviewed short papers reporting on work in progress is welcome, too. Besides globally relevant meetings with internationally representative program committees guaranteeing a strict peer-reviewing and paper selection process, conferences run by societies or of high regional or national relevance are also considered for publication.

Topics
The topical scope of CCIS spans the entire spectrum of informatics ranging from foundational topics in the theory of computing to information and communications science and technology and a broad variety of interdisciplinary application fields.

Information for Volume Editors and Authors
Publication in CCIS is free of charge. No royalties are paid, however, we offer registered conference participants temporary free access to the online version of the conference proceedings on SpringerLink (http://link.springer.com) by means of an http referrer from the conference website and/or a number of complimentary printed copies, as specified in the official acceptance email of the event.

CCIS proceedings can be published in time for distribution at conferences or as post-proceedings, and delivered in the form of printed books and/or electronically as USBs and/or e-content licenses for accessing proceedings at SpringerLink. Furthermore, CCIS proceedings are included in the CCIS electronic book series hosted in the SpringerLink digital library at http://link.springer.com/bookseries/7899. Conferences publishing in CCIS are allowed to use Online Conference Service (OCS) for managing the whole proceedings lifecycle (from submission and reviewing to preparing for publication) free of charge.

Publication process
The language of publication is exclusively English. Authors publishing in CCIS have to sign the Springer CCIS copyright transfer form, however, they are free to use their material published in CCIS for substantially changed, more elaborate subsequent publications elsewhere. For the preparation of the camera-ready papers/files, authors have to strictly adhere to the Springer CCIS Authors' Instructions and are strongly encouraged to use the CCIS LaTeX style files or templates.

Abstracting/Indexing
CCIS is abstracted/indexed in DBLP, Google Scholar, EI-Compendex, Mathematical Reviews, SCImago, Scopus. CCIS volumes are also submitted for the inclusion in ISI Proceedings.

How to start
To start the evaluation of your proposal for inclusion in the CCIS series, please send an e-mail to ccis@springer.com.

A. Mirzazadeh · Zohreh Molamohamadi ·
Efran Babaee Tirkolaee ·
Gerhard-Wilhelm Weber · Janny Leung
Editors

Optimization and Data Science in Industrial Engineering

First International Conference, ODSIE 2023
Istanbul, Turkey, November 16–17, 2023
Proceedings, Part I

Editors
A. Mirzazadeh
Kharazmi University
Tehran, Iran

Efran Babaee Tirkolaee
Istinye University
Istanbul, Türkiye

Janny Leung
University of Macau
Macao, China

Zohreh Molamohamadi
Kharazmi University
Tehran, Iran

Gerhard-Wilhelm Weber
Poznań University of Technology
Poznań, Poland

ISSN 1865-0929 ISSN 1865-0937 (electronic)
Communications in Computer and Information Science
ISBN 978-3-031-81454-9 ISBN 978-3-031-81455-6 (eBook)
https://doi.org/10.1007/978-3-031-81455-6

© The Editor(s) (if applicable) and The Author(s), under exclusive license
to Springer Nature Switzerland AG 2025

This work is subject to copyright. All rights are solely and exclusively licensed by the Publisher, whether the whole or part of the material is concerned, specifically the rights of translation, reprinting, reuse of illustrations, recitation, broadcasting, reproduction on microfilms or in any other physical way, and transmission or information storage and retrieval, electronic adaptation, computer software, or by similar or dissimilar methodology now known or hereafter developed.
The use of general descriptive names, registered names, trademarks, service marks, etc. in this publication does not imply, even in the absence of a specific statement, that such names are exempt from the relevant protective laws and regulations and therefore free for general use.
The publisher, the authors and the editors are safe to assume that the advice and information in this book are believed to be true and accurate at the date of publication. Neither the publisher nor the authors or the editors give a warranty, expressed or implied, with respect to the material contained herein or for any errors or omissions that may have been made. The publisher remains neutral with regard to jurisdictional claims in published maps and institutional affiliations.

This Springer imprint is published by the registered company Springer Nature Switzerland AG
The registered company address is: Gewerbestrasse 11, 6330 Cham, Switzerland

If disposing of this product, please recycle the paper.

Preface

The International Conference on Optimization and Data Science in Industrial Engineering (ODSIE 2023) was held virtually in Istanbul, Turkey on Nov. 16–17, 2023 by Istinye University. It provided an energetic knowledge-transferring atmosphere for participants (as several comments revealed).

ODSIE 2023 attracted the attention of students and professionals internationally. The subjects covered included, but were not limited to, "Industry 4.0, IoT and smart manufacturing", "Digital twin and virtual commissioning", "Sustainable and smart cities", "Artificial intelligence and expert systems", "Metaheuristic algorithms with applications in IE", "Machine learning algorithms", "Big data analytics and Data mining", "Robotic process automation", "Decision support systems", "E-Government, E-Commerce and E-Learning", "Supply chain design and logistics", "Optimization- and Data Science-based case studies of manufacturing/service industries", "Quantitative finance and risk modelling", and "Other fields of study related to Optimization and Data Science with applications in IE".

ODSIE 2023 was honored to be enriched by outstanding keynote speakers and workshop organizers from UK, Tunisia, Saudi Arabia, India, Pakistan, and Iran. In this event, fifteen universities from USA, UK, Czech Republic, Tunisia, Malaysia, Algeria, India, Morocco, and Turkey were present as scientific sponsors.

The conference included participation of 50 countries. The geographical diversity of international scientific committee members was from 27 countries.

The conference team received 311 English and Turkish papers, which were reviewed by at least three international reviewers (single-blind reviews). Considering the reviewers' comments in the first round of review, the papers were reviewed once more with stricter criteria to select the most appropriate ones for Springer's publication. Finally, the two-round review process resulted in the selection of 35 papers (around 11%) for Springer.

The review criteria in this step were: content; originality; relevance; contribution to the professional literature; significance and potential impact of the paper; language accuracy; study validity; accuracy of methodology and analysis; paper organization, required relevant data, citations and references; adequate reference to background information; consistency of references, symbols and units throughout the paper; quality and clarity of tables and figures.

The papers in ODSIE 2023 were presented in 21 panel sessions: "Optimization Problems Related to Logistics, Transportation and Manufacturing Systems", "Operations Research in Optimization", "Artificial Intelligence and Expert Systems", "Sustainable Business", "From Data to Knowledge - Analysis for Strategies and Challenges", "Experiment and Optimization in Research", "Industry 4.0, IoT and Smart Manufacturing", "Computational Thinking and Artificial Intelligence in Industrial Engineering", "Inventory Planning, Production and Scheduling", "Decision Support Systems",

"Other Fields of Study Related to Optimization and Data Science", "Machine Learning Solutions for Engineering Problems", "Data Analytics in Intelligent Transportation Systems", "Project Management and Fintech", "Advancing Innovation and Collaboration in Science, Engineering and Information Technology", "Artificial Intelligence for Biomedicine and Healthcare", "Machine Learning Algorithms", "Advancing Operations Management through Optimization and Data Science", "E-Government, E-Commerce and E-Learning" and "Analytics for Sustainable Society".

ODSIE 2023 included five applied workshops by outstanding lecturers with good-sized audiences. The workshop subjects were very well received by the participants, and were entitled: "From Sustainability to Circularity: Transforming Supply Chains for an Eco-centric Economy"; "Unlocking the Potential of Drones: Exploring Diverse Applications"; "A Multidisciplinary Landscape of Data Sciences"; "Indispensable Arbitration of Leaders Is Truly Conducive for the Utmost Clarity of an Organization"; and "Leading the Charge: Optimizing Sustainable Finance Strategies Through Data-Driven Decision-Making in Industrial Engineering".

December 2023

A. Mirzazadeh
Zohreh Molamohamadi
Efran Babaee Tirkolaee
Gerhard-Wilhelm Weber
Janny Leung

Organization

Conference Chairs

Abolfazl Mirzazadeh	Kharazmi University, Iran
Erfan Babaee Tirkolaee	Istinye University, Istanbul, Turkey
Nadi Serhan Aydın (Co-chair)	Istinye University, Istanbul, Turkey

Istinye University

Erkan Ibiş	Rector
Hatice Gülen	Vice Rector
Mehmet Alper Tunga	Dean of the Faculty of Engineering and Natural Sciences

Conference Coordinators

Leila Chehreghani (Planning Manager)	Ulster University, UK
Zohreh Molamohamadi (Scientific Coordinator)	Kharazmi University, Iran

Program Committee Chairs

Efran Babaee Tirkolaee	Istinye University, Turkey
Janny Leung	University of Macao, China President of IFORS
Zohreh Molamohamadi	Kharazmi University, Iran
A. Mirzazadeh	Kharazmi University, Iran
Gerhard-Wilhelm Weber	Poznań University of Technology, Poland

Editorial Committee Members

Alexandre Dolgui	IMT Atlantique, France
Arpan Kumar Kar	Indian Institute of Technology Delhi, India

Safa Bhar Layeb	University of Tunis El Manar, Tunisia
Kathryn Stecke	University of Texas at Dallas, Naveen Jindal School of Management, USA
Kok Lay Teo	JIMO Editor in Chief, Sunway University, Malaysia, Curtin University, Australia
Chefi Triki	University of Kent, UK

Steering Committee Members

Erfan Babaee Tirkolaee	Istinye University, Turkey
Babek Erdebilli	Yildirim Beyazit University, Turkey
Josef Jablonsky	Prague University of Economics and Business, Czech Republic
Janny M. Y. Leung	University of Macao, China President of IFORS
Taicir Moalla Loukil	University of Sfax, Tunisia President of The Tunisian Operational Research Society
A. Mirzazadeh	Kharazmi University, Iran
Mehmet Alper Tunga	Istinye University, Turkey
Gerhard-Wilhelm Weber	Poznań University of Technology, Poland

PhD Thesis Competition Jury

Saliha Karadayı Üsta (Chair)	Istinye University, Turkey
Emre Çakmak (Co-chair)	Istinye University, Turkey
Tatiana Tchemisova (Member)	University of Aveiro, Portugal

Scientific Committee Members

Adeyinka Peter Ajayi	Redeemer's University, Nigeria
Sadia Samar Ali	King Abdulaziz University, Saudi Arabia
Tofigh AllahViranloo	Istinye University, Turkey
Bernardo Almada-Lobo	Porto University, Portugal
Sırma Zeynep Alparslan Gök	Suleyman Demirel University, Turkey
Fayçal Belkaid	Abou Bekr Belkaid University of Tlemcen, Algeria
Sanjib Biswas	Calcutta Business School, India

Marilisa Botte	University of Naples, Italy
Jaouad Boukachour	University of Le Havre-Normandy, France
Emre Çakmak	Istinye University, Turkey
Yavuz Can	Friedrich-Alexander University, Germany
Leopoldo Eduardo Cárdenas-Barrón	Tecnológico de Monterrey, Mexico
Aybike Özyüksel Çiftçioğlu	Manisa Celal Bayar University, Turkey
Dursun Delen	Oklahoma State University, USA
Elif Kılıç Delice	Ataturk University, Turkey
Souhail Dhouib	University of Sfax, Tunisia
Serap Ergun	Isparta University of Applied Sciences, Turkey
Paulina Golinska	Poznań University of Technology, Poland
Ömer Faruk Görçün	Kadir Has University, Turkey
Sylwia Gwoździewicz	Jacob of Paradies University, Poland
Mostafa Hajiaghaei-Keshteli	Tecnológico de Monterrey, Mexico
Sarfaraz Hashemkhani Zolfani	Northern Catholic University, Chile
Marwa Hasni	University of Sfax, Tunisia
Dinh Tran Ngoc Huy	International University of Japan, Japan
Saliha Karadayi Üsta	Istinye University, Turkey
Michael G. Kay	North Carolina State University, USA
Selçuk Korucuk	Giresun University, Turkey
Eloisa Macedo	University of Aveiro, Portugal
Beata Mrugalska	Poznań University of Technology, Poland
Ammar Odeh	King Hussein School for Computing Sciences, Jordan
Fabiana Lucena Oliveira	Amazonas State University (UEA), Brazil
Mustapha Oudani	International University of Rabat, Morocco
Yavuz Selim Ozdemir	Ankara Science University, Turkey
Çağrı Özgün Kibiroğlu	Halic University, Turkey
Dragan Pamučar	University of Belgrade, Serbia
Stefan Pickl	Bundeswehr University, Germany
Sankar Kumar Roy	Vidyasagar University, India
Rubén Ruiz Garcia	Polytechnic University of Valencia, Spain
İlkay Saraçoğlu	Halic University, Turkey
Yaroslav Sergeyev	University of Calabria, Italy
Vladimir Simic	University of Belgrade, Serbia
Esra Sipahi Dungul	Aksaray University, Turkey
Renata Sotirov	Tilburg University, The Netherlands
Majid Tavana	La Salle University, USA
Tatiana Tchemisova	University of Aveiro, Portugal
Hadi Rezaei Vandchali	University of Tasmania, Australia
Ibrahim Yilmaz	Yildirim Beyazit University, Turkey

Keynote Speakers

Chefi Triki	University of Kent, UK
Safa Bhar Layeb	LR-OASIS, National Engineering School of Tunis, University of Tunis El Manar, Tunisia

Workshop Organizers

Alireza Goli	University of Isfahan, Iran
Sadia Samar Ali	King Abdulaziz University, Saudi Arabia
Maria Anjum	Lahore College for Women University, Pakistan
Rudrarup Gupta	Tagore School of Rural Development and Agriculture Management, Kalyani University, India
Sameera Fernandes	Century Financial, UAE and Garden City University, India

Executive Committee Members

Kutay Alada	Istinye University, Turkey
Anderson Chukwudi Oki	Istinye University, Turkey
Gizem Karatepe	Istinye University, Turkey
Alper Koçak	Istinye University, Turkey
Muhammet Ali Şencan	Istinye University, Turkey
Sadaf Soltani	IT Expert in DevOps University, Iran
Cansu Tayar	Istinye University, Turkey

Reviewers

Otman Abdoun	University of Abdelmalek Essaadi, Morocco
Mohamed Abdelgadir	University of Technology Malaysia, Malaysia
Salma Abid	University of Rabat, Morocco
Mohamed Nezar Abourraja	Le Havre Normandy University, France
Negar Afzali Behbahani	Islamic Azad University, Iran
Maher Agi	Rennes School of Business, France
Hadi Ahmed	Cairo University, Egypt
Sara Ahmed	Concordia University, Canada
Ghada Alhudhud	King Saud University, Saudi Arabia
Aylin Alkaya	Nevsehir Haci Bektas Veli University, Turkey

Richard Allmendinger	University of Manchester, UK
Juscelino Almeida Dias	Technical University of Lisbon, Portugal
Shabnam Amirnezhad Barough	Yıldırım Beyazit University, Turkey
El-Saed Ammar	Tanta University, Egypt
Le Thi Diep Anh	National Economics University, Vietnam
Hichem Aouag	Batna University, Algeria
Mst. Anjuman Ara	Bangladesh Army University of Science and Technology, Bangladesh
Selcen Aslan Ozsahin	TOBB University of Economics and Technology, Turkey
Mamoon Atout	Dubai Electricity and Water Authority, UAE
Nadi Serhan Aydin	Istinye University, Turkey
Salih Aytar	Suleyman Demirel University, Turkey
Youssef Baddi	Chouaib Doukkali University, Morocco
Rahmi Baki	Aksaray University, Turkey
Srinivasan Balan	North Carolina State University, USA
Igor Barahona	King Fahd University of Petroleum & Minerals, Saudi Arabia
Haripriya Barman	Vidyasagar University, India
Faycal Belkaid	University of Lorraine, France
Adil Bellabdaoui	National School of Computer Science and Systems Analysis, Morocco
Mohamed Amine Ben Rabia	National School of Computer Science and Systems Analysis, Morocco
Abderaouf Benghalia	Algiers I University, Algeria
Jamal Benhra	National School of Computer Science and Systems Analysis, Morocco
Abdelaziz Berrado	Mohammed V University, Morocco
Subir Bhattacharya	National Institute of Technology Durgapur, India
Satyajit Bhunia	Midnapore City College, India
Papatya Sevgin Bicakci	Başkent University, Turkey
Shazia Bilal	COMSATS Institute of Information Technology, Pakistan
Sanjib Biswas	Calcutta Business School, India
Bonaventure Boniface	Sabah University, Malaysia
Eleonora Bottani	University of Parma, Italy
Marilisa Botte	University of Naples, Italy
Jaouad Boukachour	Le Havre Normandy University, France
Gercek Budak	Yıldırım Beyazıt University, Turkey
Victor Camargo	Federal University of São Carlos, Brazil
Patricia Cano-Olivos	Universidad Popular Autónoma del Estado de Puebla, Mexico

Nancy-Paulina Carren	Universidad Popular Autónoma del Estado de Puebla, Mexico
Michal Cerny	Prague University of Economics and Business, Czech Republic
Ying-Hua Chang	Tamkang University, Taiwan
Maxime Chassaing	University of Skövde, Sweden
Duong Cuong	Hanoi University of Science and Technology, Vietnam
Soumen Kumar Das	Vidyasagar University, India
Manoranjan De	Vidyasagar University, India
Bikash Koli Dey	Hongik University, South Korea
Oshmita Dey	Techno India University, India
Luis C. Dias	University of Coimbra, Portugal
Oleksandr Dluhopolskyi	West Ukrainian National University, Ukraine
Aman Dua	NIFTEM University, India
Mustafa Ekici	Çanakkale Onsekiz Mart University, Turkey
Abdellatif El Afia	Mohammed V University, Morocco
Karim El Bouyahyiouy	Mohammed V University, Morocco
Adiba El Bouzekri el Idrissi	Doukkali University, Morocco
Nizar El Hachemi	Mohammed VI Polytechnic University, Morocco
Mamdouh El Haj Assad	University of Sharjah, UAE
Mohamed El Merouani	Abdelmalek Essaâdi University, Morocco
Moulay Driss El Ouadghiri	Moulay Ismail University, Morocco
Ali Emrouznejad	Surrey Business School, UK
Ergun Eraslan	Yildirim Beyazit University, Turkey
Nuh Erdogan	Robert Gordon University, UK
Serap Ergun	Isparta University of Applied Sciences, Turkey
Petr Fiala	Prague University of Economics and Business, Czech Republic
Sukono Firman	Universitas Padjadjaran, Indonesia
Martin Flegl	Tecnológico de Monterrey, Mexico
Vijay Gahlawat	National Institute of Food Technology Entrepreneurship and Management, India
Askar Garad	Muhammadiyah Yogyakarta University, Indonesia
Rakesh Garg	Amity University, Noida, India
Michel Gendreau	University of Montreal, Canada
Peiman Ghasemi	University of Calgary, Canada
Shyamali Ghosh	Vidyasagar University, India
Bibhas C. Giri	Jadavpur University, India
Beata Glinkowska-Krauze	University of Łódź, Poland
Alireza Goli	University of Isfahan, Iran
Paulina Golinska	Poznań University of Technology, Poland

Kristens Gudfinnsson	University of Skövde, Sweden
Mete Gundogan	Ankara Yildirim Beyazit University, Turkey
Shunsheng Guo	Wuhan University of Technology, China
Seyyed Ali Haddad Sisakht	Iowa State University, USA
Achraf Haibi	Moulay Ismail University, Morocco
Sondes Hammami	University of Carthage, Tunisia
Sepehr Hanjani	University of Putra, Malaysia
Amit V. Hans	Guru Gobind Singh Indraprastha University, India
Khandaker Hasan	United International University, Bangladesh
Gholamreza Haseli	Monterrey Institute of Technology, Mexico
Muhammad Hoque	Management College of Southern Africa, South Africa
Tarak Housein	Libya, Libya
Olena Hrechyshkina	Polessky State University, Belarus
Chang-Ling Hsu	Ming Chuan University, Taiwan
Jianwen Huang	China Three Gorges University, China
Qimin Huang	Case Western Reserve University, USA
Dina M. Ibrahim	Qassim University, Saudi Arabia
Hamid Reza Irani	University of Tehran, Iran
M. Y. Jaber	Ryerson University, Canada
Ajay Jain	GGSIP University, India
Madhu Jain	IIT Roorkee, India
Suresh Jakhar	Indian Institute of Management Lucknow, India
Jishu Jana	Vidyasagar University, India
Nsikan John	Covenant University, Nigeria
Sudhanshu Joshi	Doon University, India
Deepshikha Kalra	Management Education and Research Institute, India
Mohamad Amin Kaviani	University of Cyprus, Cyprus
Michael Kay	North Carolina State University, USA
Mohd Khairol Anuar Ariffin	Putra University, Malaysia
Hamiden Khalifa	Cairo University, Egypt
Soheyl Khalilpourazari	Kharazmi University, Iran
Manjeet Kharub	Institute of Management Technology, Ghaziabad, India
Mazdak Khodadadi-Karimvand	University of Science and Culture, Iran
Reza Kiani Mavi	Edith Cowan University, Australia
Iryna Kramarenko	National University of Shipbuilding, Ukraine
Aalok Kumar	Jaipuria Institute of Management, India
Pavan Kumar	VIT Bhopal University, India
Martina Kuncova	Prague University of Economics and Business, Czech Republic

Tanmoy Kundu	National University of Singapore, Singapore
Flevy Lasrado	University of Wollongong in Dubai, UAE
Mohamed Lazaar	Mohammed V University, Morocco
Emily Lee	California State University, USA
Ramesh Lekurwale	Vidyasagar University, India
Viraj Lele	DHL, USA
Viacheslav Liashenko	Academy of Economic Science of Ukraine, Ukraine
Sunil Luthra	SIET Jhajjar, India
Vinayak Madiwale	Precast India Infrastructures Private Ltd., India
Gour Mahata	Sidho Kanho Birsha University, India
Hamidreza Mahboobinejad	Wollongong University, Australia
S. Maheswaram	University of Peradeniya, Sri Lanka
Arunava Majumder	Lovely Professional University, India
Dragana Makajic-Nikolic	Belgrade University, Serbia
Paulo Manrique	National University of Mexico, Mexico
Malek Masmoudi	University of Sharjah, UAE
Tafadzwa Matiza	North-West University, South Africa
Abu Md Mashud	Hajee Mohammad Danesh Science and Technology University, Bangladesh
Jan Medlock	Oregon State University, USA
Adeel Mehmood	COMSATS University, Pakistan
Fatima-Zahra Mhada	Mohammed V University, Morocco
Sudipta Midya	Vidyasagar University, India
Shashi Mishra	Banaras Hindu University, India
Froilan Mobo	Philippine Merchant Marine Academy, Philippines
Nanees Mohamed	Helwan University, Egypt
Essaaidi Mohammed	Mohammed V University, Morocco
Arijit Mondal	Vidyasagar University, India
Rahul Mor	National Institute of Food Technology Entrepreneurship and Management, India
Chanicha Moryadee	Suan Sunandha Rajabhat University, Thailand
Regaieg Mouna	University of Sfax, Tunisia
Mahabub Musa	Yusuf Maitama Sule University, Kano, Nigeria
Mahantesh M. Nadakatti	KLS Gogte Institute of Technology, India
Peter Nadeem	University of Derby, UK
Ahmed Nait Sidi Moh	Jean Monnet University, France
Mehdi Najib	International University of Rabat, Morocco
Behnam Nakhai	Millersville University of Pennsylvania, USA
Amir Hossein Nasiri	University of Putra, Malaysia
Luka Neralic	University of Zagreb, Croatia

Luan-Thanh Nguyen	HUFLIT-Ho Chi Minh City University of Foreign Languages-Information Technology, Vietnam
Lewis Njualem	Texas Tech University, USA
Mohammed Nusari	Lincoln University College, Malaysia
Mehmet Onur Olgun	Suleyman Demirel University, Turkey
Khaoula Ouaddi	National School for Computer Science, Morocco
Rachid Oucheikh	Norwegian University of Science and Technology, Norway
Mustapha Oudani	International University of Rabat, Morocco
Kenza Oufaska	International University of Rabat, Morocco
Amar Oukil	Sultan Qaboos University, Oman
Fatima Ouzayd	National School for Computer Science, Morocco
Mohamed Ouzineb	National Institute of Statistics and Applied Economics, Morocco
İsmail Ozcan	Suleyman Demirel University, Turkey
Manvinder Pahwa	Manipal University Jaipur, India
Ash Pain	UMT, Pakistan
Brojeswar Pal	University of Burdwan, India
Osman Palanci	Suleyman Demirel University, Turkey
Iztok Palcic	University of Maribor, Slovenia
Anupama Panghal	National Institute of Food Technology Entrepreneurship and Management, India
Sarla Pareek	Banasthali Vidyapith, India
Hiren Patel	Ganpat University, India
Asim Paul	Vidyasagar University, India
Sanjoy Paul	University of Technology Sydney, Australia
Dina Pereira	University of Beira Interior, Portugal
Magfura Pervin	Vidyasagar University, India
Jaroslav Pushak	Lviv State University of Internal Affairs, Ukraine
Juanjuan Qin	Tianjin University of Finance and Economics, China
Kumar Rahul	National Institute of Food Technology Entrepreneurship and Management, India
Shivani Rana	Hoshiarpur S.D. College, India
Shalendra Rao	Mohanlal Sukhadia University, India
Svetlana Rastvortseva	National Research University Higher School of Economics, Russian Federation
D. Raut	Indian Institute of Management Mumbai, India
Goncalo Reis Figueira	INESC TEC, FEUP, Portugal
Naoufal Rouky	Le Havre Normandy University, France
Darin Rungklin	Suratthani Rajabhat University, India
Sahara Sahara	Bogor Agricultural University, Indonesia
Tkatek Said	University Ibn Tofail, Morocco

Luis Salinas	New Jersey Institute of Technology, USA
Guruprasad Samanta	Indian Institute of Engineering Science and Technology, India
Sushant Samir	PEC University of Technology, India
Shib Sankar Sana	Kishore Bharati Bhagini Nivedita College, India
Thomy Saputro	Porto University, Portugal
Fanny Saruchera	University of the Witwatersrand, South Africa
Seyyed Hadi Seifi	Mississippi State University, USA
Marc Sevaux	University of South Brittany, France
Bhavin Shah	IIM Sirmaur, India
Janmejai Shah	Graphic Era University, India
Nita Shah	Gujarat University, India
Asadullah Shaikh	Najran University, Saudi Arabia
Manu Sharma	Graphic Era University, India
Ali Akbar Sheikh	University of Burdwan, India
Arun Kumar Shettigar	National Institute of Technology Karnataka, India
Om Ji Shukla	National Institute of Technology Patna, India
Parag Siddique	University of Louisville, USA
Vladimir Simic	University of Belgrade, Serbia
Avanish Singh Chauhan	Manipal University, India
Mustafa Soba	Usak University, Turkey
Roya Soltani	Khatam University, Iran
Karan Sukhija	Panjab University, India
Aneerav Sukhoo	Ministry of ITC and Operations, Mauritius
Yulin Sun	Southwestern University of Finance and Economics, China
Faustino Taderera	National University of Science and Technology, Oman
Nezih Tayyar	Usak University, Turkey
Ali Tehci	Ordu University, Turkey
Stefania Testa	University of Genoa, Italy
Vikas Thakur	Norwegian University of Science and Technology, Norway
Sunil Tiwari	National University of Singapore, Singapore
Mehdi Toloo	Technical University of Ostrava, Czech Republic
Achraf Touil	Hassan 1st University, Morocco
Rakesh Tripathi	Dr. APJ Abdul Kalam Technical University, India
Pinar Usta	Isparta University of Applied Science, Turkey
Tatapudi Vasista	Srinivas University Mangalore, India
Agnes Nalini Vincent	University of Third Age, Mauritius
Gongming Wang	Tsinghua University, China

Wenqing Wu	Southwest University of Science and Technology, China
Liangping Wu	Sichuan Normal University, China
Mehmet Yetim	Suleyman Demirel University, Turkey
Abdullah Yildizbasi	Ankara Yıldırım Beyazıt University, Turkey
Nurullah Yılmaz	Suleyman Demirel University, Turkey
Ramadan Zenedean	Cairo University, Egypt
Jian Zhang	Sichuan Normal University, China
Haining Zheng	Massachusetts Institute of Technology, USA
Karim Zkik	ESAIP School of Engineers, France
Shakib Zohrehvandi	Technical University of Kosice, Slovakia

Scientific Sponsors

International University of Rabat, Morocco

The University of Texas at Dallas, USA

Prague University of Economics and Business, Czech Republic

Kent Business School, UK

University of Sfax, Tunisia
Faculty of Economics and Management

University of Sfax, Tunisia

Sunway University, Malaysia

Ankara Science University, Turkey

Manisa Celal Bayar University, Turkey

Manufacturing Engineering Laboratory of Tlemcen, Algeria

Abou Bekr Belkaid Tlemcen University, Algeria

Global Academy, Turkey

Suleyman Demirel University, Turkey

Calcutta Business School, India

Halic University, Turkey

Contents – Part I

Smart and Intelligent Transportation Systems

Optimizing Traffic Flow: A Multi-agent Approach to Dynamic Signal Control Accounting for Vehicle Types 3
Serap Ergün

Estimating Parking Lot Occupancy Based on Traffic Congestion for Route Planning 19
Uğur Güven Adar, Osman Çayli, and Atınç Yilmaz

A Framework for Simulating the Optimal Allocation of Shared E-Scooters 32
Halil Ibrahim Ayaz and Bilal Ervural

Performance Enhancement of Two-Way DF Relayed Cooperative NOMA Vehicular Network with Outdated CSI and Imperfect SIC 46
Potula Sravani and Ijjada Sreenivasa Rao

Machine/Deep/Reinforcement Learning in Industries

Integration of Artificial Intelligence for the Analysis and Monitoring of Projects Within Companies 67
Ouissem Mougari, Sarra Bouzid, and Sihem Saadi

Multi-criteria Inventory Classification with Machine Learning Algorithms in the Manufacturing Industry 86
Özge Albayrak Ünal and Burak Erkayman

Analyzing Results of Business Process Automation with Machine Learning Methods 104
Elif Yigit and Seda Özmutlu

Predictive Analysis of Surface Defects in Engineering Structures Using Machine Learning Technologies 121
Roman Mysiuk, Iryna Mysiuk, Volodymyr Yuzevych, Roman Shuvar, Svyatoslav Tsyuh, and Nataliia Pavlenchyk

Enhancing Autonomous Industrial Navigation: Deep Reinforcement Learning for Obstacle Avoidance in Challenging Environments 133
Richa Mohta, Vaishnavi Desai, Tanvi Khurana, M. R. Rahul, and Shital Chiddarwar

Fast and Accurate Right-Hand Detection Based on YOLOv8
from the Egocentric Vision Dataset 153
    Van-Dinh Do, Trung-Minh Bui, and Van-Hung Le

Fine-Tuning of 3D Hand Pose Estimation on HOI4D Dataset
by Convolutional Neural Networks 171
    Van-Dinh Do and Van-Hung Le

Prediction of Ethereum Prices Based on Blockchain Information
in an Industrial Finance System Using Machine Learning Techniques 189
    Syrine Ben Romdhane, Fahmi Ben Rejab, and Khadija Mnasri

Email Classification of Text Data Using Machine Learning and Natural
Language Processing Technique 212
    Oluwaseyi Ijogun, Hayden Wimmer, and Carl Rebman Jr.

The Fake News Detection Model Explanation and Infrastructure Aspects 237
    Maksym Lupei and Myroslav Shliakhta

Advances of Artificial Intelligence/Operational Research Tools in Healthcare

An Advanced Approach to COVID-19 Detection Using Deep Learning
and X-ray Imaging 257
    Hela Limam and Wided Oueslati

Exploring Factors that Affect the User Intention to Take Covid
Vaccine Dose 274
    Pradipta Patra and Arpita Ghosh

Unsupervised Incremental-Decremental Attribute Learning Healthcare
Application Based Feature Selection 292
    Siwar Gorrab, Fahmi Ben Rejab, and Kaouther Nouira

On Solving the Physicians Scheduling Problem at an Emergency
Department: A Case Study from Canada 311
    Ghada Yakoubi, Chahid Ahabchane, and Safa Bhar Layeb

Temporal Emotional and Thematic Progression (TETP): A Novel Analysis
of Mental Health Discussions on Social Platforms 336
    Sharath Kumar Jagannathan and Gulhan Bizel

Author Index 355

Contents – Part II

Technology, Learning and Analytics in Intelligent Systems

Key Factors of Mobile Applications in Government Services (MG-App) Adoption in Indonesia: An Empirical Study Based on Netnography and Text Mining Approach 3
   Nur Rahmah Syah Ramdani and Zulkarnain Zulkarnain

Assessing Operation of Mobile Banking Applications that Support Customers to Access Banking Services Remotely 20
   Daniel Okari Orucho, Fredrick Mzee Awuor, and Collins Oduor

Analyzing the USA Housing Complaints to Score the County Problems 36
   Sharath Kumar Jagannathan, Gulhan Bizel, Vijay Kumar Voddi, and J. V. Thomas Abraham

LEADOW: An Empowering Indoor Navigation System for Individuals with Visual Impairments Using Bluetooth Beacons and Audio Guidance 52
   Suhail Odeh, Murad Al Rajab, Jannat Natsheh, Isra' Zahran, Khader Ballout, and Mahmoud Obaid

The Role of Banks in Financial Literacy and Increasing the Use of Financial Technology: An Exploratory Study of a Sample of Private Bank Managers in Baghdad 67
   Alaa Abdulkareem Ghaleb Almado and Raghad Mohammed Najm Algburi

SecureCloudX: An Innovative Approach to Enhance Data Security Through Advanced File Encryption 77
   Shashwat Kumar, Anannya Chuli, Shivam Raj, Aditi Jain, and D. Aju

Nasdaq-100 Companies' Hiring Insights: A Topic-Based Classification Approach to the Labor Market 98
   Seyed Mohammad Ali Jafari and Ehsan Chitsaz

Expert Systems, Decision Analysis, and Advanced Optimization

Hybrid Reasoning Based Intelligent Decision Support System for Maintenance Management: A Boiler Combustion System Case Study 117
   Bakhta Nachet, Djamila Bouhalouan, and Abdelkader Adla

The Mediating Role of Artificial Intelligence Strategy on Strategic
Thinking Skills and Digital Leadership at Baghdad's General Company
for Electrical and Electronic Industries 137
   *Saba Noori Alhamdany, Mohammed Noori Alhamdany,
   and Saad Noori Alhamdany*

A Two-Phase Time-Based Robust Shift Genetic Algorithm
for Multi-resource Flexible Project Scheduling Under Uncertainties 154
   Do Hong Nhat and Nguyen Van Hop

Artificial Intelligence in Education: A Comprehensive Examination
of Integration, Impact, and Future Implications 182
   *Kishori Kasat, Uday Sinha, Sakshi Juneja, Anurupa Ghatge,
   Nikhil Thorat, and Naim Shaikh*

LEO Satellites: Enhancing Connectivity and Data Collection Across
Industries 199
   Nimet Selen Yesilkaya and Irem Bayraktar

Exploring Robotic Arm Dynamics in Mobile Platforms for Space
Industrial Applications 214
   Cahit Taslicali and Abdullah Demiray

Digital Transformation of Supply Chain and Logistics Systems

Blockchain Based Privacy Preservation and Misbehavior Analysis
in Financial Supply Chain 231
   M. R. Sumalatha, Aditya Kumar, Nethra Janardhanan, and S. Abhinash

Analysis of Blockchain Barriers and Enablers for Improving
the Transparency of Agriculture Supply Chain 249
   Houda Dahbi, Abla Chaouni Benabdellah, and Amine Belhadi

Investigating Critical Success Factors in IoT-Based Marketing Systems 259
   Hamed Nozari, Hamid Reza Irani, and Maryam Rahmaty

Author Index 273

Smart and Intelligent Transportation Systems

Optimizing Traffic Flow: A Multi-agent Approach to Dynamic Signal Control Accounting for Vehicle Types

Serap Ergün[✉]

Department of Computer Engineering, Isparta University of Applied Sciences, Isparta, Turkey
serapbakioglu@isparta.edu.tr

Abstract. This study addresses the critical need for effective traffic signal control in reducing traffic congestion, enhancing road safety, and minimizing environmental impacts. It proposes a real-time dynamic traffic signal control system that incorporates multi-agent systems and considers vehicle type differences. Through comprehensive evaluation experiments, the method proves highly effective in reducing delay time. Experiment 1 optimizes the split control method parameter α in a straightforward setting, while Experiment 2 extends these findings to a complex 4 × 4 intersection, emphasizing the impact of vehicle type differences. Experiment 3 assesses physical decentralization for multiple intersections, showcasing effective split calculations despite occasional execution time fluctuations. The results demonstrate the proposed method's substantial advantages over static control and methods neglecting vehicle type, exhibiting a notable 5.8% reduction in average delay time and a 12.9% decrease in the average number of waiting times. Despite occasional execution time fluctuations in Experiment 3, the method underscores its adaptability to diverse traffic conditions and configurations, offering promising prospects for advancing traffic signal control strategies.

Keywords: Traffic Signal Optimization · Multi-Agent Traffic Control · Vehicle Type Considerations · Real-time Signal Timing · Urban Traffic Management

1 Introduction

Traffic congestion, acknowledged as a grave concern in urban areas worldwide, leads to diverse issues including economic losses, heightened air pollution, and an upsurge in traffic accidents. The operation of traffic lights at intersections, a significant contributor to congestion, necessitates the use of appropriate parameters. Given the substantial variations in traffic flow across different times and days, such as morning and evening rush hours, daytime, night-time, and holidays, this research focuses on dynamically adjusting traffic light control parameters in response to the evolving traffic conditions (Li et al., 2021).

The regulation of traffic in urban areas through the use of traffic signals plays a crucial role in ensuring the safety of both vehicles and pedestrians at intersections. These signals allocate time among conflicting traffic flows, thereby mitigating the most significant

© The Author(s), under exclusive license to Springer Nature Switzerland AG 2025
A. Mirzazadeh et al. (Eds.): ODSIE 2023, CCIS 2204, pp. 3–18, 2025.
https://doi.org/10.1007/978-3-031-81455-6_1

conflicts. This primary motivation led to the installation of signals at intersections during the early 1900s (Zhu et al., 2019).

Various urban traffic control (UTC) systems are present globally, ranging from basic fixed-time plans to the latest generations that enhance control flexibility.

The CRONOS algorithm, a real-time UTC system originated at INRETS in France (Boillot et al., 2006), determines the upcoming traffic light states (green, amber, red) at junctions for the next several seconds through optimized traffic criteria. Employing the rolling horizon concept, CRONOS possesses three distinctive features: firstly, it can simultaneously control a zone comprising a few junctions (typically 5 or 10 junctions), facilitated by the favorable properties of the modified Box algorithm version used for optimization; secondly, it offers flexibility in selecting traffic light states, as CRONOS is acyclic, and no predefined stages are established. Each junction is defined by a set of safety constraints on the traffic light groups, specifying forbidden durations and state correlations; and thirdly, CRONOS has the capability to incorporate video-based measurements, such as queue lengths along the links or spatial occupancy in the inner junction storage zones (Boillot et al., 2006). To assess the real-world performance of CRONOS, an 8-month experiment was conducted on a single junction following the establishment of a comprehensive control chain.

Its responsiveness to sudden changes, such as accidents and disasters, is limited. Furthermore, the manual setting of control areas results in inflexibility in expansion and contraction, leading to broader control ranges and increased calculation costs. To address these challenges, Ikidid et al. (2021b) proposed an autonomous distributed signal control system, enhancing responsiveness and offering flexibility in adjusting control areas. This study extends their approach, incorporating vehicle types like large vehicles and regular vehicles, and validates the model's efficacy.

This paper makes several significant contributions to the field of traffic signal control, differentiating itself from previous studies in the following ways:

Integration of Multi-agent Systems and Vehicle Type Considerations: Unlike conventional traffic signal control systems, which often rely on simplistic actuation based on traffic volume or predetermined schedules, our proposed method integrates multi-agent systems. This enables a more dynamic and responsive approach, considering various vehicle types and their distinct traffic characteristics. This holistic consideration of diverse vehicles contributes to a more adaptive and efficient traffic management strategy.

Real-Time Dynamic Traffic Signal Control: The paper introduces a real-time dynamic traffic signal control system, emphasizing the importance of adjusting signal timing on-the-fly. This real-time adaptability is a departure from static control methods, allowing the system to respond dynamically to changing traffic conditions. The incorporation of real-time adjustments enhances overall traffic flow efficiency and reduces delays.

Optimization of Control Parameters: Experiment 1 of our study focuses on optimizing the split control method parameter α. This fine-tuning is crucial for achieving optimal performance in different scenarios, particularly when considering various vehicle types. The study extends this optimization approach to Experiment 2, highlighting its effectiveness in diverse and complex traffic environments.

Consideration of Environmental and Economic Factors: While addressing traffic congestion and safety concerns, our study goes beyond by considering environmental and economic impacts. The evaluation experiments assess the method's effectiveness in reducing delay times and waiting periods, contributing to potential reductions in fuel consumption and emissions. This broader consideration aligns with contemporary urban planning goals that prioritize sustainability and economic efficiency.

Analysis of Physical Decentralization: Experiment 3 explores the impact of physical decentralization on the proposed system's performance. This investigation into the decentralization of control for multiple intersections provides valuable insights into the scalability and adaptability of the method. The study also acknowledges and discusses potential limitations, such as occasional execution time fluctuations, providing a realistic perspective for future implementation.

In-depth Evaluation of Real-World Applicability: Responding to the need for practical insights, the study discusses the real-world applicability of the proposed method. By addressing scalability, integration with existing urban traffic systems, and potential challenges associated with implementation, the paper provides guidance for urban planners and policymakers. This practical focus distinguishes the study from purely theoretical contributions.

In summary, this paper contributes to the field by presenting an innovative approach to traffic signal control that incorporates multi-agent systems, adapts to real-time conditions, optimizes control parameters, considers diverse vehicle types, and evaluates the broader impacts on the environment and economy. The in-depth exploration of real-world applicability and acknowledgment of limitations further enhance the paper's contribution to the advancement of traffic management strategies.

The organizational of the paper is structured as follows: The main perspective of the study which is the Traffic Light Control is presented briefly in Sect. 2. The proposal techniques with preliminaries are given in Sect. 3. Evaluation experiment on considering vehicle type differences are presented in Sect. 4 and evaluation experiment on physical decentralization is proposed in Sect. 5. Finally, this paper ends with Conclusion part in Sect. 6.

2 Traffic Light Control

Traffic signal control is a critical aspect of urban infrastructure management, playing a pivotal role in regulating vehicular and pedestrian movement at intersections. Efficient traffic signal control involves a nuanced understanding of various parameters that influence display changes, ultimately optimizing the overall flow of traffic. The term "present" refers to the simultaneous right-of-way granted to specific traffic flows, accommodating both vehicles and pedestrians within the intersection dynamics (Abbracciavento et al, 2023).

To delve deeper into the intricacies of traffic signal control, several key parameters come into play (Chen et al., 2020). These parameters include "cycle length," "split," and "offset," each contributing significantly to the orchestration of traffic signal changes. The "cycle length" represents the total time it takes for the traffic signal to complete one full sequence of displays, including green, yellow, and red phases for each direction. Meanwhile, the "split" denotes the portion of the cycle assigned to a particular movement, such as the duration of the green light for vehicles traveling in a specific direction.

Optimizing traffic flow is a multifaceted challenge that demands a delicate balance among these control parameters. The research at hand is dedicated to advancing our understanding of how adjusting the "split" among these parameters can positively impact traffic flow. By strategically managing the allocation of time for different traffic movements, we aim to enhance overall efficiency and reduce congestion at intersections.

This study employs advanced modeling techniques and real-time data analysis to identify optimal configurations for traffic signal control. Machine learning algorithms are leveraged to adaptively adjust signal parameters based on current traffic conditions, time of day, and other relevant factors. The overarching goal is to create a dynamic and responsive traffic signal control system that can adapt to the ever-changing demands of urban traffic.

Traffic signal control involves managing parameters that influence display changes, optimizing traffic flow. The term "present" denotes the simultaneous right of way granted to specific traffic flows, encompassing pedestrians, at an intersection. Signal control parameters, including "cycle length," "split," and "offset," contribute to display switching. This research aims to enhance traffic flow by controlling the split among these parameters.

3 Proposal Techniques

The proposed technique is applied by the authors to optimize traffic flow, utilizing a decentralized approach where autonomous agents at each intersection cooperate. This method builds upon existing models and aims to improve traffic management by considering different vehicle types. The planned demonstration experiment validates the model's functionality in real-world scenarios. By gathering data from various sources and enabling communication between agents and a central database, informed decisions are aimed to be made and collaboration among traffic lights and intersections facilitated.

In this research, traffic lights at each intersection autonomously and decentrally cooperate. A multi-agent model (Ge et al., 2021), distributing autonomous agents at each traffic light, determines signal control parameters, optimizing the overall environment. Each agent manages the splitting of traffic lights at its intersection, building upon the split control proposed by Ikidid et al. (2021a) with enhancements for vehicle types. The study anticipates a demonstration experiment, physically deploying multiple agents to validate the model's functionality.

Each agent possesses information about its own intersection, including cycle length, preset parameters, green light start time, and the number of lanes in each direction. Additionally, agents can gather data on vehicle ID, lane, speed, and length within the sensor range through radio radar (Guo et al., 2021). Communication with a database allows access to information such as total vehicles entering adjacent intersections, offset values, and green hour start times.

3.1 Traffic Simulator

The research harnesses the power of the Simulation of Urban MObility (SUMO) traffic simulator, a sophisticated tool developed by the German Aerospace Center's Transport Systems Research Institute (DLR). SUMO, having been an open-source simulator since its inception in 2002, stands as a pivotal component in the realm of traffic engineering, providing researchers and urban planners with a robust platform for simulating and analyzing complex traffic scenarios (Behrisch et al., 2011).

SUMO's versatility is a key asset, offering the capability to create intricate vehicle routes, conduct detailed simulations, and integrate online functionalities for real-time analysis. The simulator's open-source nature fosters collaboration and innovation within the research community, allowing for continuous improvements and adaptations to meet the evolving challenges of urban mobility.

One notable feature of SUMO is its ability to model a wide array of transportation modes, including not only conventional vehicles but also pedestrians, cyclists, and public transportation (Behrisch et al., 2011). This comprehensive modeling capability enables researchers to explore and optimize traffic management strategies in diverse urban settings where various modes of transportation coexist.

The simulator's user-friendly interface, coupled with its powerful backend capabilities, facilitates the development and testing of complex traffic signal control algorithms. Researchers can define specific traffic scenarios, set parameters such as vehicle types, and assess the impact of different control strategies under varying conditions. This level of detail is crucial for understanding the nuanced interactions within traffic systems and refining control methods for enhanced real-world applicability.

Moreover, SUMO supports the integration of real-world geographical data, allowing researchers to simulate traffic conditions on actual road networks. This realistic simulation environment enhances the validity of research findings and aids in the development of traffic signal control strategies that can seamlessly integrate with existing urban infrastructures.

As urban environments continue to evolve, SUMO remains at the forefront of traffic simulation technology, providing a reliable and scalable platform for researchers to push the boundaries of traffic management research. The collaboration between the research community and the open-source SUMO project exemplifies a dynamic synergy, driving advancements in the field of urban mobility and contributing to the development of smarter, more efficient transportation systems for the cities of tomorrow.

3.2 Split Control for Each Intersection

Agents placed at intersections calculate the split for each cycle by applying traffic volume to the balance equation of the split model, based on Ikidid et al.'s split model (2021b). This study introduces new aspects, detailed in Sect. 3.2.1.

Definition of Traffic Volume. The split value in this approach is determined by utilizing the traffic volume q from each direction. To establish a foundational understanding, let this be defined first.

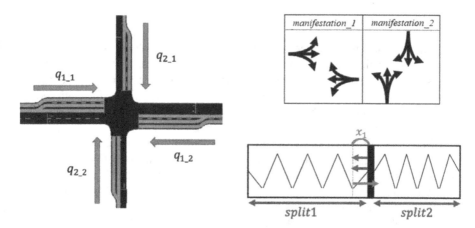

Fig. 1. The schematic diagram of the split models with two visible intersections

Traffic volume, at its core, constitutes the sum of inflows and the residual number of vehicles at the conclusion of each display. This can be succinctly expressed in the following formula:

$$q = \frac{n_{inflow}}{n_{lane}} + a^{\frac{n_{res}}{n_{lane}}} \quad (1)$$

At this juncture, the inflow amount is denoted as n_{inflow}, the number of remaining vehicles as n_{res} and the lane as n_{lane}. The exponential growth of this term, achieved by setting α as a constant greater than 1, occurs as the number of vehicles awaiting sorting increases. The definitions of n_{inflow} and n_{res} in this context are as follows:

$$n_{inflow} = n_{all\_inflow} + \alpha n_{large\_inflow} \quad (2)$$

$$n_{res} = n_{all\_res} + \alpha n_{large\_res} \quad (3)$$

The term n_{all_inflow} designates the total number of inflows from all vehicles, while n_{large_inflow} denotes the inflows specifically from large vehicles. Correspondingly, n_{all_res} signifies the overall count of remaining vehicles, encompassing all types, and n_{large_res} represents the count of remaining large vehicles subsequent to the conclusion of the display. These definitions establish a framework wherein large vehicles are accounted for "$1 + \alpha$" times more than regular vehicles.

Split Calculation. In Fig. 1, a schematic illustration of the split model featuring two simultaneous intersections is presented. For an intersection embodying dual manifestations, a model is assumed in which two splits are interlinked, as depicted in Fig. 1. The split values for manifestation 1 and manifestation 2 are denoted as *split*1 and *split*2, respectively.

The split values, *split*1 and *split*2, are derived by incorporating the difference d_1 between the traffic volume of display 1 and the traffic volume of display 2 onto the connecting surface of the split. This adjustment is realized through the split balance equation. In cases where the number of displayed splits expands, each split can be determined by augmenting the number of splits and executing the same calculation process.

$$split1 = \frac{1}{2} + \frac{d_1}{2k} \qquad (4)$$

$$split2 = \frac{1}{2} - \frac{d_1}{2k} \qquad (5)$$

Right Turn Only Sign Influence. A schematic depiction of the split model determining the split for right-turn only indications is presented in Fig. 2. In instances where a right-turn only sign is in effect, a model is presumed wherein the split derived in Sect. 3.2.2 connects a split for an indication that turns green in all directions and a split exclusively activating for right turns, as illustrated in Fig. 2.

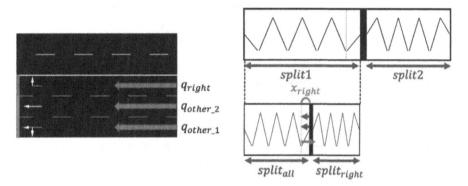

Fig. 2. The schematic diagram of a split model considering right turn only indication

The proportions of the omnidirectional green display and the right-turn-only display are denoted as $split_{all}$ and $split_{right}$ respectively. The derivation of $split_{all}$ and $split_{right}$ is facilitated by introducing "dright," representing the discrepancy between the number of waiting cars in the left-turn straight lane and the right-turn lane, to the split connection surface. This addition enables the determination of $split_{all}$ and $split_{right}$ through the split balance equation.

$$split_{all} = \frac{1}{2} + \frac{d_{right}}{2k} \qquad (6)$$

$$split_{right} = \frac{1}{2} - \frac{d_{right}}{2k} \qquad (7)$$

3.3 Physical Decentralization

In the implementation of the demonstration experiment detailed in Sect. 3.2, a novel approach is taken to enhance the physical realism of the simulation. To achieve this, agents

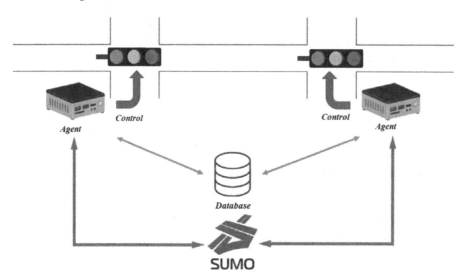

Fig. 3. The overview of simulation experiment

are strategically dispersed at each intersection, simulating their physical presence in a distributed manner. The acquisition of crucial vehicle information within the sensor range is facilitated through the innovative utilization of radio radar during the demonstration experiment. This allows the agents to dynamically collect real-time data, contributing to a more accurate representation of the traffic dynamics within the simulation.

It's particularly noteworthy that the integration of the SUMO online connection function, as expounded in Sect. 3.1, plays a pivotal role in extracting essential vehicle information from the SUMO server. This function, depicted in Fig. 3, acts as a communication bridge, enabling seamless data exchange between the various entities involved in the simulation.

Building on the methodology employed by Ikidid et al. (2021a), where all entities operated on a single PC, this present study takes a significant stride towards a more sophisticated simulation environment. In contrast, a distinct and innovative approach is adopted here, with each agent and the SUMO simulator executing on separate PCs, thereby introducing physical separation. This departure from the conventional single PC setup is motivated by a desire to enhance scalability, simulate realistic communication constraints, and mirror the distributed nature of traffic management systems.

The procedural details of this distributed setup are elucidated as follows: the initiation of the SUMO server occurs on the first PC, and concurrently, the database is installed on the same PC to ensure seamless data access. Subsequently, each agent responsible for governing traffic lights is connected to the SUMO server, initiating the simulation across the distributed network. This approach necessitates the use of multiple PCs, with the total number required being equal to the sum of the intersections plus one.

By adopting this distributed architecture, the study introduces a level of sophistication that aligns more closely with real-world scenarios. This approach not only enhances the scalability and realism of the simulation but also provides a foundation for exploring the intricacies of communication delays and network effects within the context of dynamic traffic signal control. This innovative methodology represents a significant step forward in the pursuit of more accurate and applicable traffic management strategies in the ever-evolving landscape of urban mobility.

4 Evaluation Experiment on Considering Vehicle Type Differences

The split control method, incorporating considerations for vehicle type as detailed in Sect. 3.2, undergoes evaluation in this experiment.

4.1 Overview of the Experiment

The standardized parameters across all experiments encompass a yellow light time and red light time of 3 steps each, a cycle length of 100 steps, and a simulation period set at 14400 steps. It is imperative to note that in SUMO, the unit of time is referred to as a step, with 1 step equating to 1 s. Table 1 delineates the vehicle settings for both large vehicles and general vehicles in the simulation.

Table 1. Experimental vehicle settings.

Setting items	Large vehicle	General vehicle
Vehicle length	7.1 m	4.3 m
Acceleration	1.3 m/s^2	2.9 m/s^2
Deceleration	4.0 m/s^2	7.5 m/s^2

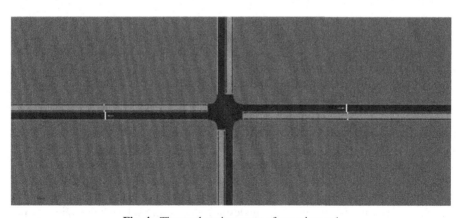

Fig. 4. The road environment of experiment 1

For comparative analysis, a static control method with a fixed split set at 0.5 is employed. Additionally, Ikidid et al.'s control (2021b), which does not account for vehicle type, serves as the benchmark method.

4.2 Experiment 1

In Experiment 1, the utility of the traffic volume definition, considering vehicle type as expounded in Sect. 3.2.1, is affirmed. Simultaneously, the search for the optimal value of the parameter α is undertaken. The experimental setting assumes the simplest environment, with only one intersection and one lane, as delineated in Fig. 4. To maintain simplicity and prevent complications in the search process, especially with the absence of dedicated right-turn lanes, the experiment incorporates a scenario where right-turning vehicles may potentially occupy the front of the queue, leading to substantial traffic congestion.

In Experiment 1, a careful balance is maintained by establishing a ratio of left-turning vehicles to right-turning vehicles to straight-going vehicles at intersections, set at 1:0:9. This ensures a methodical evaluation of the scenario. The vehicle inflow ratio is equally divided between the east-west and north-south directions at 1:1, with the evaluation index focusing on delay time.

Under the specified conditions, simulations are conducted for both large and ordinary vehicles in the east-west direction. Three distinct patterns with vehicle ratios of 1:9 (sim1), 3:7 (sim2), and 5:5 (sim3) are meticulously prepared and executed. The outcomes of Experiment 1, as depicted in Fig. 5, provide a comparative analysis of each evaluation index.

Upon scrutinizing the results of sim1, minimal differences are observed in the outcomes. In contrast, a comparative examination of sim2 and sim3 reveals the efficacy of our method, which factors in vehicle type, in reducing delay time—particularly significant when the presence of large vehicles is substantial. This simulation reinforces the consideration that the optimal value for the rate parameter α stands at 0.3.

4.3 Experiment 2

In Experiment 2, the optimal parameter α identified in Experiment 1 is employed to showcase the efficacy of defining traffic volume, considering large vehicles, in diverse environments. The road configuration features a 4×4 intersection, delineated in Fig. 6. Each inflow road spans a length of 600 m, and the distance between intersections is 1200 m. The second road from the bottom functions as the main road, boasting two lanes, while the remaining roads are general roads. Notably, a right-turn lane is positioned 50 m before the intersection on all main roads, and non-arterial roads are equipped with a single lane, incorporating a right-turn lane 50 m before the intersection with the main road.

However, dedicated right-turn lanes are absent at intersections between non-arterial roads. Consequently, right-turn-only signs are implemented at intersections along arterial roads, and the determination of the split for these signs adheres to the method outlined in Sect. 3.2.3. The ratio of left-turning vehicles to right-turning vehicles to straight-going vehicles at each intersection is standardized at 1:1:8. The evaluation criteria encompass delay time and the average number of waiting times.

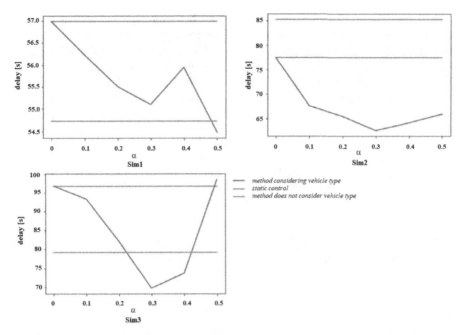

Fig. 5. Comparison of average delay times in Experiment 1

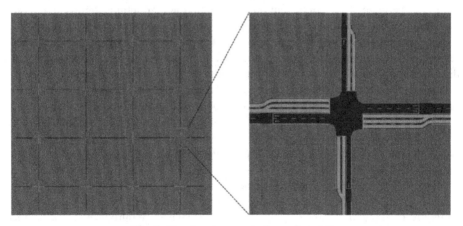

Fig. 6. Road environment of experiment 2

4.4 Discussion

The outcomes of Experiment 1 affirm the efficacy of the method, considering vehicle type. Furthermore, the simulation highlights the optimum value for the parameter α, standing at 0.3. Experiment 2 extends this understanding by showcasing the effectiveness of considering vehicle type in a more intricate environment.

Differences in acceleration and vehicle length between large and regular vehicles emerge as influential factors in demonstrating the efficacy of methods that account for these vehicle types. As indicated in Table 1, large vehicles exert a more substantial impact on congestion, boasting a vehicle length 1.65 times greater and acceleration 0.45 times that of regular vehicles. Consequently, by augmenting the influence of large vehicles through a multiplication factor of $1 + \alpha$, there exists the potential for improvements in average delay time and average number of waiting times compared to methods neglecting vehicle type.

Table 2. Comparison of average delay time and average number of waits in Experiment 2.

Control method	Average delay time [s]	Average number of waiting times [times]
Static control	760.120	6.569
Method that does not consider vehicle type	719.885	5.828
Method considering vehicle type	716.382	5.723

Conversely, the effectiveness of this method and the parameter α optimization pursued in Experiment 1 are contingent on the assumption that optimization is attainable using prior research methods, specifically under conditions involving only regular vehicles. For instance, if the previous research method were applied in a scenario where only regular vehicles are present, and congestion is more probable in the east-west direction than in the north-south direction, the introduction of large vehicles might inadvertently exacerbate the situation. Adopting this method could potentially lead to an increase in the split in the east-west direction, where large vehicles are more predominant, thus potentially worsening the outcomes. While significant effectiveness was demonstrated under simplified conditions like Experiment 1, the lower efficacy observed in more complex situations like Experiment 2 may be attributed to factors beyond the purview of this method.

Experiment 2 navigates an environment where traffic conditions fluctuate, highlighting the adaptability of the method. Changes in the volume of vehicles per hour and variations in the ratio of large vehicles to general vehicles are introduced to assess the method's responsiveness to these alterations.

The results in Table 2 unveil the method's efficacy in considering vehicle type, showcasing a 5.8% reduction in average delay time and a 12.9% reduction in the average number of waiting times when compared to static control. However, when compared to the method of prior research that neglects vehicle type, the average delay time sees a marginal reduction of 0.5%, and the average number of waiting times experiences a modest decrease of 1.8%. Though improvements are discernible, they are relatively modest in comparison.

Table 3. Results of experiment 3.

Evaluation index	sim1	sim2	sim3
Split correct answer rate	100%	100%	100%
Average execution time per step [ms]	15.236	28.240	67.158
Maximum execution time per step [ms]	799.000	667.000	1078.000

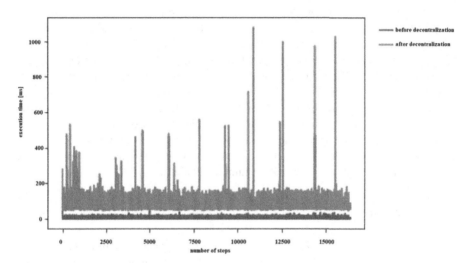

Fig. 7. The changes in execution time per step in Experiment 3 (sim3).

5 Evaluation Experiment on Physical Decentralization

5.1 Overview of the Experiment

In this experiment, the proper functioning of the physical decentralization outlined in Sect. 3.3 is verified. The evaluation is based on two key aspects: achieving a 100% correct answer rate for splitting and ensuring an execution time per step of less than 1 s, even with decentralization. The crucial observation that the correct split answer rate reaches 100% affirms that split calculations can be executed precisely in the same manner as when the simulation is performed on a single PC without decentralization.

5.2 Experiment 3

Experiment 3 aims to validate the appropriate physical distribution of control for multiple intersections. Additionally, it explores the impact of delay time resulting from an increased number of controlled traffic lights and the corresponding rise in required PCs. Simulations are conducted with configurations involving 1 (sim1), 2 (sim2), and 4 (sim3) controlled traffic lights. The results in Table 3 reveal that sim1 and sim2 meet the two-point evaluation criteria. In sim3, with a total of 4 controlled traffic lights, there are instances where the maximum execution time per step exceeds 1 s, particularly evident in Fig. 7, where occasional spikes in execution time are observed.

5.3 Discussion

All simulations confirm the fulfilment of the first evaluation point, affirming that split calculations can be seamlessly executed after decentralization, identical to the pre-decentralization scenario. Regarding the second evaluation point, ensuring execution within 1 s per step post-decentralization is demonstrated in sim1 and sim2, where the number of controlled traffic lights is 2 or fewer. In sim3, encompassing 4 controlled traffic lights and 4 steps, some steps surpass the 1-s threshold. However, the majority of steps maintain an execution time below 100 ms, albeit occasional fluctuations. Though there is a potential for sudden increases in execution time for sim1 and sim2, this is recognized as a limitation.

The issue appears more related to the instability of the communication environment than the increment in the number of controlled traffic lights. Consequently, significant alterations to this method are deemed unnecessary. Even if execution time occasionally exceeds 1 s, it remains feasible to conduct a demonstration experiment by implementing measures such as substituting alternative numerical values.

6 Conclusion

In conclusion, the comprehensive exploration of the proposed method through a series of experiments has provided valuable insights into its efficacy, limitations, and adaptability in addressing the complexities of traffic management. The proposed method utilizes a decentralized approach where autonomous agents at each intersection cooperate to optimize traffic flow.

Experiment 1 established the effectiveness of the method, emphasizing the critical role of considering different vehicle types and optimizing the parameter α for optimal performance. Experiment 2 extended this understanding by navigating a more intricate environment, highlighting the need to account for variations in acceleration and vehicle length between large and regular vehicles.

The discussion illuminated the influence of vehicle characteristics on congestion, underscoring the substantial impact of large vehicles and the necessity of incorporating these factors into traffic management strategies. Experiment 2's comparison with static control methods and approaches neglecting vehicle type showcased the superior performance of the proposed method, albeit with a nuanced level of improvement.

Contingency on prior research methods, as discussed, raises questions about the assumption of optimization in scenarios involving only regular vehicles. The introduction of large vehicles may disrupt the effectiveness of established methods, prompting careful consideration in real-world applications. Experiment 2 further demonstrated the adaptability of the method to fluctuating traffic conditions, affirming its responsiveness to changes in vehicle volume and the ratio of large vehicles to general vehicles.

Experiment 3 provided additional insights into the method's stability and split decision accuracy under varying simulation scenarios. While the method exhibited a high split correct answer rate, the comparison with a prior research method in Table 2 revealed modest gains in average delay time and waiting times, prompting a nuanced evaluation of its overall impact.

While the proposed method exhibits efficacy in addressing traffic congestion, enhancing road safety, and minimizing environmental impacts, several limitations should be acknowledged. The study recognizes the method's limited responsiveness to sudden changes such as accidents and disasters. Additionally, the manual setting of control areas introduces inflexibility in expansion and contraction, leading to broader control ranges and increased calculation costs. Despite its adaptability to diverse traffic conditions, the method's limitations are evident in its potential for occasional increases in execution time, particularly in scenarios involving a higher number of controlled traffic lights.

In summary, the proposed method demonstrates efficacy, particularly in considering vehicle type and adapting to changing traffic conditions. However, the discussion has highlighted the nuanced nature of its effectiveness, acknowledging trade-offs and limitations compared to prior research methods. These findings underscore the importance of carefully weighing the benefits and drawbacks of the proposed approach for its optimal application in real-world traffic management scenarios. Future research should delve deeper into refining the method, addressing identified limitations, and exploring avenues for enhanced performance in diverse and dynamic urban environments.

References

Abbracciavento, F., Zinnari, F., Formentin, S., Bianchessi, A.G., Savaresi, S.M.: Multi-intersection traffic signal control: a decentralized MPC-based approach. IFAC J. Syst. Control **23**, 100214 (2023)

Behrisch, M., Bieker, L., Erdmann, J., Krajzewicz, D.: SUMO–simulation of urban mobility: an overview. In: Proceedings of SIMUL 2011, The Third International Conference on Advances in System Simulation. ThinkMind (2011)

Boillot, F., Midenet, S., Pierrelee, J.C.: The real-time urban traffic control system CRONOS: algorithm and experiments. Transp. Res. Part C Emerg. Technol. **14**(1), 18–38 (2006)

Chen, C., et al.: Toward a thousand lights: decentralized deep reinforcement learning for large-scale traffic signal control. Proc. AAAI Conf. Artif. Intell. **34**(04), 3414–3421 (2020)

Ge, H., et al.: Multi-agent transfer reinforcement learning with multi-view encoder for adaptive traffic signal control. IEEE Trans. Intell. Transp. Syst. **23**(8), 12572–12587 (2021)

Guo, Q., Li, L., Ban, X.J.: Urban traffic signal control with connected and automated vehicles: a survey. Transp. Res. Part C Emerg. Technol. **101**, 313–334 (2019)

Guo, X., et al.: Urban traffic light control via active multi-agent communication and supply-demand modeling. IEEE Trans. Knowl. Data Eng. (2021)

Ikidid, A., El Fazziki, A., Sadgal, M.: A fuzzy logic supported multi-agent system for urban traffic and priority link control. JUCS: J. of Univ. Comput. Sci. **27**(10) (2021a)

Ikidid, A., El Fazziki, A., Sadgal, M.: A multi-agent framework for dynamic traffic management considering priority link. Int. J. Commun. Netw. Inf. Secr. **13**(2), 324–330 (2021)

Li, Z., Yu, H., Zhang, G., Dong, S., Xu, C.Z.: Network-wide traffic signal control optimization using a multi-agent deep reinforcement learning. Transp. Res. Part C Emerg. Technol. **125**, 103059 (2021)

Soon, K.L., Lim, J.M.Y., Parthiban, R.: Coordinated traffic light control in cooperative green vehicle routing for pheromone-based multi-agent systems. Appl. Soft Comput. **81**, 105486 (2019)

Wang, T., Cao, J., Hussain, A.: Adaptive traffic signal control for large-scale scenario with cooperative group-based multi-agent reinforcement learning. Transp. Res. Part C Emerg. Technol. **125**, 103046 (2021)

Wang, Y., Xu, T., Niu, X., Tan, C., Chen, E., Xiong, H.: STMARL: a spatio-temporal multi-agent reinforcement learning approach for cooperative traffic light control. IEEE Trans. Mob. Comput. **21**(6), 2228–2242 (2020)

Wang, Z., Bian, Y., Shladover, S.E., Wu, G., Li, S.E., Barth, M.J.: A survey on cooperative longitudinal motion control of multiple connected and automated vehicles. IEEE Intell. Transp. Syst. Mag. **12**(1), 4–24 (2019)

Wu, T., et al.: Multi-agent deep reinforcement learning for urban traffic light control in vehicular networks. IEEE Trans. Veh. Technol. **69**(8), 8243–8256 (2020)

Xu, M., An, K., Vu, L.H., Ye, Z., Feng, J., Chen, E.: Optimizing multi-agent based urban traffic signal control system. J. Intell. Transp. Syst. **23**(4), 357–369 (2019)

Yang, S.: Hierarchical graph multi-agent reinforcement learning for traffic signal control. Inf. Sci. **634**, 55–72 (2023)

Zhang, H., et al.: Cityflow: a multi-agent reinforcement learning environment for large scale city traffic scenario. In: The world Wide Web Conference, May 2019, pp. 3620–3624 (2019)

Zhang, X.M., et al.: Networked control systems: a survey of trends and techniques. IEEE/CAA J. Automatica Sinica **7**(1), 1–17 (2019)

Zhu, F., Lv, Y., Chen, Y., Wang, X., Xiong, G., Wang, F.Y.: Parallel transportation systems: toward IoT-enabled smart urban traffic control and management. IEEE Trans. Intell. Transp. Syst. **21**(10), 4063–4071 (2019)

Estimating Parking Lot Occupancy Based on Traffic Congestion for Route Planning

Uğur Güven Adar[1(✉)], Osman Çayli[2], and Atınç Yilmaz[2]

[1] Istanbul Beykent University, Istanbul, Turkey
ugurguvenadar@beykent.edu.tr
[2] VBT Yazılım A.Ş., Istanbul, Turkey

Abstract. Traffic congestion in large cities negatively affects daily life and is a significant concern. In crowded cities like Istanbul, not only waiting in traffic but also the availability of parking spaces at the destination is a critical issue. Predicting the occupancy of parking lots in advance can supply a significant advantage for drivers. In this study, traffic congestion in Istanbul is attempted to be estimated using artificial intelligence techniques such as K-NN, Decision Trees, and Artificial Neural Networks, with 2022 vehicle traffic data from the Istanbul Metropolitan Municipality (IBB) Open Data Portal. As a result of the tests, the decision tree method provided the best results in the dataset, estimating traffic congestion with an R^2 value of 0.8469. The occupancy status of parking lots in areas close to the traffic congestion estimation points was determined using the İspark Parking Detailed Information dataset. Based on the obtained information, it is aimed to predict the occupancy status of parking lots according to traffic congestion without the need for live data using the Dijkstra algorithm in the future.

Keywords: parking lot · traffic congestion estimation · k-nn · decision trees · artificial neural networks

1 Introduction

As the world's population continues to grow, and with increased migration to major cities, various problems are emerging in urban areas referred to as metropolises. With the increase in construction and the rise in individual well-being, a significant amount of activity is concentrated in limited spaces. This mobility brings to the forefront one of the primary challenges in metropolises: traffic congestion. Despite various attempted solutions, traffic remains a persistent problem in nearly all metropolitan cities worldwide. While these are optimistic solutions about traffic congestion, it is crucial to consider the economic, social, and cultural changes brought about by this loss of time.

The negative impact of traffic on the quality of life for people in metropolitan areas is an undeniable reality. However, another significant issue in metropolitan living caused by traffic is the need for car owners to find appropriate and safe parking spaces at their destinations, in addition to the time spent in traffic. When traveling from one location to another, finding a suitable and available parking spot at the destination is essential. This situation can be thought of as a cyclical problem. The inability to find proper parking spaces leads to incorrect parking, which exacerbates traffic congestion. Consequently, the increase in traffic directly escalates the parking problem and waiting time. Therefore, it can be inferred that the proper parking of vehicles in locations that do not disrupt traffic is directly linked to the traffic problem. In addition to parking issues being a cause of traffic in metropolises, the availability of parking spaces at destinations in individuals' travel or route planning is another separate problem to consider.

Istanbul is home to nearly 1.2 million buildings. With the city's population having grown rapidly since the 1980s, the question of how parking and traffic problems have been addressed in this construction surge remains unanswered. Turkey has seen various revisions to its parking regulations over the years. In Istanbul, the metropolitan municipality has established its own regulations based on these national parking rules. These regulations mandate that each building must have a parking facility proportional to its number of floors and apartments. However, as noted by experts, the fines for buildings without parking are relatively low cost-wise, and these penalties can be negated over time through legal loopholes such as statutes of limitations and amnesties. This situation leads to a reluctance to construct parking facilities. The failure to plan Istanbul's buildings with adequate parking, combined with the fact that parking regulations have only been enforced since 1985 and the leniency in penalizing non-compliance, results in vehicle owners resorting to roadside parking, which undeniably contributes to increased traffic congestion.

The study is guided by two main motivations within the smart cities concept. The first is to empower city dwellers to make informed decisions about their travel by predicting traffic congestion and parking lot occupancy without relying on live data. It aims to create a smart system that estimates the correlation between parking availability and traffic congestion, enabling predictions of parking lot occupancy at destination points. The second motivation is to alleviate traffic exacerbation caused by parking practices by forecasting the occupancy rates of parking lots at specific locations. By providing estimated parking fullness based on traffic congestion, the system encourages users to consider alternatives, potentially easing traffic congestion and addressing traffic issues in large cities.

This study leverages artificial intelligence techniques to identify the most effective AI method for its objectives, using datasets from the Istanbul Metropolitan Municipality (IBB) Open Data Portal. Focusing on Istanbul's 2022 daily vehicle count data from various areas, it analyzed this information with K-Nearest Neighbors (K-NN), Decision Trees, and Machine Learning techniques to determine the most accurate. The selected technique was then employed to predict parking lot occupancy based on nearby traffic congestion, correlating traffic congestion location data with İspark Locations dataset information from the IMM Open Data Portal. The aim was to develop a decision-support system to estimate parking lot fullness using existing data, circumventing the need for live data. This system is intended to facilitate route planning for future users. However, it's important to note that the study's scope is confined to Istanbul and IBB-owned İspark parking lots, with its findings limited to the insights gleaned from the AI techniques applied to the datasets in question.

The study is structured into four key sections: literature review, methodology, results, and conclusion. The literature review explores and contrasts similar studies, setting the context for this research. The methodology outlines data acquisition and processing, detailing the artificial intelligence techniques used and the implementation of the proposed system. The findings section presents the analysis results. Lastly, the conclusion reflects on the datasets, methods, and findings, offering insights for future research and potential advancements in the field.

2 Literature Review

In recent years, the issue of parking space identification and management has become a significant challenge, particularly in dense urban centers. Researchers have addressed this issue within the context of smart cities and sustainable development [1–3]. Concerns about parking in urban areas are increasingly becoming a focal point in supporting city centers. The Internet of Things (IoT) and cloud-based technologies are expanding the scope of smart vehicle parking services, offering new applications that better regulate traffic related to vehicle parking [3, 4]. Additionally, the relationship between land use, parking, and traffic congestion plays a crucial role in developing management strategies to reduce traffic congestion and emissions in urban areas [5]. These advancements highlight the importance of integrating technology and urban planning to address the growing challenges of parking and traffic in metropolitan cities.

There are studies indicating that the availability of parking spaces is closely related to land use and traffic congestion, and in this context, parking reform offers a more efficient way to alleviate congestion by coordinating the density of spaces and areas in residential regions [5]. The interactions between urban parking and traffic systems must be considered for the improvement of both systems' performance. Research in this area significantly contributes to evaluating different parking policies and identifying potential strategies for reducing traffic delays. The use of system dynamics based on urban traffic's parking-related conditions is a vital tool for effectively assessing urban traffic and parking systems on a macroscopic level [6]. This approach underscores the necessity of a holistic and integrated view of urban mobility, focusing on the synergy between parking management and overall traffic flow.

Advanced technologies like vehicular ad hoc networks (VANETs) can be utilized to predict the occupancy status of parking spaces, thereby aiding drivers in finding available parking spots at their destinations. VANETs, in particular, hold the potential to offer solutions to parking problems in cities with heavy traffic [7, 8]. Machine learning algorithms and other artificial intelligence techniques can be employed to predict the occupancy of parking spaces and their future occupancy levels. These predictions, when integrated into smart parking management systems, can assist drivers in finding available parking spots more quickly, thus helping to reduce traffic congestion [9]. This integration of technology highlights the increasing role of AI and networked systems in resolving urban mobility challenges, emphasizing a shift towards more data-driven and responsive urban planning solutions.

In this context, existing literature presents various solutions for parking space detection using decision trees [10], K-Nearest Neighbors (K-NN) [11], and Machine Learning

[12]. This study employs all of these methods to identify the most effective one for predicting parking space availability. By comparing the outcomes of these different AI techniques, the research aims to optimize the prediction accuracy and efficiency of the proposed smart parking management system. This approach reflects a comprehensive and comparative methodology, ensuring that the selected solution is best suited to address the complexities and dynamic nature of urban parking challenges.

3 Methodology and Application

3.1 Data Acquisition and Preprocessing

The study employs two datasets. The first is the "2022 Daily Vehicle Count" dataset from the IBB Open Data Portal [13]. This dataset, derived from sensor data at 706 points between January 1, 2022, and December 31, 2022, contains a total of 96,599 records organized into six columns. An example view of this dataset is illustrated in Fig. 1(a). The dataset is utilized for two main purposes in this study. Firstly, it is analyzed to calculate traffic congestion using vehicle counts, latitude and longitude data, and date-specific information. Secondly, it serves to predict the occupancy rate of parking lot capacities based on the latitude and longitude information from the İspark dataset, in relation to the traffic data from the nearest coordinates.

The second dataset used in the study is the İSPARK Locations dataset, also available on the IBB Open Data Portal [14]. This dataset contains information about parking areas provided by İSPARK. It includes data such as Park Name, Location Name, Park Type ID, Park Type Description, Park Capacity, Operating Hours, District Name, Longitude, and Latitude. Comprising a total of 708 records, an example of this dataset is illustrated in Fig. 1(b). The latitude and longitude information along with the parking capacities from the İspark Locations are utilized to estimate the occupancy rate based on traffic congestion.

After the relevant datasets were recorded, they were analyzed. The Python programming language was chosen for this study. Python's open-source nature, the diversity of its modules, and ease of use were the primary reasons for its selection. Data was read using the Pandas module [15] in Python, followed by data organization. The initial steps included adjusting for Turkish characters, removing unnecessary columns, and checking for missing data. Then, the locations in both datasets were examined using the folium module [16], allowing for a visual evaluation of the dataset's status. Maps created with the folium module showed that while the traffic data from the vehicle count dataset appeared at road locations and stations throughout Istanbul, the locations from the İspark location dataset were more concentrated in the inner parts of Istanbul. Examples of these maps are provided in Fig. 2.

3.2 Methodology and Application

In this study, the Python programming language was chosen, and relevant modules were included. The Pandas module [15] was used for organizing, cleaning, saving, and creating data frames for the datasets. Maps were generated using the Folium module [16] based on

Fig. 1. (a) Upper Image: Example View of the 2022 Daily Vehicle Count Dataset [13]. (b) Lower Image: Example View of the İspark Locations Dataset [14].

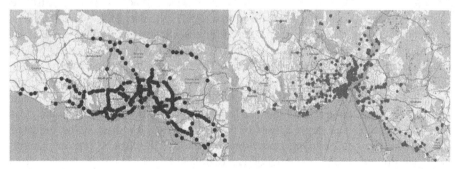

Fig. 2. Folium Module Generated Vehicle Count Locations (Left), İspark Locations (Right) are Presented.

latitude and longitude information. The primary module used for implementing artificial intelligence methods and recording results was Scikit-Learn [17]. The Matplotlib module [18] was utilized to visualize comparative graphs of the results, while the Graphviz module [19] was used for visualizing decision trees and artificial neural networks. The preference for using these modules was due to their BSD licensing, which guarantees fast and effective results.

The use of artificial intelligence methods such as K-Nearest Neighbors (K-NN), Decision Trees, and Artificial Neural Networks for Machine Learning is of significant importance for this study. These three methods were employed to determine which would yield the best results from the datasets. The study was concluded using the method that provided the most accurate outcome. In this context, a comparative analysis was conducted for traffic congestion detection, and the same method was applied to estimate İspark capacity fullness. This methodology highlights the comparative approach of the study, ensuring the selected AI techniques are robust and capable of providing reliable predictions for both traffic and parking lot occupancy.

K-NN is a machine learning algorithm that is utilized for solving classification or regression problems [20]. The K-NN algorithm classifies or makes predictions for a new data point by considering the nearest k other data points (selected from the training data). The classes or values of these nearest neighbors are used to predict the class or value of the new data point. The algorithm calculates the distance from each point in the dataset to another and determines the nearest k neighbors based on these distances. K is a hyperparameter that dictates the number of neighbors to consider, and its selection can significantly impact the algorithm's performance. Larger values of K result in less overfitting and more resistance to noise, while smaller values of K may be more prone to overfitting. Although K-NN is a straightforward algorithm, its computational cost increases with the size of the data. The performance of K-NN also depends on factors such as the scale, dimensionality, and noise level of the data [20].

Decision trees are one of many machine learning algorithms and are used to solve data classification or regression problems [21]. They segment the dataset into branches under certain conditions based on the features, creating a tree structure. The decision tree begins at a root node and conducts a test based on an attribute, continuing down the branch according to the outcome. Each branch is segmented according to an attribute and its condition. As a result, the decision tree divides the dataset into smaller, more homogenous subsets, which can then be used for classification or prediction. Decision trees are popular because they are easy to understand and interpret, making them widely used in classification or regression problems. They are also faster compared to other algorithms and can handle large datasets. Training decision trees involves using a training dataset that includes the features of the dataset and the target classes. This training data is used to construct the tree, and the tree learns to determine the conditions that best separate the features and target classes of the training data. The tree can be validated using a separate test dataset to measure its performance on the dataset [21]. An example of how decision trees work in prediction is provided in Fig. 3.

Artificial Neural Networks are inspired by the functioning of biological neurons [22]. This model consists of an input layer, one or more hidden layers, and an output layer. Each layer is made up of units called neurons. Each neuron generates an output by taking a weighted sum of the input features and applying an activation function. In this study, the MLPRegressor model from the sklearn module [17] was used. This model utilizes a feedforward ANN, which is a type of Multilayer Perceptron. In a feedforward ANN, each layer is interconnected, processing outputs from the previous layer to produce an output in the final layer. The model training is conducted using a training dataset that contains features and target outputs. The model learns to best match the input features to the target outputs. During this process, parameters such as weights and biases are

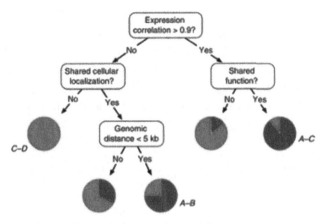

Fig. 3. Example Image for Decision Trees [21].

automatically adjusted. It is noted that this model can perform particularly well on noisy or complex datasets [17].

4 Results

4.1 Traffic Congestion Prediction

After preprocessing the 2022 Daily Vehicle Count dataset, traffic congestion predictions were made using K-NN, Decision Trees, and Artificial Neural Networks methods. The process began with the application of K-NN. The vehicle counts and date columns of the dataset were identified as independent variables, with vehicle count serving as the dependent variable. The dataset was then split into training and test sets, with 80% of the data used for training and 20% for testing. Subsequently, the K-NN Regression model was created and employed using the nearest 5 neighbors. The results yielded an MAE (Mean Absolute Error) of 13192.87, MSE (Mean Squared Error) of 499,854,115.13, and an R2 score of 0.8216. Additionally, a graph comparing the predicted values with the actual values was evaluated and presented in Fig. 4 (left). Although this method achieved relatively good results, other methods were also assessed. Upon examining the graph and results, it was observed that while the model produced accurate predictions for low traffic volumes, it deviated at higher traffic densities.

The dataset was re-analyzed using decision trees, similar to the K-NN approach. The data was divided into training and testing sets to ensure comparable performance evaluation with the other methods. The decision tree model exhibited better performance than our KNN model. The Mean Absolute Error (MAE) for the decision tree model was

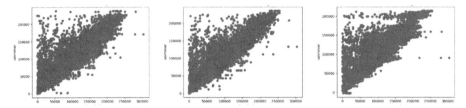

Fig. 4. On the left: K-NN Prediction Results Graph, in the middle: Decision Tree Prediction Graph, and on the right: Prediction Graph obtained with Artificial Neural Networks.

measured at 12130, the Mean Squared Error (MSE) at 428844950, and the coefficient of determination (R2 Score) at 0.8469. These results indicate that the model's predictions are closer to the actual values. The prediction graph for the decision tree model can be found in the middle of Fig. 5.

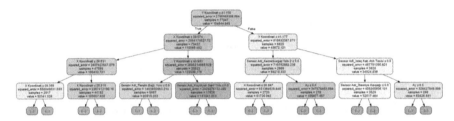

Fig. 5. A Small Section of The Decision Tree.

The dataset was similarly tested with Artificial Neural Networks, with 80% of the data used for training and 20% for testing. The neural network comprised three hidden layers, each containing 100 neurons, and the number of iterations set for the model was 500. The results obtained are as follows: the Mean Absolute Error (MAE) value is 15746 vehicles. This indicates that each prediction in the test dataset, on average, had an error of about 15746 vehicles. The Mean Squared Error (MSE) value is considerably high, suggesting that the average of the squared errors is higher due to larger individual errors. The R2 Score, which indicates the explanatory power of the model, is 0.79. This value suggests that our model has sufficient explanatory power to predict traffic congestion.

Table 1. Comparison of Artificial Intelligence Methods Used in Traffic Prediction

Model Type	MAE	MSE	R2
Artificial Neural Networks	15746.51	561492883.02	0.7996
K-NN	13192.87	499854115.13	0.8216
Decision Tree	12130.87	428844950.37	0.8470

Due to the objective of predicting traffic congestion in this part of the study, Decision Trees were chosen based on the outcomes presented in Table. 1. The rationale behind this selection is the necessity to opt for the method with the lowest MAE and MSE values, along with the highest R2 score. In this context, the trained data was used to generate the vehicle count prediction graph, which is provided in Fig. 6. Following the completion of this phase, the study progresses to the stage of predicting parking lot occupancy.

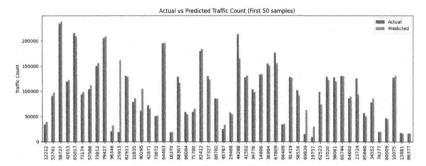

Fig. 6. Actual and Predicted Values Obtained with the Decision Tree Model.

4.2 Prediction of Parking Lot Occupancy Based on Traffic Congestion

After predicting traffic congestion, a comparison was made between the latitude and longitude information in the İspark locations dataset and the sensor locations in the 2022 Daily Vehicle Count dataset. Subsequently, the nearest İspark points were matched with traffic congestion data to create a new dataset. In this new dataset, the data was divided into 70% for training and 30% for testing. A model was then constructed using decision trees, and 5-fold cross-validation was performed to optimize the hyperparameters of this model. The hyperparameters included the depth of the decision tree and the node weight, and the "Grid Search" method was used for their optimization. Grid search tries to find the best values of hyperparameters by testing all possible combinations. The performance of the model resulted in an MSE of 18188.659, an MSE of 762461769.75, and an R2 value of 0.72. This indicates that the model's overall structure works with an accuracy of approximately 72.75%. The prediction results of the corresponding capacities are presented in Fig. 7.

Given the objective of route planning in this study, the results of İspark parking lot occupancy rates, based on traffic sensor locations, have been transferred onto a map. The map provided in Fig. 8 will be evaluated in the final part of the study. In the map, red points indicate areas where parking lots are likely to be full, while orange points suggest that the parking lots may be 50% occupied. Parking capacities that are less than 50% full are marked in green, but due to traffic congestion, a green point is only found in the Istanbul Airport area.

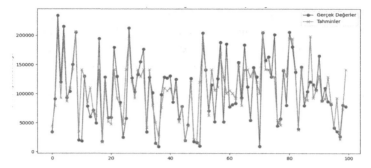

Fig. 7. İspark Capacity Occupancy Training Result Data.

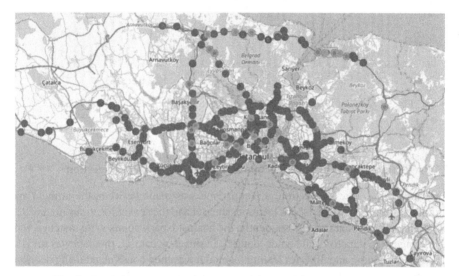

Fig. 8. Predictions of Parking Lot Occupancy Based on Traffic Congestion.

5 Conclusion

While numerous approaches have been explored to address the growing traffic problems in metropolitan cities, it is widely acknowledged that these solutions only decelerate the progression of the issue rather than completely resolving it. This study introduces a comprehensive model within the smart cities' framework, designed to alleviate traffic congestion. It achieves this by accurately predicting the capacities of parking lots at various destinations using historical data. The model encompasses an in-depth analysis of two datasets obtained from the IBB Open Data Portal. A series of meticulous preprocessing steps, coupled with advanced artificial intelligence techniques, were employed to pinpoint the most effective method for data analysis. This rigorous process was aimed at producing reliable and conclusive results. Throughout the study, there was a strong emphasis on maintaining uniformity in the parameters of the methods applied to the datasets. This consistency was vital to ensure an unbiased and fair evaluation of all the methods under consideration. Among the various techniques tested, it was observed that decision trees stood out, offering the most suitable and accurate results for the datasets used. The study achieved a notable level of precision, concluding with an impressive 72.75% accuracy rate in predicting the occupancy of parking lots. This was based on an analysis of prevailing traffic congestion patterns, demonstrating the model's efficacy in providing practical solutions to traffic management in densely populated urban areas.

The study's accuracy, though ideal, may not be entirely sufficient. Surprisingly, decision tree regression outperformed other machine learning methods. The results suggest higher accuracy at less congested sensor points and lower effectiveness in areas with more traffic. Exploring various artificial intelligence methods could enhance the study's accuracy.

The İspark Location dataset used in the study only represents paid parking areas owned by the IBB. The absence of comprehensive data on other parking options such as free parking areas, shopping mall parking's, and private parking lots, as well as parking facilities at individuals' residences or workplaces, is a limitation of the study. This limitation led to the assumption that all vehicles would park in İspark facilities for the purpose of this research. This aspect can be considered a weakness of the study. However, determining the total number of parking spaces in Istanbul, including all potential parking areas, is a challenging task that was not addressed in this research and is difficult to ascertain due to data availability constraints. Therefore, making predictions based on the available data was the most feasible approach, and the proposed system accomplishes this by estimating parking occupancy based on traffic congestion patterns.

The study acknowledges that each vehicle in traffic can move between locations, leading to potential double-counting in traffic data. To address this, the study used the average vehicle counts for model training. Despite this limitation, the location-based daily vehicle count data proved effective for predicting traffic congestion, demonstrating the model's applicability in real-world scenarios.

During dataset preprocessing, several errors emerged, such as duplicate counts for the same location under different names and inconsistent listings for İspark parking facilities. Despite the IBB Open Data Portal's usefulness, it highlighted the necessity for rigorous data checking and preprocessing. This process is vital to eliminate empty entries and duplicates, ensuring data integrity and reliability.

The occupancy of parking lots may not solely depend on traffic congestion. In some cases, vehicles may remain in a parking lot for more than a day. Therefore, this study was limited to the scope of traffic congestion only. Due to the unavailability of other variable data, traffic congestion was the primary focus. While reports from the IBB contain information about İspark's revenues, they do not provide data on the number of vehicles in a parking lot at any given time throughout the day. Attempts were made to research this information, but due to its unavailability, it could not be included in the study.

The study effectively determined the best artificial intelligence method for the datasets, thoroughly preprocessed the data, and developed a parking lot occupancy prediction model. It sets a benchmark in early research in this area. Future research could integrate systems like Google Maps API or OpenStreetMap for route optimization using the Dijkstra algorithm, marking this research as a foundational step for similar future studies. In conclusion, having an ample number of parking spaces in metropolitan cities and directing individuals to these parking lots can prevent improper parking of vehicles and, to some extent, reduce traffic congestion. Similarly, within the scope of smart cities, residents of metropolises can enhance their quality of life by predicting in advance how full parking lots in their desired destination might be, based on traffic congestion. This study underscores the potential of using data-driven approaches and AI to improve urban mobility and parking management, contributing to the broader goal of enhancing urban living in smart cities.

References

1. Gandhi, B.K., Rao, M.K.: A prototype for IoT based car parking management system for smart cities. Indian J. Sci. Technol. **9**(17), 1–6 (2016)
2. Ji, Z., Ganchev, I., O'Droma, M., Zhao, L., Zhang, X.: A cloud-based car parking middleware for IoT-based smart cities: design and implementation. Sensors **14**(12), 22372–22393 (2014)
3. Sadhukhan, P.: An IoT-based E-parking system for smart cities. In: 2017 International Conference on Advances in Computing, Communications, and Informatics (ICACCI), pp. 1062–1066. IEEE (September 2017)
4. Hans, V., Sethi, P.S., Kinra, J.: An approach to IoT based car parking and reservation system on Cloud. In: 2015 International Conference on Green Computing and Internet of Things (ICGCIoT), pp. 352–354. IEEE (October 2015)
5. Shen, T., Hong, Y., Thompson, M.M., Liu, J., Huo, X., Wu, L.: How does parking availability interplay with the land use and affect traffic congestion in urban areas? The case study of Xi'an. China. Sustain. Cities Soc. **57**, 102126 (2020)
6. Dogaroglu, B., Caliskanelli, S.P.: Investigation of car park preference by intelligent system guidance. Res. Transp. Bus. Manag. **37**, 100567 (2020)
7. Tu, S., Ayaz, M., Arshad, A., Iftikhar, U., Harrath, Y., Waqas, M.: Cloud-based smart parking system using internet of things. In: 2023 International Wireless Communications and Mobile Computing (IWCMC), pp. 1377–1382. IEEE (June 2023)
8. Abdelhamid, S., Jabeur, K.: Intelligent traffic congestion prediction model for VANET-based smart cities. Procedia Comput. Sci. **110**, 408–415 (2017)
9. Barker, J., ur Rehman, S.: Investigating the use of machine learning for smart parking applications. In: 2019 11th International Conference on Knowledge and Systems Engineering (KSE), pp. 1–5. IEEE (October 2019)
10. Chou, C.H., Chen, C.Y.: A decision tree-based parking space availability prediction system. In: 2016 10th International Conference on Innovative Mobile and Internet Services in Ubiquitous Computing (IMIS), pp. 15–19 (2016)
11. Li, M., Liu, Y., Dai, Y.: A parking space availability prediction algorithm based on K-nearest neighbor regression. In: 2019 IEEE 3rd Information Technology, Networking, Electronic and Automation Control Conference (ITNEC), pp. 1538–1542 (2019)
12. Hussein, M., Hammad, M.: A machine learning approach for predicting parking space availability. Int. J. Comput. Sci. Softw. Eng. **6**(12), 246–252 (2017)
13. Istanbul Metropolitan Municipality (IBB): Daily Vehicle Count for the Year 2022 (2022). https://data.ibb.gov.tr/dataset/gunluk-arac-sayimi/resource/5358a86b-93b0-4e85-bbc8-95780b4515ca Accessed 7 May 2023
14. Istanbul Metropolitan Municipality (IBB): İSPARK Locations (2023). https://data.ibb.gov.tr/dataset/ispark-otopark-bilgileri/resource/f4f56e58-5210-4f17-b852-effe356a890c. Accessed 7 May 2023

15. Pandas: Documentation (2023). htttps://pandas.pydata.org/. Accessed 7 May 2023
16. Story, R.: Folium (2013). https://python-visualization.github.io/folium/. Accessed 7 May 2023
17. Scikit-Learn: sklearn.neural_network.MLPRegressor (2023). https://scikit-learn.org/stable/. Accessed 7 May 2023
18. Matplotlib: Matplotlib (2023). https://matplotlib.org/. Accessed 7 May 2023
19. The Graphviz: The Graphviz Documentation (2023). https://graphviz.org/. Accessed 7 May 2023
20. Ali, M., Jung, L.T., Abdel-Aty, A.-H., Abubakar, M.Y., Elhoseny, M., Ali, I.: Semantic-k-NN algorithm: an enhanced version of traditional k-NN algorithm. Expert Syst. Appl. **151**, 113374 (2020). https://doi.org/10.1016/j.eswa.2020.113374
21. Kingsford, C., Salzberg, S.L.: What are decision trees. Nat. Biotechnol. **26**(9), 1011–1013 (2008). https://doi.org/10.1038/nbt0908-1011
22. Picton, P., Picton, P.: What is a Neural Network? Macmillan Education UK, pp. 1–12 (1994)

A Framework for Simulating the Optimal Allocation of Shared E-Scooters

Halil Ibrahim Ayaz(✉) and Bilal Ervural

Necmettin Erbakan University, 42090 Konya, Turkey
{hiayaz,bervural}@erbakan.edu.tr

Abstract. The rising popularity of shared e-scooters has created a need for an efficient way to match riders with available vehicles in urban areas. This study proposes a new approach that uses simulation to tackle this challenge. Unlike previous research that has focused on allocating infrastructure, our study prioritizes optimizing the allocation of e-scooters to riders, which is an essential gap in the literature. We conducted simulation experiments using three scenarios, considering factors such as total travel time, rider walking distance, and total distance traveled by vehicles. Our findings provide insights into the optimal allocation method for shared e-scooters, which can help urban planners, transportation authorities, and shared mobility service providers make informed decisions to improve rider satisfaction, minimize environmental impact, and ensure operational efficiency in modern urban environments.

Keywords: Shared E-Scooters · Simulation-Based Optimization · Urban Mobility

1 Introduction

Shared electric scooters, often referred to as e-scooters, have gained significant popularity as an urban transportation mode in recent years [1, 2]. These compact, electric-powered scooters are typically made available for short-term rental through a network of docking stations or app-based systems [3]. Users can easily locate, unlock, and ride these scooters for short distances within urban areas.

Several factors contribute to the growing popularity of shared e-scooters [4, 5]. First, they offer a convenient and eco-friendly alternative for short trips, reducing the reliance on traditional modes of transportation and contributing to a greener urban environment [6, 7]. Second, the accessibility and ease of use provided by smartphone apps make them a convenient option for last-mile connectivity, bridging the gap between public transportation and final destinations [8–10]. Additionally, the cost-effectiveness and flexibility of e-scooter rentals appeal to a broad demographic, from commuters to tourists, making them an attractive option for urban mobility [11]. However, the increasing use of shared e-scooters has also raised concerns about safety, regulation, and integration with existing transportation systems [12]. As cities continue to grapple with these challenges, the popularity of shared e-scooters underscores the ongoing evolution of urban transportation and the demand for innovative, sustainable solutions [13].

While existing literature extensively covers various aspects of shared mobility, such as ridesharing and bicycle sharing [5, 14, 15], scant attention has been given to the critical issue of optimizing rider-vehicle matching within the context of e-scooter sharing systems. A review of studies in the literature shows that they generally focus on the optimal allocation or assignment of e-scooters to charging stations [16, 17], optimal route selection [18], or the positioning of charging stations [19]. To our knowledge, there are no studies on allocating scooters to riders. This study aims to bridge this gap by introducing an innovative approach grounded in simulation-based modeling. By shifting a focus from infrastructure allocation to rider-scooter matching, our research offers a fresh perspective on enhancing the efficiency and effectiveness of shared e-scooter systems. In this study, we focus on developing an allocation model, or in other words, a recommendation system, to guide e-scooter users to the most suitable scooter for their needs. Through comprehensive analysis and evaluation of three distinct scenarios, we provide valuable insights into the multifaceted considerations involved in optimizing the shared mobility experience. Our study not only fills a significant void in the existing literature but also lays the groundwork for informed decision-making by urban planners, transportation authorities, and shared mobility service providers. We m to contribute to the discussion on sustainable urban transportation and promote a more efficient and eco-friendly mobility ecosystem for all stakeholders by providing practical suggestions for optimizing the process of allocating riders with vehicles.

The rest of this paper is structured as follows. Section 2 presents an extensive literature review. Section 3 introduces the simulation model and scenarios used in this study. Section 4 describes the empirical research used as a case study in detail, and the insights for each plan are discussed along with the analytical results. The study concludes with a discussion in Sect. 5.

2 Literature Review

There is a considerable body of literature on planning and optimizing electric scooters. However, as stated in the introduction, no study directly presents a model for driver-scooter matching. This section addresses the literature in two parts: studies conducted on e-scooters and related studies using simulation models for assignment problems.

Firstly, we examined studies that specifically investigate the operational dynamics and challenges associated with electric scooters. These works encompass a spectrum of topics, ranging from ridership patterns and urban mobility dynamics to charging infrastructure optimization. By examining these studies, we aim to compile insights that contribute to the overarching goal of refining the driver-scooter matching process. Masoud et al. [17] aimed to find the optimal tours for all chargers to pick up e-scooters in the form of routes, such that each route contains one charger, and each e-scooter is visited only once by the set of routes. The study developed a mathematical model for assigning e-scooters to freelance chargers and adapted a simulated annealing metaheuristic to determine a near-optimal solution. Kim et al. [20] introduced a rebalancing algorithm to address undermining the aesthetics of the city and obstructing the passage of pedestrians and put forth an efficient operational strategy. A comprehensive relocation was executed to align the demand and supply, optimize operations, and minimize the surplus of shared e-scooters. The optimal rebalancing algorithm, tailored to the characteristics of e-scooters, was formulated using genetic algorithms and then implemented in real-world scenarios. Masoud et al. [21] proposed a mixed integer linear programming model to solve the e-scooter-chargers allocation problem. The proposed model allocates the e-scooters to the chargers, emphasizing minimizing the chargers' average traveled distance to collect the e-scooters. The study by Gao et al. [22] addressed the matching problem in the community ride-sharing setting, where drivers and riders have distinct personal preferences for their matched counterparts. Gao et al. [14] introduced a comprehensive dispatching framework designed to match drivers with riders in ride-hailing systems. The objective is to formulate matching solutions that optimize social welfare, benefiting drivers and riders. The overarching aim is to ensure the sustainable growth of the ride-hailing system. Karamanis et al. [15] devised a novel algorithm for shared-ride assignment and pricing using combinatorial double auctions. Lin et al. [23] constructed an optimized allocation model for grid-based scooter battery swapping stations. This model addresses the challenge associated with deploying battery-operated scooters. Zakhem and Smith-Colin [24] introduced a tool crafted to assist shared e-scooter riders in navigating from a specified starting point to a designated destination along routes that adhere to city regulations and ordinances, ensuring safety and convenience. Additionally, the tool aids in identifying the closest shared micro-mobility parking facility near the rider's destination and issues alerts to avoid prohibited infrastructures. With advances in battery and charging technology, the strategic placement of charging stations to maximize the use of e-scooters poses a significant challenge. Akova et al. [25] focused on identifying optimal locations for cost-effective charging stations for e-scooters, emphasizing the critical importance of precise energy consumption calculations. To address this, they presented a mixed integer linear programming formulation. A similar study is centered on identifying optimal locations for electric scooter charging stations, specifically focusing on sustainability, an integral aspect of smart city initiatives [26]. A novel three-stage Pythagorean fuzzy group decision-making methodology is introduced to address this challenge. Initially, experts are assessed, accounting for variations in their experience and knowledge level. Subsequently, criteria are established based on expert opinions and literature, and their weights are computed using the Pythagorean Fuzzy Stepwise Weight Assessment Ratio Analysis (PF-SWARA) method. Finally, the Pythagorean Fuzzy Combinative Distance-based Assessment (PF-CODAS) is proposed to evaluate alternative locations. A practical case study is conducted in Istanbul, providing both policy and managerial implications for decision-makers.

In addition to the direct studies into electric scooters, we explored related studies that employ simulation models to address assignment problems in various contexts. These simulations may not be explicitly tailored to electric scooters, but their methodologies and outcomes could offer valuable perspectives and methods for developing a robust matching model. By drawing parallels with these broader simulation studies, we aim to extract transferable principles and techniques that can be adapted to the unique challenges posed by the electric scooter ecosystem. Sebastiani et al. [16] proposed an optimized method for allocating electric charging stations based on a simplified traffic model for urban mobility and vehicles' energy consumption. For this purpose, they introduced a discrete event simulation created with Arena software. They used a simulation model because it considers stochastic information that is difficult to characterize for particular scenarios. Muñoz-Villamizar et al. [27] aimed to optimize private transportation services by defining the optimal fleet allocation to routes. In this context, they proposed a simulation-based optimization model to allocate vehicles that minimize user waiting times. Mclean et al. [19] introduced a simulation model to explore design issues and management strategies for a shared micro-mobility program based on e-scooters. The detailed workload model, developed by analyzing empirical e-scooter data, encompasses e-scooter trip characteristics, trip volume, and geospatial distribution within the downtown area. The subsequent application of the simulation model allows for experiments that examine the performance of the e-scooter service, considering factors such as e-scooter fleet size, parking infrastructure location and capacity, and battery re-charging infrastructure. In another study using simulation. Lebeau et al. [28] aimed to evaluate the difference in performance between a center using diesel fuel and a center using electric vehicles. For this purpose, the operations of an urban distribution center were modeled in a discrete event simulation, and different scenarios were evaluated. Rodriguez et al. [29] developed a model using a simulation-optimization approach to improve emergency response by accurately positioning fire stations and optimizing the assignment of vehicles. Li et al. [30] introduced a vehicle relocation strategy for Shared Automated Electric Vehicles (SAEVs) utilizing an agent-based simulation modeling approach.

Agent-based models (ABMs) simulate traffic operations in large urban areas and road networks. These models integrate into existing frameworks to simulate the performance and effects of innovative transport modes, including shared autonomous vehicles, demand-responsive buses, electric scooters, and electric taxis. Tzouras et al. [31] present a detailed review and a qualitative evaluation to analyze the current ABMs and bridge the gap between ABMs and e-scooter-sharing services.

Through a comprehensive exploration of the existing literature, we identify a significant gap in research: while shared mobility has been extensively covered, with a focus on ridesharing, little attention has been paid to the critical issue of optimizing the allocation of riders to vehicles in the context of e-scooter sharing systems. Existing studies typically concentrate on the optimal allocation or assignment of e-scooters to charging stations, the selection of optimal routes, or the placement of charging stations. As far as we know, no research has been conducted on allocating scooters to riders. Our study seeks to fill this gap by proposing an innovative approach based on simulation-based modeling.

3 Simulation Model

In this section, we detail the general framework and performance metrics that underpin our simulation model. Building a robust simulation model is crucial to gaining valuable insights into the complexity of the driver-scooter matching process. By explaining our model's complexity, we aim to provide a comprehensive understanding of its architecture and the criteria by which its performance will be evaluated.

3.1 Simulation Framework

The simulation model diagram for the considered system is depicted in Fig. 1. The diagram illustrates the system's architecture and outlines the flow for the scenarios addressed in the study. The primary focus of the investigation is rider-scooter matching, and three distinct scenarios are explored, each highlighting different scooter recommendation strategies. These scenarios emphasize proximity, battery charge, and overall travel time as critical factors in the decision-making process. The selection of optimal alternatives among these scenarios depends on various factors, including riders' specific priorities and preferences, as well as the operational goals of the scooter-sharing service. Riders may prioritize quick access, reliable battery levels, or an efficient overall journey based on their individual needs. Simultaneously, the service provider must balance these considerations to meet user expectations and operational efficiency.

In the first scenario, the recommendation system places a premium on proximity and availability, striving to offer riders the most convenient option. The system suggests the scooter nearest to the rider's current location is also available for immediate use. This approach is designed to facilitate a quick and efficient start to the rider's journey, aligning with the dynamic nature of urban mobility where prompt access to transportation is crucial.

Shifting the focus to the second scenario, the recommendation system centers on evaluating the battery status of available scooters in close proximity to the rider. The system suggests the scooter with the highest battery level among the options, aiming to enhance the rider's experience by mitigating the risk of the scooter running out of charge during the trip. This strategy addresses the common concern of range anxiety, providing riders with a more reliable and uninterrupted journey.

The third scenario introduces a comprehensive approach to the recommendation system by considering both scooter access time and travel time to the destination. The system calculates and combines the required time from accessing the scooter to reaching the rider's destination. By prioritizing the minimization of this combined time, the recommendation system aims to provide an efficient and time-effective solution for the rider's journey. This approach acknowledges that riders value quick access to scooters and prioritize a streamlined and expedited overall travel experience.

Each scenario reflects a distinct emphasis within the rider-scooter matching process. The first scenario prioritizes convenience and immediacy, the second addresses concerns about battery life, and the third takes a holistic approach by optimizing access and travel times. These scenarios cater to different rider preferences and operational goals, demonstrating the versatility of the recommendation system in accommodating a range of user needs in the context of e-scooter sharing.

3.2 Performance Metrics

The performance evaluation of the scenarios in Sect. 3.1 involves considering several key performance metrics. These metrics are designed to provide a comprehensive assessment of each scenario's effectiveness and efficiency. To calculate these metrics, rectilinear distance is used between rider and scooter and scooter to destination. A sample for locations of riders and scooters is given in Fig. 2.

A Framework for Simulating the Optimal Allocation of Shared e-Scooters

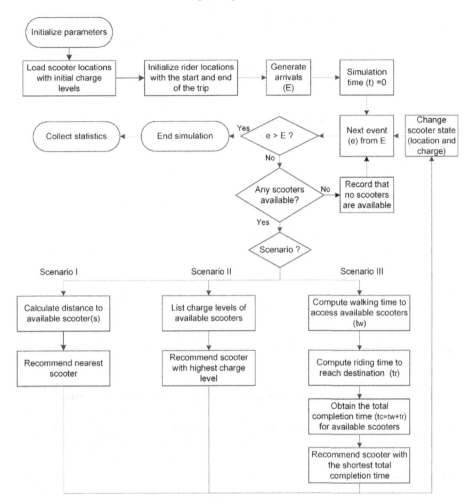

Fig. 1. Structure of simulation model

1. Average Scooter Access Time (Walking Time):

The time travelers take to access a suitable scooter is isolated and measured separately. This metric focuses on the efficiency of scooter availability and accessibility, shedding light on the convenience for users in acquiring a scooter. Walking time (WT) or access time to a scooter can be calculated as division of the total distance to the scooter (TDS) by the average walking speed (V_w) of a rider. It can be calculated using Eq. (1).

$$WT = \frac{TDS}{V_w} \tag{1}$$

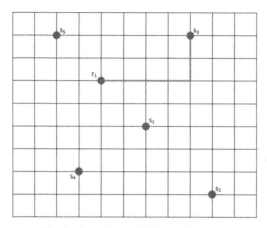

Fig. 2. Locations of riders and scooters

Walking time can be rewritten as Eq. (2) considering the coordinates of the rider and scooter.

$$WT = \frac{|x_r - x_s| + |y_r - y_s|}{V_w} \quad (2)$$

where the coordinates of the rider and the coordinates of the scooter.

2. Average Riding Time:

This metric explicitly captures the time spent riding the scooter to the destination. It provides insights into the efficiency of the transportation mode in terms of travel time from pick-up to drop-off. Similarly, riding time (RT) is calculated by dividing the distance between scooter location and destination (distance to scooter to destination, DSD) of riders by the average scooter speed (V_s). The formula is given in Eq. (3) to calculate riding time.

$$RT = \frac{DSD}{V_s} \quad (3)$$

Considering the coordinates of the scooter and destination, Eq. (3) can be rewritten as Eq. (4)

$$RT = \frac{|x_d - x_s| + |y_d - y_s|}{V_s} \quad (4)$$

where (x_d, y_d) coordinates of destination.

3. Total Travel Time (Completion Time):

This metric encompasses the entire journey, from the traveler reaching a suitable scooter to the driving time to the destination. It offers a holistic view of each scenario's overall time efficiency. Total travel time (TTT) is the summation of walking time (WT) and riding time (RT). It can be calculated using Eq. (5).

$$TTT = WT + RT \quad (5)$$

4. Percentage of Unserved Travelers:

This indicator quantifies the rate at which incoming demand is satisfied. It represents the percentage of travelers who cannot be served, highlighting the scenarios' ability to meet the overall demand and serve the users effectively. There are two reasons for unserved rides. The first one, walking time from rider location to destinations (distance of rider location to destination, DRD), is less than the total travel time. In this situation, riders do not prefer to use scooters. The formula for the number of unserved riders is given in Eq. (6).

$$U_{i,1} = \begin{cases} 1 & \frac{DRD}{V_s} \geq TTT \\ 0 & o.w. \end{cases} \quad (6)$$

where is whether the rider is unserved or not. On the other hand, riders may not find an available scooter when they demand to use one. This case is given in Eq. (7).

$$U_{i,2} = \begin{cases} 1 & \text{Available scooter not exist} \\ 0 & o.w. \end{cases} \quad (7)$$

The number of unserved riders can be calculated using Eq. (8).

$$\sum_{i=1}^{n} (U_{i,1} + U_{i,2}) \quad (8)$$

where n is the total number of riders.

5. Balance of Charge Utilization:

This metric involves analyzing the charging balance under each scenario. It provides insights into the effective utilization and distribution of scooter charging, contributing to the overall sustainability and reliability of the e-scooter sharing system.

By employing diverse performance metrics, the study aims to offer a nuanced understanding of each scenario's strengths and weaknesses. Considering both time-related metrics (total travel time, average scooter access time, average riding time) and demand-related metrics (percentage of unserved travelers) ensures a well-rounded assessment of the scenarios' impact on user experience and system efficiency. Additionally, the analysis of the balance of charge utilization introduces a sustainability perspective, reflecting the scenarios' impact on the long-term operational aspects of the e-scooter sharing system. Overall, this multi-faceted approach to performance evaluation contributes to a comprehensive and informed comparison of the scenarios presented in Sect. 3.1.

4 Results and Analysis

The empirical study described in this section provides a structured approach to analyzing and comparing the results of three scenarios in the context of an e-scooter-sharing system. Several key components of the study methodology are outlined below:

- Spatial Framework and Demand Generation

The study utilizes a 100 x 100 coordinate grid as the spatial framework, providing a structured environment for the simulation. The arrival of scooter demands ($r = 1,..., R$) is modeled using an exponential distribution of inter-arrival times, capturing the stochastic nature of real-world demand patterns. Initial e-scooter positions ($s = 1,..., S$) are randomly distributed across the grid, reflecting the variability in scooter availability and placement.

– Distance Measurement

Rectilinear distances are employed to measure travel distances between demand points and scooter locations. This choice of distance measurement is common in spatial analyses and helps simulate realistic travel scenarios.

– Experimental Factors:

The study conducts experiments at three different levels, varying the number of scooters to 25, 50, and 100. This multi-level approach allows for a comprehensive understanding of how system performance scales with the number of scooters.

– Computational Environment:

All computational experiments are conducted on a system featuring an Intel Core i7 processor and 16 GB RAM. This information is crucial for understanding the computational resources used in the simulation. MATLAB 9.1 is employed as the simulation tool, indicating the software framework used to implement the simulation model. MATLAB is widely used for simulation and numerical analysis, providing a robust platform for this type of study.

– Replication and Duration:

The pilot study involves 100 replications, each with a duration of 180 min. This replication process is essential for achieving robust statistical analysis by capturing variability and trends across multiple runs. The 180-min duration chosen for each replication aligns with a realistic timeframe, allowing the simulation to capture the dynamics of e-scooter demand and system performance over a meaningful period.

Overall, the study's methodological details provide transparency and reliability in the analysis. The use of a spatial framework, realistic demand generation, varying levels of scooter availability, and the employment of a well-established simulation tool contribute to the validity of the findings. Including multiple replications and a sufficiently long duration ensures that the results are statistically sound and representative of diverse scenarios, enhancing the study's overall credibility and applicability.

When the results of the experimental study given in Table 1 are analyzed, it is observed that to ensure customer satisfaction, scenarios 1 and 3 emerge as preferable from the rider's perspective. This is based on assigning the scooter closest to the rider in terms of average travel time and time to reach the e-scooter. However, Scenario 2 presents a superior plan due to a balanced utilization of scooters, ensuring a continuous scooter charging cycle.

In Scenario 1, prioritizing proximity and immediate availability demonstrates a commitment to providing riders with a convenient and timely solution. Quick access to a

Table 1. The results of the different scenarios are based on the performance criteria.

Performance metrics	Scenarios	# of scooters		
		25	50	100
Avg. Scooter access time	S1	1.53	1.41	1.08
	S2	1.69	1.64	1.44
	S3	1.51	1.40	1.11
Avg. Riding time	S1	2.62	2.64	2.62
	S2	2.62	2.65	2.62
	S3	2.57	2.57	2.52
Total completion time	S1	4.15	4.05	3.70
	S2	4.31	4.29	4.06
	S3	4.08	3.97	3.63
Unserved demand (%)	S1	39.7	23.9	3.9
	S2	39.2	23.2	3.5
	S3	41.1	25.3	5.1

nearby scooter aligns with the on-the-go nature of urban mobility and ensures an efficient start to the rider's journey. This strategy mainly benefits individuals seeking prompt transportation without prolonged waiting times. The results are shown in Table 1. Comparing the results of Scenario 1 with Scenario 3, it can be seen that Scenario 1 is highly competitive in time-related metrics, while it is better in unserved demand percentages (See Table 1). The positive correlation between the increase in the number of scooters and the decrease in both times and unserved demand rates further underscores the scalability and effectiveness of this scenario.

Scenario 2 introduces a nuanced perspective by incorporating battery charge levels into the recommendation process. This approach addresses the common concern of range anxiety by suggesting the scooter with the highest charge among the closest options. The emphasis on a fully charged vehicle enhances the reliability of the service, minimizing the risk of disruptions due to insufficient battery capacity during the ride. Despite a higher total completion time compared to other scenarios, Scenario 2 excels in minimizing unserved demand through a more balanced charging utilization, as depicted in Table 1. The box plot in Fig. 3 visually reinforces the superiority of Scenario 2 in achieving a charging balance, emphasizing its potential to provide a robust and dependable e-scooter-sharing service.

The comprehensive strategy presented in Scenario 3, considering both scooter access time and travel time to the destination, underscores the importance of efficiency in the overall journey. By optimizing the total travel time, this scenario aligns with the broader goal of shared mobility services to provide not just accessible but also time-effective transportation solutions. This approach offers a balance between convenience and time efficiency to riders who prioritize. The efficiency in completion time, as demonstrated in Table 1, positions Scenario 3 as a strong contender for providing an overall effective and efficient rider experience. However, it is essential to note that it falls behind in terms of the percentage of unserved demand, suggesting a trade-off between time efficiency and meeting all user demands.

Including a box plot (Fig. 3) further supports the findings, highlighting the differences in charge usage amounts. Figure 3 summarizes the results from 100 independent runs. It visually confirms that Scenario 2 surpasses the other scenarios in achieving a superior charging balance. This visual representation provides a clear insight into the effectiveness of Scenario 2 in managing scooter charging cycles, reinforcing its potential to enhance the overall reliability and availability of the e-scooter sharing system.

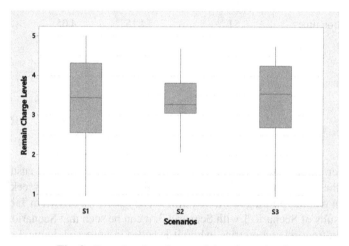

Fig. 3. Box-plot chart for remaining charge levels

The experimental study analyzed three scenarios to determine optimal approaches for ensuring customer satisfaction in scooter-sharing systems. Scenario 1, prioritizing proximity and immediate availability, proves highly competitive in time-related metrics and unserved demand percentages. Scenario 2 introduces a balanced utilization of scooters, emphasizing fully charged vehicles to address range anxiety and minimize disruptions. Although Scenario 2 has a higher total completion time, it provides a more balanced charging cycle. Scenario 3, considering both scooter access time and travel time, optimizes total travel time, aligning with the broader goal of shared mobility services. While Scenario 3 is the best for completion time, it lags in unserved demand percentages. The choice between these scenarios depends on whether riders prioritize immediate access, balanced charging, or overall travel efficiency.

5 Conclusion

In conclusion, this study has aimed to understand and optimize shared e-scooter systems by exploring three distinct assignment scenarios. By analyzing the implications of these scenarios on various performance metrics, including total travel time, rider's walking distance, and vehicle utilization, we have provided valuable insights into the challenges and opportunities associated with rider-scooter matching. Our findings underscore the importance of considering diverse factors in designing and implementing shared mobility systems to ensure customer satisfaction and operational efficiency. Moreover, by explaining the trade-offs between different assignment strategies, we have empowered stakeholders, including urban planners, transportation authorities, and shared mobility service providers, to make informed decisions that balance the competing objectives of immediacy, efficiency, and sustainability. Our research provides a relevant perspective on the current conversation around sustainable mobility, as e-scooters are becoming increasingly popular in urban areas. The study highlights the potential for simulation-based modeling to bring about positive changes in modern urban environments.

Furthermore, our study fills a significant gap in the literature by shifting the focus from infrastructure development to rider-scooter allocation, thereby contributing to a more comprehensive understanding of sustainable urban transportation. The practical implications of our findings are significant for urban planners, authorities, and service providers. By optimizing rider-scooter matching, stakeholders can enhance rider satisfaction, minimize environmental impact, and improve operational efficiency. This leads to more sustainable and accessible shared e-scooter systems, benefiting both users and the broader urban community. While the findings are valuable, we should consider a few factors as we interpret the results. The strategies analyzed in our research are limited in scope, and future studies could explore additional allocation strategies to optimize rider-scooter matching further. Additionally, future studies could incorporate real-world data to refine and validate the effectiveness of different allocation approaches. Lastly, additional performance metrics can be considered for future studies. For example, waiting times or arrival times of riders will be essential performance metrics.

References

1. Latinopoulos, C., Patrier, A., Sivakumar, A.: Planning for e-scooter use in metropolitan cities: A case study for Paris. Transp. Res. Part D Transp. Environ. **100**, 103037 (2021). https://doi.org/10.1016/j.trd.2021.103037
2. Vosooghi, R., Puchinger, J., Jankovic, M., Vouillon, A.: Shared autonomous vehicle simulation and service design. Transp. Res. Part C Emerg. Technol. **107**, 15–33 (2019). https://doi.org/10.1016/j.trc.2019.08.006
3. Bai, S., Jiao, J., Chen, Y., Guo, J.: The relationship between E-scooter travels and daily leisure activities in Austin, Texas. Transp. Res. Part D Transp. Environ. **95**, 102844 (2021). https://doi.org/10.1016/j.trd.2021.102844
4. Nikiforiadis, A., Paschalidis, E., Stamatiadis, N., Raptopoulou, A., Kostareli, A., Basbas, S.: Analysis of attitudes and engagement of shared e-scooter users. Transp. Res. Part D Transp. Environ. **94**, 102790 (2021). https://doi.org/10.1016/J.TRD.2021.102790
5. Lee, H., Baek, K., Chung, J.H., Kim, J.: Factors affecting heterogeneity in willingness to use e-scooter sharing services. Transp. Res. Part D Transp. Environ. **92**, 102751 (2021). https://doi.org/10.1016/j.trd.2021.102751
6. Badia, H., Jenelius, E.: Shared e-scooter micromobility: review of use patterns, perceptions and environmental impacts. Transp. Rev. **43**, 811–837 (2023). https://doi.org/10.1080/01441647.2023.2171500

7. Hosseinzadeh, A., Algomaiah, M., Kluger, R., Li, Z.: E-scooters and sustainability: investigating the relationship between the density of E-scooter trips and characteristics of sustainable urban development. Sustain. Cities Soc. **66**, 102624 (2021). https://doi.org/10.1016/j.scs.2020.102624
8. Weschke, J.: Scooting when the metro arrives — Estimating the impact of public transport stations on shared e-scooter demand. Transp. Res. Part A Policy Pract. **178**, 103868 (2023). https://doi.org/10.1016/j.tra.2023.103868
9. Guo, Z., Liu, J., Zhao, P., Li, A., Liu, X.: Spatiotemporal heterogeneity of the shared e-scooter–public transport relationships in Stockholm and Helsinki. Transp. Res. Part D Transp. Environ. **122**, 103880 (2023). https://doi.org/10.1016/j.trd.2023.103880
10. Aman, J.J.C., Smith-Colin, J., Zhang, W.: Listen to E-scooter riders: mining rider satisfaction factors from app store reviews. Transp. Res. Part D Transp. Environ. **95**, 102856 (2021). https://doi.org/10.1016/j.trd.2021.102856
11. Mitropoulos, L., Stavropoulou, E., Tzouras, P., Karolemeas, C., Kepaptsoglou, K.: E-scooter micromobility systems: review of attributes and impacts. Transp. Res. Interdiscip. Perspect. **21**, 100888 (2023). https://doi.org/10.1016/j.trip.2023.100888
12. McKenzie, G.: Spatiotemporal comparative analysis of scooter-share and bike-share usage patterns in Washington, D.C. J. Transp. Geogr. **78**, 19–28 (2019). https://doi.org/10.1016/J.JTRANGEO.2019.05.007
13. Gössling, S.: Integrating e-scooters in urban transportation: problems, policies, and the prospect of system change. Transp. Res. Part D Transp. Environ. **79**, 102230 (2020). https://doi.org/10.1016/j.trd.2020.102230
14. Gao, J., Li, X., Wang, C., Huang, X.: BM-DDPG: an integrated dispatching framework for ride-hailing systems. IEEE Trans. Intell. Transp. Syst. **23**, 11666–11676 (2022). https://doi.org/10.1109/TITS.2021.3106243
15. Karamanis, R., Anastasiadis, E., Angeloudis, P., Stettler, M.: Assignment and pricing of shared rides in ride-sourcing using combinatorial double auctions. IEEE Trans. Intell. Transp. Syst. **22**, 5648–5659 (2021). https://doi.org/10.1109/TITS.2020.2988356
16. Sebastiani, M.T., Luders, R., Fonseca, K.V.O.: Allocation of charging stations in an electric vehicle network using simulation optimization. In: Proceedings - Winter Simulation Conference 2015-Janua, pp. 1073–1083 (2015). https://doi.org/10.1109/WSC.2014.7019966
17. Masoud, M., Elhenawy, M., Liu, S.Q., Almannaa, M., Glaser, S., Alhajyaseen, W.: A simulated annealing for optimizing assignment of e-scooters to freelance chargers. Sustainability. **15**, 1869 (2023). https://doi.org/10.3390/su15031869
18. Zhang, W., Buehler, R., Broaddus, A., Sweeney, T.: What type of infrastructures do e-scooter riders prefer? A route choice model. Transp. Res. Part D Transp. Environ. **94**, 102761 (2021). https://doi.org/10.1016/j.trd.2021.102761
19. Mclean, R., Williamson, C., Kattan, L.: Simulation modeling of Urban E-Scooter mobility. In: 2021 29th International Symposium on Modeling, Analysis, and Simulation of Computer and Telecommunication Systems (MASCOTS), pp. 1–8. IEEE (2021). https://doi.org/10.1109/MASCOTS53633.2021.9614305
20. Kim, S., Lee, G., Choo, S.: Optimal rebalancing strategy for shared e-scooter using genetic algorithm. J. Adv. Transp. **2023**, 1–13 (2023). https://doi.org/10.1155/2023/2696651
21. Masoud, M., Elhenawy, M., Almannaa, M.H., Liu, S.Q., Glaser, S., Rakotonirainy, A.: Heuristic approaches to solve e-scooter assignment problem. IEEE Access. **7**, 175093–175105 (2019). https://doi.org/10.1109/ACCESS.2019.2957303
22. Gao, J., Wong, T., Selim, B., Wang, C.: VOMA: a privacy-preserving matching mechanism design for community ride-sharing. IEEE Trans. Intell. Transp. Syst. **23**, 23963–23975 (2022). https://doi.org/10.1109/TITS.2022.3197990

23. Lin, M.D., Liu, P.Y., Yang, M.D., Lin, Y.H.: Optimized allocation of scooter battery swapping station under demand uncertainty. Sustain. Cities Soc. **71**, 102963 (2021). https://doi.org/10.1016/j.scs.2021.102963
24. Zakhem, M., Smith-Colin, J.: An E-scooter route assignment framework to improve user safety, comfort and compliance with city rules and regulations. Transp. Res. Part A Policy Pract. **179**, 103930 (2024). https://doi.org/10.1016/J.TRA.2023.103930
25. Akova, H., Hulagu, S., Celikoglu, H.B.: Effects of energy consumption on cost optimal recharging station locations for e-scooters. In: 2021 7th International Conference on Models and Technologies for Intelligent Transportation Systems (MT-ITS) 2021, pp. 1–6 (2021). https://doi.org/10.1109/MT-ITS49943.2021.9529282
26. Ayyildiz, E.: A novel pythagorean fuzzy multi-criteria decision-making methodology for e-scooter charging station location-selection. Transp. Res. Part D Transp. Environ. **111**, 103459 (2022). https://doi.org/10.1016/j.trd.2022.103459
27. Muñoz-Villamizar, A., Montoya-Torres, J.R., Moreno-Camacho, C.A.: Simulation-based optimization approach for vehicle allocation in a private transport service: a case study. Manag. Sci. Lett. **9**, 193–204 (2019). https://doi.org/10.5267/j.msl.2018.12.003
28. Lebeau, P., Macharis, C., van Mierlo, J., Maes, G.: Implementing electric vehicles in urban distribution: a discrete event simulation. World Electr. Veh. J. **6**, 38–47 (2013). https://doi.org/10.3390/WEVJ6010038
29. Rodriguez, S.A., De la Fuente, R.A., Aguayo, M.M.: A simulation-optimization approach for the facility location and vehicle assignment problem for firefighters using a loosely coupled spatio-temporal arrival process. Comput. Ind. Eng. **157**, 107242 (2021). https://doi.org/10.1016/j.cie.2021.107242
30. Li, L., Lin, D.C., Pantelidis, T., Chow, J., Jabari, S.E.: An agent-based simulation for shared automated electric vehicles with vehicle relocation. In: 2019 IEEE Intelligent Transportation Systems Conference (ITSC), 2019. pp. 3308–3313 (2019). https://doi.org/10.1109/ITSC.2019.8917253
31. Tzouras, P.G., et al.: Agent-based models for simulating e-scooter sharing services: a review and a qualitative assessment. Int. J. Transp. Sci. Technol. **12**, 71–85 (2023). https://doi.org/10.1016/J.IJTST.2022.02.001

Performance Enhancement of Two-Way DF Relayed Cooperative NOMA Vehicular Network with Outdated CSI and Imperfect SIC

Potula Sravani[✉] and Ijjada Sreenivasa Rao

ECE Department, GITAM Deemed to be University, Visakhapatanam, AP, India
sravaniphd2020@gmail.com, sijjada@gitam.edu

Abstract. In the realm of 5G-enabled Intelligent Transportation networks, despite persistent unresolved challenges exploration of new technological paradigms is underway. Regardless of the new explorations in Vehicular communications, dynamic nature of vehicle networks and rapidly fluctuating channel conditions make it grim to achieve comprehensive recompenses. Many researchers had focused on mitigating the effects of fading channels, interference and throughput under dynamic channel conditions. Very less attention was made to address continuous time varying channels and its implications over the channel sate information. This work proposes a novel approach employing two-way half-duplex decode-and-forward relaying with Non-orthogonal multiple access to counteract the impact of outdated channel state information and imperfect successive interference cancellation resulting from dynamic channel behaviour. This is achieved through the utilization of Self-proclaiming relay selection and multi-phase transmission strategies. Explicit closed-form equations are developed and sum outage probabilities are analysed based on the location of decodable relay nodes between the source and destination vehicles. The performance of the proposed strategy is compared with two-way amplify-and-forward relaying and two-way decode and forward relaying with and without network coding. Furthermore, an analysis of the effect on Packet error rate is conducted across various sub transmission phases, while also examining the trade-offs associated with determining the optimal phase value during the data transmission phase and relay selection phase.

Keywords: Orthogonal Multiple Access · Non-Orthogonal Multiple Access · Imperfect Successive Interference Cancellation · Outdated Cannel State Information and Vehicular Networks

1 Introduction

Intelligent Transportation System (ITS) has been substituted by vehicle automations with the merge of new technological advances in communications and improvements in the Internet of Things (IOT) [1, 2]. One of the hard and interesting study areas in 5G communications is vehicular network communication involving moving vehicles, roadside equipment, and pedestrians having communication capabilities. The inherent dynamic character of vehicle communication owing to mobility, constantly changing channels, and irregular connection made system design more complicated [3–6]. Many obstacles still needed attention and just dimly understood [7–10].

The principles and advantages of cooperative relaying networks in conjunction with Non-Orthogonal Multiple Access (C-NOMA) has explored eclectic popularity in 5G and other wireless communications [11, 12]. This novel paradigm is being used to increase spectral efficiency, reliability, interference cancellation, linkage dependability, and network latency [13–15]. Orthogonal multiple access (OMA) has proved in limiting the fading effects and in-spite of it technological advantages it suffers with limitations of inadequate resources. NOMA implements non-orthogonal resource sharing, superposition coding, power domain multiplexing, and successive interference cancellation (SIC) [16–19]. In disparity to OMA, NOMA outpaces OMA in terms of accommodating many users, channel capacity and reduced complexity [20–22] based on information theory.

In automotive networks, intermediate vehicles in the networks are designated as cooperative relay nodes to dissipate the information to destination with their available resources. In most of the earlier related works and proposals, the same appraised channel response was exploited for selection of relay and information delivery. Nevertheless, in practical scenario due to dynamic behaviour of the time varying channels, it will be unrealistic to consider same channel response for choosing relay and data transmission. In [23], In this paper, we consider relay selection for turbo coded cooperative networks subject to Nakagami-m fading when the channel state information (CSI) is known at the receiver but is not necessarily ideal. A closed-form expression for the exact outage probability is derived as well as its asymptotic expression in the high signal-to-noise (SNR) regime. Moreover, upper bounds on the bit-error rate (BER) are presented and a study of the diversity order reveals that for ideal CSI, full diversity in the number of relays and fading parameters m is achieved as opposed to outdated CSI. The system performance of the proposed system is also investigated through the pairwise error probability (PEP) analysis. A closed-form expression of the PEP for all values of the fading parameter m. In [24] authors examined the deterioration in performance with obsolete CSI and presented an outage optimum relay technique to reduce the provisional probability for specified relay transmitted power. In [25, 26] A two-stage relay selection was investigated for cognitive radio scenarios that consists of primary user and secondary user. The limitation of decoding requirements and its complexity in DF, a relay is selected from a group of AF relays to forward the message to end users. The spatial randomness of the relays are examined in this work by implementing stochastic geometric tools. The impact of the power allocation was also studied in the work. The AF outperformed over DF in terms of the outage probability and ergodic rate. In [27], NOMA-AF relaying was investigated and analysed in a cooperative relay system. The evaluation of C-NOMA performance gain was validated using dedicated full-duplex relaying. The work did not consider the delay constraints, which effect the throughput and link reliability. The relay was selected based the intended data rates of the far end user. The work did not consider the relaying protocols and the mobility factor was missing. In [28] examined the outage behaviours of cooperative relaying protocols with NOMA for cell-edge users. The outage probability and ergodic rate of a two-way NOMA system for cellular communication were investigated in [29, 30]. A two-way relay non-orthogonal multiple access system was investigated, where users exchange messages with the aid of one HD-DF relay. Since the SINRs of NOMA signals mainly depend on effective successive interference cancellation (SIC) schemes, imperfect SIC (ipSIC) and perfect SIC (pSIC) were taken into account. In performance is characterized and derive closed-form expressions for both exact and asymptotic outage probabilities of NOMA users' signals with ipSIC/pSIC. Based on the derived results, the diversity order and throughput of the system are examined. They studied the ergodic rates of users' signals by providing the asymptotic analysis in high SNR regimes.

In [31, 32], authors projected the performance of C-NOMA with fixed DF relaying over the Nakagami fading with outage probability and aggregate rate. The relaying protocols were evaluated using power domain NOMA and network coding. In [33], the outage performance of two-way/one-way full-duplex/half-duplex fixed-gain AF relay systems are investigated over Rayleigh fading channels in the presence of residual loop-interference (LI). A unified end-to-end SINR expression is obtained. A new unified exact outage probability expression for all fixed-gain AF relay systems was derived in a closed-form by using cumulative distribution function approach. They projected that two-way full-duplex fixed-gain AF relay systems provides better outage performance as long as the quality of LI cancellation improves and/or the bandwidth efficiency increases. Where as in [34], a two-way relay with multi-relays were considered, with the joint decoding (JD), decoding by parts (DP), bit-level exclusive-or (XOR), and superposition coding (SC) schemes for the two-way DF. They focused on the outage performance analysis of the relay selection strategy using the aforementioned coding and decoding methods. Based on an approximate analysis of the outage performance in the high SNR regime the diversity-multiplexing trade-off for the considered decoding and coding schemes is evaluated.

Most of the previous works cited above considered either implicitly or explicitly block fading for their simplicity, but the fading effect will be constant during data transmission and fluctuates at the time of selection of relay. This variation in channel response will have severe effect in performance if not considered while designing. The bulk of vehicular network research has not engrossed on mobility effects under time-varying channel conditions. Hence dynamically changing channel status information should be considered while selecting relays and transmitting data, particularly in scenarios such as vehicle communication. In this work to mitigate the consequence of outdated CSI in relay selection and data transmission, the transmission process is divided into subphases and the relay selection is minimized on the metric of updated CSI estimation.

The remaining work is structured as follows. Section II presents two-way-DF C-NOMA relaying approach in Vehicular Networking. Section III presents the cluster formation and decodable relay selection with self-proclaiming approach. In section IV, the influence of a continuous time-varying channel model on propose model with NOMA-SIC approach and also projecting expression for packet error rate (PER). Section V projects closed form expressions for the outage probability at sum outage probability based on the location of selected decodable relay. presents simulation results assessing PER performance for different M values and analysis of sum outage probability based on location of decodable relay and average Signal to Noise Ratio are presented in Section VI. Summary of the proposed work is depicted in Section VII.

2 Network Modelling

Figure 1, portray network modelling of C-NOMA for vehicular networks. To increase the spectral efficiency, the proposed model is investigated with relay aided NOMA with two-way half-duplex approach with decode-and-forward relaying by assuming there is no line of sight between the vehicle nodes. The complete relaying procedure is split into two phases: multiple access (MA) phase and broadcasting phase (BC). The notations under consideration are articulated as reference in Table 1.

Table 1. Description of parameters with notations

Notation	Parameter
S	Source
D	Destination
x_1, x_2	Transmit information from S and D
r	Decodable relay vehicle
P_t	Transmit power
P_r	Retransmit power
P	Transmit power ($P_t = P_r = P$)
$h_{xy}(t)$	Time varying channel response between X and Y
g_{xy}	Channel Strength
σ_{xy}^2	Variance
α	Pathloss exponent
λ	Carrier wavelength
d_0	Reference distance
d_{xy}	Distance between X and Y
$F_{g_{ij}}(x)$	Cumulative Distributive Function
$f_{g_{ij}}(x)$	Probability Density Function
$d(r,k)$	Euclidean distance between vehicle r and k
K	Set of neighbor vehicles
R	Total number of vehicles in single-hop cluster
$|\Delta V|$	Vehicle velocity difference between k and r
r_a	Achievable data rate
R_t	Threshold data rate at decodable relay
$CH(r)$	Self-proclaiming cluster head
y_r	Received signal at CH(r)

(*continued*)

Table 1. (*continued*)

Notation	Parameter
η_r	Additive White Gaussian Noise (AWGN)
$h_{xy}(t_n) \approx h_{xy}^{(n)}$	Channel response of nth symbol from X and Y
$H_{xy}(f)$	Normalized Doppler Spectrum
f_{xy}	Frequency shift due to relative velocity between two nodes
ρ_r	Correlation coefficient
τ	Relay selection duration
$J_0(\cdot)$	Zero order Bessel function of first kind
SNR	Signal to Noise Ratio
SINR	Signal to Interference Noise Ratio
γ	Transmit SNR
N_0	Noise Variance
Δ_{MA}	Time ratio of the multiple access in first timeslot for information exchange
Δ_{BC}	Time ratio of the broadcast phase
K	Joint PDF of the channel strength
$R_t(r)$	Achievable threshold data rate at r
R_t	Threshold Achievable data rate at S and D
θ	Threshold SNR
σ_1, σ_2	Time varying channel response of $g_{Sr}^{(1)}$ and $g_{Dr}^{(1)}$
$\frac{E_b}{N_0}$	Energy per bit with noise variance

In MA phase the *S* and *D* vehicles transmit their x_1 and x_2 information to designated decodable relay vehicle (*r*). The *r* detects x_1 and x_2 through NOMA-SIC technique where Imperfect successive Interference Cancellation (ipSIC) is considered in our modelling. Though perfect SIC consideration in the design will be more idealistic, but practically SIC cannot completely mitigate the effects of interference, hence for realistic approach we considered ipSIC [21]. In order to analyse the proposed model with more practical scenarios, the imperfect SIC (ipSIC) condition is assumed. The transmit and retransmit powers are assumed to be similar i.e., $P_t = P_r = P$ and the channel distribution between *S*, *r* and *D* is assumed to be independent and identically distributed Nakagami-*m* fading channels. The time varying channel response from *S* to *r* and *D* to *r* is represented as $h_{Sr}(t)$ and $h_{rD}(t)$, and strength of the channel is exponentially distributed as $g_{xy} = |h_{xy}(t)|^2$ with an average value $\bar{g}_{xy} = \sigma_{xy}^2$. The variance can be expressed as:

$$\sigma_{xy}^2 = (\lambda/4\pi d_0)^2 (d_0/d_{xy})^\alpha \quad (1)$$

where, α: pathloss exponent, λ: carrier wavelength, d_0: reference distance and d_{xy}: distance between *X* and *Y*. $F_{g_{ij}}(x) = 1 - e^{-\sigma_j x}$ depicts Cumulative Distributive Function

(CDF) and $f_{g_{ij}}(x) = \sigma_j e^{-\sigma_j x}$ represents Probability Density Function (PDF) for the random variable, where $i \in \{1, 2\}$.

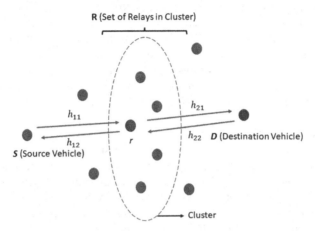

Fig. 1. Two-way C-NOMA Vehicular Network with decodable HD Relay.

The transmission phase under C-NOMA is presumed to be driven into M multiple phases to mitigate the effect of outdated CSI [23]. As depicted in Fig. 2, the sub-phase is further sub-split as relay selection phase and data forward phase. In relay selection phase the decodable relay is selected from the cluster based on the multiple metrics as specified in section III along with strongest channel response. This multiphase splitting will allow to mitigate the performance degradation of outdated CSI by allowing updation of CSI in each phase till considerable deviation in CSI correlation is observed between two subphases. This approach reduces the complexity of relay selection multiple times with considerable tolerance of deviation in CSI.

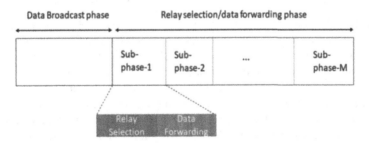

Fig. 2. Multiphase for Relay Selection and data transmission

3 Cluster Formation and Cluster Head Selection

We considered the effective cluster formation with the knowledge of location and direction of the vehicle's movement. The proposed algorithm uses the Euclidean distance and to further improve the stability, adaptivity is explored with cluster time span due to the limitations of IEEE802.11P DSRC (Dedicated short-range communication) coverage radio specifications. Each vehicle node broadcasts its present kinematic information like position, speed and direction of movement to create consciousness of its existence to all other vehicles in single-hop transmission. If the distance between intended vehicles is considerably high, multi-hop clustering can be considered which is out of this work scope.

With the aid of cooperative awareness message broadcasted by the vehicles, each vehicle with this aggregated information develops its single-hop neighbouring vehicles list. With this hysteresis and empirically deduced awareness message will help in getting the information of the vehicles density in the cluster dynamically and trigger for adaptive changes in reselection of Cluster Head (CH). Vehicle r can be self-assigned as CH, with the conditions of having maximum number of one hop neighbouring vehicle list (Neighbour Information Table, NIT), having lowest relative speed with other neighbouring vehicles present in its vicinity and minimum Euclidean distance from the other vehicles. The same can be articulated as

$$F(r) = \left(\sum_{k \in R} d(r,k)\right) \Big/ R + \left(\sum_{k \in R} |\Delta V|\right) \Big/ R - r + \Pr(r_a \geq R_t) \quad (2)$$

where, $d(r,k)$ signifies Euclidean distance between Vehicle r and k, R depicts the total number of vehicles in the single-hop cluster. $|\Delta V| = |v_k - v_r|$ explore vehicle velocity difference between k and r, r_a achievable data rate and R_t Threshold data rate at decodable relay.

The self-proclaiming Cluster Head (r) will be selected if it satisfies:

$$CH(r) = \langle r | F(r) = \min(F(k), \forall k \in R) \rangle \quad (3)$$

After the selection of decodable relay (r), it broadcasts beacon messages to the neighbours in its vicinity about its essence. Due to mobility and direction of movement of vehicles the number of vehicles in cluster vary dynamically. New vehicle can join the cluster only after receiving Join response from CH as an acknowledgment to its Join Request message. The implementation of the cluster selection can be described with a state diagram as presented in Fig. 3. In the Initial state every vehicle will receive beacon signals from all other neighbouring vehicles. The vehicle remains in same state till it completes In-TIMER period in which the aggregation of information and updation of NIT will be pursued. When the timer runs out, the vehicle in cluster enters the Selection state. If the CH Condition is met, it will immediately enter the CH state and begin sending its CH beacon messages. If a vehicle in the Selection state gets a CH beacon transmission, it will change to the other's state. The shortest number of connection hops till reaching the CH will be preferred. If not, it will return to the Selection State. The vehicle r will remain in CH state until the CH-TIMER is completed; if no cars are linked to it, it will return to Selection state.

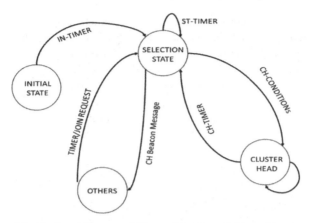

Fig. 3. State Diagram of Cluster Head Decodable Relay (r)

4 Continuous Time Varying Channel Modelling with NOMA

According to the system paradigm, the decodable relay r receives the signal from the source (S) and destination (D) similar to NOMA uplink. The selected r (CH) is presumed to be operating in HD decode-and forwarding (DF) protocol with C-NOMA-SIC scheme. It is presumed that S, r and D utilize comparable transmit power (P). y_r depicts the received signal at CH (r) and is expressed as:

$$y_r = \sqrt{P} h_{Sr}^{(1)} x_1 + \sqrt{P} h_{Dr}^{(1)} x_2 + \eta_r \qquad (4)$$

where x_1, x_2 represent transmitted information from S and D respectively and η_r symbolizes AWGN. Owing to the relative velocity between the V-V and V-I, the channel is assumed to be persistent inside symbol sequence and fluctuates amid symbols owing to minimal symbol interval for relay selection and data transmission phases. The channel response of the nth symbol from X to Y can be expressed as $h_{xy}(t_n) = h_{xy}^{(n)}$. Similarly, for N symbols the channel response of time-varying channels are articulated as $\left[h_{xy}^{(1)}, h_{xy}^{(2)}, h_{xy}^{(3)}, ..., h_{xy}^{(N)}\right] \left[g_{xy}^{(1)}, g_{xy}^{(2)}, g_{xy}^{(3)}, ..., g_{xy}^{(N)}\right]$ depicts square channel strength vector. The Doppler frequency response between any two nodes with considerable relative velocity is specified as $\sqrt{H_{xy}(f)}$, $H_{xy}(f)$ depicts is normalized Doppler spectrum. With Jakes channel model is represented as:

$$H_{xy}(f) = \frac{1}{\pi f_{xy} \sqrt{1 - (f/f_{xy})^2}} \qquad (5)$$

The frequency shift due to relative velocity between two nodes is denoted as f_{xy}. The dynamic time-varying channels persuade complexity and relay selection channel response will be obsolete to the channel response experienced during data transmission. It can be symbolized as $h_{rD}^{(0)}$ when compared to the first symbol channel response $h_{rD}^{(1)}$. Hence, $h_{rD}^{(1)}$ conditioned over $h_{rD}^{(0)}$ can be articulated in Gaussian distribution as:

$$h_{rD}^{(1)} | h_{rD}^{(0)} \sim CN\left(\rho_r h_{rD}^{(0)}, \left(1 - \rho_r^2\right)\sigma_{rD}^2\right) \qquad (6)$$

$\rho_r = J_0(2\pi f_{rD}\tau)$, ρ_r is correlation coefficient. τ depicts relay selection duration and $J_0(\bullet)$ signifies zero order Bessel function of first kind. For decoding the messages from S and D, the received SINR at the decodable relay is expressed as

$$SINR_{Sr}^{(1)} = \frac{P\left|h_{sr}^{(1)}\right|^2}{P\left|h_{Dr}^{(1)}\right|^2 + N_0} = \frac{\gamma g_{Sr}^{(1)}}{\gamma g_{Dr}^{(1)} + 1} \tag{7}$$

$$SINR_{Dr}^{(1)} = \frac{P\left|h_{Dr}^{(1)}\right|^2}{N_0} = \gamma g_{Dr}^{(1)} \tag{8}$$

wherever, γ symbolizes transmit SNR, $\gamma = \frac{P}{N_0}$.

After receiving signals from S and D, the r with the aid of NOMA, may decode the received information either with Joint Decoding, Phase detection or with SIC.

A. **Joint Decoding:** The decodable relay (r) employs Maximum likelihood (ML) criteria to jointly decode the two information signals modulated from the constellation space Q from the received signals as in Eq. (4).

$$(\hat{x}_1, \hat{x}_2) = \arg\min_{(x_1,x_2)\in Q}\left|y_r - \sqrt{P}h_{Sr}^{(1)}x_1 - \sqrt{P}h_{Dr}^{(1)}x_2\right|^2 \tag{9}$$

B. **Phase Detection:** In this scheme, the relay employs ML criteria for recovery of one information and other information is treated as interference.

$$\hat{x}_1 = \arg\min_{x_1\in Q}\left|y_r - \sqrt{P}h_{Sr}^{(1)}x_1\right|^2$$

The r employs SIC, $\hat{y}_r = y_r - \sqrt{P}h_{Sr}^{(1)}\hat{x}_1$. The remaining information $\hat{x}_2$ is extracted as:

$$\hat{x}_2 = \arg\min_{x_2\in Q}\left|\hat{y}_r - \sqrt{P}h_{Dr}^{(1)}x_2\right|^2$$

C. Applying the NOMA-SIC scheme, the r first decodes x_1 information by virtue of treating x_2 as interreference, hence SINR at r to detect x_1 is given by

$$SINR_{Sr\rightarrow x_1} = \frac{\gamma_{Sr}^{(1)}\left|h_{Sr}^{(1)}\right|^2}{\gamma_{Dr}^{(1)}\left|h_{Dr}^{(1)}\right|^2 + 1}$$

After decoding x_1 at r with NOMA-SIC approach, the received SINR at r to detect x_2 is articulated as:

$$SINR_{Dr\rightarrow x_2} = \frac{\gamma_{Dr}^{(1)}\left|h_{Dr}^{(1)}\right|^2}{\omega\gamma_{Dr}^{(1)}|g|^2 + \gamma_{Sr}^{(1)}\left|h_{Sr}^{(1)}\right|^2 + 1}$$

where ω represents imperfect SIC (ipSIC) factor and modelled as Rayleigh fading channel model signified as g.

In the next time slot, the S and D receive the superimposed information with the virtue of decodable relay (r) similar to the NOMA downlink transmission. For simplified analysis we initially assumed $P_t = P_r = P$, in this regard the power allocations coefficients in the downlink power domain NOMA remain same. In practical scenario based on the location of the S and D, the influence of dynamic power allocation coefficients on the performance are analysed intensively in many literature works. After extracting the demodulated information from S and D, r re-encodes and modulate the information sequence as x_r. Where, r uses Bit-Level Exclusive-OR ($x_r = x_1 \oplus x_2$) and retransmit to S an D. The received signal at S or D i.e., $i = \{S, D\}$ is expressed as:

$$y_i = \sqrt{P} h_{ri} x_r + n_i \tag{10}$$

The S and D estimates the received information $\hat{x}_r$ using ML detection as given in Eq. (11).

$$\hat{x}_r = \arg \min_{x_r \in Q} \left| y_i - \sqrt{P} h_{ri} x_r \right|^2 \tag{11}$$

The information can be extracted by performing Bit-Level Exclusive-OR ($x_r \oplus x_i$) at S and D. In other words, the S can extract the D information by performing Bit-Level-Exclusive-OR over superimposed information with its own transmitted information. In the first timeslot i.e., in MA phase the following two conditions must be gratified to decode correctly by the r:

$$\Delta_{MA} = \log_2 \left(1 + \frac{\gamma_{Sr}^{(1)} \left| h_{Sr}^{(1)} \right|^2}{\gamma_{Dr}^{(1)} \left| h_{Dr}^{(1)} \right|^2 + 1} \right) \geq R_t(r)$$

$$\Delta_{MA} = \log_2 \left(1 + \frac{\gamma_{Dr}^{(1)} \left| h_{Dr}^{(1)} \right|^2}{\omega \gamma_{Dr}^{(1)} |g|^2 + \gamma_{Sr}^{(1)} \left| h_{Sr}^{(1)} \right|^2 + 1} \right) \geq R_t(r)$$

where Δ_{MA} depicts the time ratio of the multiple access in first timeslot for information exchange and $R_t(r)$ represents achievable threshold data rate at r. Similarly in the broadcast phase, to extract the information at S and D, it should satisfy $\Delta_{BC} = \log_2 \left(1 + \gamma_{rS} |h_{rS}|^2 \right) \geq R_t$ and $\Delta_{BC} = \log_2 \left(1 + \gamma_{rD} |h_{rD}|^2 \right) \geq R_t$ respectively, where R_t depicts the threshold achievable data rate at S and D.

Packet Error rate: PER is a performance metric evaluated to find the percentage of the data that could not be retrieved due to channel impairments caused due to varying channel conditions.

$$PER = \int_0^{+\infty}\int_0^{+\infty}\int_0^{+\infty} \left\{1 - \Pi(1 - \overset{N}{\underset{n=1}{SER}}(x_n))\right\} \\ \times \underbrace{f_{g_{xy}^{(1)}, g_{xy}^{(2)}, \ldots, g_{xy}^{(N)}}(x_1, x_2, \ldots, x_N)}_{K} dx_1 dx_2 \ldots dx_N \quad (12)$$

where, K represent the joint PDF of the channel strength between any two nodes for various time intervals. In selection of relay with outdated CSI, PER is assessed with multiple integrals and joint PDF. The PER performance of outdated CSI in cooperative networks was projected intensively in many research works, but the PER performance of the outdated CSI for the proposed model is analyzed for the various M Phase values. From the simulations, it is observed that outdated CSI effect can be compensated with the M phase transmission and performance can be brought close to the block fading transmission by decreasing the multiple relay selections within short duration of data transmission. On the other side of observation, with linear rise of M value further leads to initiation of relay selection within short interval and may lead to further complexity in practical scenarios.

5 Outage Probability Based on Location of Decodable Relay

For the proposed work, two conditions of channel gain are assumed to calculate Outage probability at S and D to evaluate the influence of dynamic time-varying channels.

A. If $g_{Sr}^{(1)} > g_{Dr}^{(1)}$.

When the channel gain between the S and r is more than the D and r, and when the achievable data rates of both links are less than the threshold R_t, the Outage probability at node D is articulated as:

$$P_{out\_D^{(1)}} = \underbrace{\Pr\left[R_{Sr}^{(1)} < R_t\right]}_{a} + \underbrace{\Pr\left[R_{Sr}^{(1)} \geq R_t, R_{rD}^{(1)} < R_t\right]}_{b} \quad (13)$$

where, $R_{Sr}^{(1)} = \frac{1}{2}\log_2\left(1 + SNR_{Sr}^{(1)}\right) = \frac{1}{2}\log_2\left(1 + \frac{\gamma g_{Sr}^{(1)}}{\gamma g_{Dr}^{(1)} + 1}\right)$ and
$R_{rD}^{(1)} = \frac{1}{2}\log_2\left(1 + SNR_{rD}^{(1)}\right) = \frac{1}{2}\log_2\left(1 + \gamma g_{rD}^{(1)}\right).$
a in Eq. (13) is can be articulated as

$$a = \left[g_{Sr}^{(1)} > g_{Dr}^{(1)}, g_{Sr}^{(1)} < \theta g_{Dr}^{(1)} + \theta/\gamma\right]$$
$$= \Pr\left[g_{Dr}^{(1)} < g_{Sr}^{(1)} < \theta g_{Dr}^{(1)} + \theta/\gamma\right]$$
$$= \begin{cases} \int_0^\infty f_{g_{Dr}^{(1)}}(x)\left[-F_{g_{Sr}^{(1)}}(x) + F_{g_{Sr}^{(1)}}(\theta x + \theta/\gamma)\right], & 1-\theta < 0 \\ \int_0^\omega f_{g_{Dr}^{(1)}}(x)\left[-F_{g_{Sr}^{(1)}}(x) + F_{g_{Sr}^{(1)}}(\theta x + \theta/\gamma)\right], & 1-\theta > 0 \end{cases} \quad (13a)$$

where, $\theta = 2^{2R_t} - 1$, $\omega = \frac{\theta}{(1-\theta)\gamma}$. If σ_1 and σ_2 symbolizes the exponential distribution of channel gains $g_{Sr}^{(1)}$ and $g_{Dr}^{(1)}$ respectively. The closed form expression can be derived by applying PDF of $g_{Dr}^{(1)}$ and CDF of $g_{Sr}^{(1)}$.

$$a = \begin{cases} \frac{\sigma_2}{\sigma_1+\sigma_2} - \frac{\sigma_2 e^{-\sigma_1\theta/\gamma}}{\sigma_2+\sigma_1\theta}, & 1-\theta < 0 \\ \frac{\sigma_2}{\sigma_1+\sigma_2}\left(1 - e^{-(\sigma_1+\sigma_2)\omega}\right) - \frac{\sigma_2 e^{-\sigma_1\theta/\gamma - (\sigma_2+\sigma_1\theta)\omega}}{\sigma_2+\sigma_1\theta} \times \left(e^{(\sigma_2+\sigma_1\theta)\omega} - 1\right), & 1-\theta > 0 \end{cases} \quad (13b)$$

b in Eq. (13) is similarly articulated as

$$b = \Pr\left[g_{Sr}^{(1)} > g_{Dr}^{(1)}, g_{Sr}^{(1)} \geq \theta g_{Dr}^{(1)} + \theta\big/\gamma, g_{rD}^{(1)} < \theta/\gamma\right]$$

The destination's outage probability with DF relaying under the NOMA-SIC scheme may be articulated using the probabilities in Eq. (13b) and Eq. (14).

$$= \begin{cases} \Pr\left[g_{Sr}^{(1)} \geq \theta g_{Dr}^{(1)} + \theta/\gamma, g_{rD}^{(1)} < \theta/\gamma\right], & 1-\theta < 0 \\ \Pr\left[g_{Sr}^{(1)} \geq \theta g_{Dr}^{(1)} + \theta/\gamma, g_{Dr}^{(1)} < \theta g_{Dr}^{(1)} + \theta/\gamma, g_{rD}^{(1)} < \theta/\gamma\right], & 1-\theta > 0 \\ + \Pr\left[g_{Dr}^{(1)} \geq \theta g_{Dr}^{(1)} + \theta/\gamma, g_{Sr}^{(1)} > g_{Dr}^{(1)}, g_{rD}^{(1)} < \theta/\gamma\right] \end{cases}$$

$$= \begin{cases} \int_0^\infty \left(f_{g_{Dr}^{(1)}}(x)\left[1 - F_{g_{Sr}^{(1)}}(\theta x + \theta/\gamma)\right] \times F_{g_{rD}^{(1)}}(\theta/\gamma)dx\right), & 1-\theta < 0 \\ \int_0^\omega \left(f_{g_{Dr}^{(1)}}(x)\left[1 - F_{g_{Sr}^{(1)}}(\theta x + \theta/\gamma)\right] \times F_{g_{rD}^{(1)}}(\theta/\gamma)dx\right) \\ + \int_0^\infty \left(f_{g_{Dr}^{(1)}}(x)\left[1 - F_{g_{Sr}^{(1)}}(x)\right] \times F_{g_{rD}^{(1)}}(\theta/\gamma)dx\right), & 1-\theta > 0 \end{cases}$$

$$= \begin{cases} \left(1 - e^{\sigma_2\theta/\gamma}\right)\frac{\sigma_2 e^{-\sigma_1\theta/\gamma}}{\sigma_2+\sigma_1\theta}, & 1 < \theta \\ \sigma_2\left(1 - e^{-\sigma_2\theta/\gamma}\right) \times \left\{\left(\frac{e^{-\sigma_1\theta/\gamma - \omega\sigma_2 - \omega\sigma_1\theta}}{\sigma_2+\sigma_1\theta}\right) \times \left(e^{(\sigma_2+\sigma_1\theta)\omega} - 1\right) + \frac{e^{-(\sigma_2+\sigma_1)\omega}}{\sigma_1+\sigma_2}\right\}, & 1 > \theta \end{cases} \quad (14)$$

Similarly, in the proposed scheme, the outage probability is conditioned by $g_{Sr}^{(1)} > g_{Dr}^{(1)}$ and given as

$$P_{out\_S^{(1)}} = \underbrace{\Pr\left[R_{Dr}^{(1)} < R_t\right]}_{a^1} + \underbrace{\Pr\left[R_{Dr}^{(1)} \geq R_t, R_{rS}^{(1)} < R_t\right]}_{b^1} \quad (15)$$

where, $R_{rD}^{(1)} = \frac{1}{2}\log_2\left(1 + \gamma g_{rD}^{(1)}\right)$ and $R_{rS}^{(1)} = \frac{1}{2}\log_2\left(1 + \gamma g_{rS}^{(1)}\right)$

Hence,

$$a^1 = \Pr\left[g_{Sr}^{(1)} > g_{Dr}^{(1)}, g_{Dr}^{(1)} < \theta/\gamma\right]$$

$$= \int_0^{\theta/\gamma} f_{g_{Dr}^{(1)}}(x)\left(1 - F_{g_{Sr}^{(1)}}(x)\right)dx \qquad (16)$$

$$= \frac{\sigma_2}{\sigma_1 + \sigma_2}\left(1 - e^{-\theta/\gamma(\sigma_1+\sigma_2)}\right)$$

$$b^1 = \Pr\left[g_{Sr}^{(1)} > g_{Dr}^{(1)}, g_{Dr}^{(1)} > \theta/\gamma, g_{rS}^{(1)} < \theta/\gamma\right]$$

$$= F_{g_{rS}^{(1)}}(\theta/\gamma) \times \int_{\theta/\gamma}^{\infty} f_{g_{Dr}^{(1)}}(x)\left(1 - F_{g_{Sr}^{(1)}}(x)\right)dx \qquad (17)$$

$$= \left(1 - e^{-\sigma_1\theta/\gamma}\right)\frac{\sigma_2}{\sigma_2 + \sigma_1}e^{-(\sigma_1+\sigma_2)\theta/\gamma}$$

B. If $g_{Sr}^{(1)} > g_{Dr}^{(1)} g_{Sr}^{(1)} > g_{Dr}^{(1)}$

With a time-varying channel response σ_1 and σ_2 parameters, we can calculate the probability of failure for S and D. The possibilities of an interruption are expressed as $P_{out\_S^{(2)}} = a^2 + b^2$ and $P_{out\_D^{(2)}} = a^3 + b^3$.

$$a^2 = \begin{cases} \frac{\sigma_1}{\sigma_1+\sigma_2} - \frac{\sigma_1 e^{-\sigma_2\theta/\gamma}}{\sigma_1+\sigma_2\theta} & , 1-\theta < 0 \\ \frac{\sigma_1}{\sigma_1+\sigma_2}\left(1 - e^{-(\sigma_1+\sigma_2)\omega}\right) - \frac{\sigma_1 e^{-\sigma_2\theta/\gamma - (\sigma_1+\sigma_2)\omega}}{\sigma_1+\sigma_2\theta} \times \left(e^{(\sigma_1+\sigma_2\theta)\omega} - 1\right) & , 1-\theta > 0 \end{cases} \qquad (18)$$

$$b^2 = \begin{cases} \left(1 - e^{-\sigma_1\theta/\gamma}\right)\frac{\sigma_1 e^{-\sigma_2\theta/\gamma}}{\sigma_1+\sigma_2\theta} & , 1-\theta < 0 \\ \sigma_1\left(1 - e^{-\sigma_1\theta/\gamma}\right)\left(\frac{e^{-\sigma_2\theta/\gamma - \omega\sigma_1 - \omega\sigma_2\theta}}{\sigma_1+\sigma_2\theta}\right) \times \left(e^{(\sigma_1+\sigma_2\theta)\omega} - 1\right) & , 1-\theta > 0 \\ +\sigma_1\left(1 - e^{-\sigma_1\theta/\gamma}\right)\left(\frac{e^{-(\sigma_2+\sigma_1)\omega}}{\sigma_1+\sigma_2}\right) & \end{cases}$$

$$(19)$$

$$a^3 = \frac{\sigma_1}{\sigma_1+\sigma_2}\left(1 - e^{-\theta/\gamma(\sigma_1+\sigma_2)}\right) \qquad (20)$$

$$b^3 = \left(1 - e^{-\sigma_2\theta/\gamma}\right)\frac{\sigma_1}{\sigma_1+\sigma_2}e^{-(\sigma_2+\sigma_1)\theta/\gamma} \qquad (21)$$

The combined outage probabilities at S and D in the proposed model is articulated as

$$P_{out\_Sum} = P_{out\_S^{(1)}} + P_{out\_D^{(1)}} + P_{out\_S^{(2)}} + P_{out\_D^{(2)}}$$

$$= 1 - \frac{1}{\sigma_1 + \sigma_2} \left[\sigma_2 \left(1 - e^{-(2\sigma_1 + \sigma_2)\theta/\gamma} \right) + \sigma_1 \left(1 - e^{-(\sigma_1 + 2\sigma_2)\theta/\gamma} \right) \right]$$

$$+ \begin{cases} 1 - \frac{\sigma_2 e^{-(\sigma_1+\sigma_2)\theta/\gamma}}{\sigma_2+\sigma_1\theta} - \frac{\sigma_1 e^{-(\sigma_1+\sigma_2)\theta/\gamma}}{\sigma_2+\sigma_1\theta} \\ \left(1 - e^{-(\sigma_1+\sigma_2)\omega}\right) - \frac{\sigma_2 e^{-\sigma_1\theta/\gamma - (\sigma_2+\sigma_1\theta)\omega}}{\sigma_2+\sigma_1\theta}\left(e^{(\sigma_2+\sigma_1\theta)\omega} - 1\right) - \frac{\sigma_1 e^{-\sigma_2\theta/\gamma - (\sigma_1+\sigma_2\theta)\omega}}{\sigma_1+\sigma_2\theta}\left(e^{(\sigma_1+\sigma_2\theta)\omega}\right) &, 1-\theta < 0 \\ \sigma_2\left(1 - e^{-\sigma_2\theta/\gamma}\right) \left\{ \left(\frac{e^{-\sigma_1\theta/\gamma - \omega\sigma_2 - \omega\sigma_1\theta}}{\sigma_2+\sigma_1\theta}\right)\left(e^{(\sigma_2+\sigma_1\theta)\omega} - 1\right) + \frac{e^{-(\sigma_2+\sigma_1)\omega}}{\sigma_1+\sigma_2} \right\} + \\ \sigma_1\left(1 - e^{-\sigma_1\theta/\gamma}\right) \left\{ \left(\frac{e^{-\sigma_2\theta/\gamma - \omega\sigma_1 - \omega\sigma_2\theta}}{\sigma_1+\sigma_2\theta}\right)\left(e^{(\sigma_1+\sigma_2\theta)\omega} - 1\right) + \frac{e^{-(\sigma_2+\sigma_1)\omega}}{\sigma_1+\sigma_2} \right\} &, 1-\theta > 0 \end{cases}$$

(22)

6 Simulation Results

The projected Two-Way DF C-NOMA in vehicular networks under dynamic channel varying conditions is assessed in terms of PER and cumulative outage probability. From the simulation results projected in Fig. (4), it is worth noted that, PER decreased with the increasing in the multiple data phases (M) and average SNR. With the splitting of data transmission phase, the amount of data transmitted in each phase decreases and the probability of outdated CSI effect over the transmission will be minimal. This finding is valid for restricted data phases because as with the increases of more phases, the estimation of channel correlation induce complexity, latency and multiple initiation for relay selection.

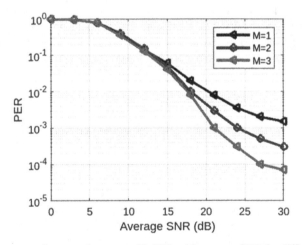

Fig. 4. Performance of proposed system with PER with average SNR for different M values

Sum outage probability metric is evaluated as a performance metric for the proposed system model for different locations of decodable relay r between source and destination vehicles. For the simulation a two-dimensional plan with S, r and D are assumed in the range of 10 m to 90 m. According to Fig. 5, the aggregate outage probability is optimum at 30 m and 70 m distances between S and D because the position of the r at the source and destination has an influence on performance. The performance improvement is evident due to the SIC and power allocation in the NOMA. The propose model with TW-DF NOMA performance is observed to have better performance compared to the TW-AF scheme [33], TW-DF with and without network coding [34].

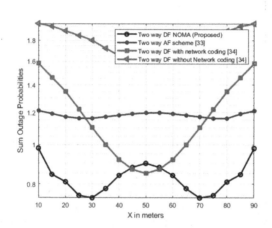

Fig. 5. Performance of Sum Outage Probabilities with r location.

The simulation findings shown in Fig. 6 depict that the cumulative outage probability enhances with transmitted power and the received SNR. The proposed system with two-way HD DF C-NOMA with SIC has better reduction outages, improved channel capacity compared to the proposal TW-AF, TW-DF with and without network coding. The Table 2 depict sum outage probabilities of the proposed system in comparison with Ref. [33, 34].

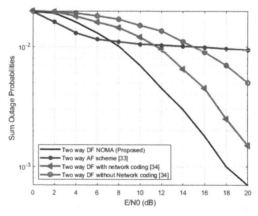

Fig. 6. Performance with Sum Outage Probabilities for different E/N_0 with exponential surge in transmit power (γ).

Table 2. Contains the aggregate outage probability from the simulation findings shown in Figs. 5 and 6

$\frac{E}{N_0}$ (dB)	Position of r (meters)	TW-HD DF-NOMA-SIC (Proposed)		TW-AF Approach [33]		TW- DF with network coding [34]		TW- DF without Network coding [34]	
		Figure 5	Figure 6	Figure 5	Figure 6	Figure 5	Figure 6	Figure 5	Figure 6
0	0	0.98	1.95	1.21	1.95	1.58	1.98	1.98	1.98
20	2	0.81	1.9	1.17	1.6	1.35	1.95	1.9	1.95
30	4	0.74	1.6	1.16	1.3	1.1	1.79	1.8	1.9
40	6	0.84	1.3	1.18	1.15	0.92	1.6	1.65	1.8
50	8	0.9	1.01	1.195	1.1	0.85	1.44	1.6	1.69
60	10	0.84	0.7	1.19	1.05	0.92	1.2	1.65	1.5
70	12	0.74	0.45	1.17	1.03	1.1	0.95	1.8	1.35
80	14	0.81	0.3	1.16	1.01	1.35	0.65	1.9	1.1
90	16	0.98	0.18	1.21	0.99	1.58	0.45	1.98	0.9
–	18	–	0.1	–	0.96	–	0.25	–	0.7
–	20	–	0.07	–	0.94	–	0.15	–	0.5

7 Conclusion

In this work, we introduced a novel system model for vehicular networks, employing TW-HD decode-and-forward (DF) relaying with Non-orthogonal multiple access (NOMA) and successive interference cancellation (SIC) over Nakagami-m fading channels. We investigated the trade-off in packet error rate (PER) performance by dividing the transmission phase into M sub-phases. Our analysis revealed that while increasing M initially leads to a reduction in PER, further subdivision introduces complexities, heightened latency, and unnecessary relay selections, thus diminishing system efficiency. Nonetheless, this segmentation of the transmission phase was found effective in mitigating the impact of time-varying channels.

We derived closed-form expressions to evaluate the aggregate outage probabilities at the source and destination. Our simulations confirmed that the selection of the decodable relay within the cluster significantly influences outage probabilities, particularly as transmit power increases. Notably, optimal outage performance was observed when the selected relay was positioned at distances of 30 m and 70 m.

Comparative analysis against traditional methods demonstrated the superiority of our proposed system. Specifically, the total outage probability was significantly lower, with the sum of outage probabilities at the source and destination showing notable improvements. In particular, the two-way NOMA with half-duplex DF relay outperformed existing schemes by 31.20%, 25.87%, and 52.82% compared to the two-way amplify-and-forward scheme, two-way DF with network coding, and two-way DF without network coding, respectively. These findings underscore the effectiveness and efficiency of our proposed system in enhancing communication reliability and performance in vehicular networks.

References

1. Sjoberg, K., Andres, P., Buburuzan, T., Brakemeier, A.: Cooperative intelligent transport systems in europe: current deployment status and outlook. IEEE Veh. Technol. Mag. **12**, 89–97, (2017).
2. Brincat, A.A., Pacifici, F., Martinaglia, S., Mazzola, F.: The Internet of Things for Intelligent Transportation Systems in Real Smart Cities Scenarios, 2019 IEEE 5th World Forum on Internet of Things (WF-IoT), pp. 128–132. Limerick, Ireland (2019)
3. Liang, L., Peng, H., Li, G.Y., Shen, X.: Vehicular communications: a physical layer perspective. IEEE Trans. Veh. Technol. **66**(12), 10647–10659 (2017).
4. Kim, S.I., Oh, H.S., Choi, H.K.: Mid-amble aided OFDM performance analysis in high mobility vehicular channel. In: Proceedings 2008 IEEE Intelligent Vehicles Symposium, pp. 751–754, (Jun 2008)
5. Cortes, J.A., Aguayo-Torres, M.C., Fransisco.: Vehicular channels: characteristics, models and implications on communication systems design. Wirel. Pers. Commun. Int. J. **106**(1), 237–260 (2019)
6. Eze, E.C., Zhang, S., Liu, E.: Vehicular ad hoc networks (VANETs): current state, challenges, potentials and way forward. In: 20th International Conference on Automation and Computing, Cranfield, pp. 176–181 (2014)
7. Zeadally, S., Javed, M.A., Hamida, E.B.: Vehicular Communications for ITS: Standardization and Challenges. IEEE Commun. Stand. Mag. **4**(1), 11–17 (2020)
8. Boban, M., d'Orey, P.M.: Exploring the practical limits of cooperative awareness in vehicular communications. IEEE Trans. Veh. Technol. **65**, 3904–3916 (2016)
9. Schwarz, S., Philosof, T., Rupp, M.: Signal processing challenges in cellular-assisted vehicular communications: efforts and developments within 3GPP LTE and beyond. IEEE Signal Process. Mag. **34**(2), 47–59 (2017)
10. Ahmed, E., Gharavi, H.: Cooperative vehicular networking: a survey. IEEE Trans. Intell. Trans. Syst. **19**(3) 996–1014 (2018)
11. Dai, L., Wang, B., Ding, Z., Wang, Z., Chen, S., Hanzo, L.: A Survey of non-orthogonal multiple access for 5G. IEEE Commun. Surv. Tutorials **20**(3), 2294–2323 (2018)
12. Di, B., Song, L., Li, Y., Han, Z.: V2X meets NOMA, Non-orthogonal multiple access for 5G-enabled vehicular networks. IEEE Wirel. Comm. **24**(6), 14–21 (2017)
13. Ding, Z., Peng, M., Poor, H.V.: Cooperative Non-Orthogonal Multiple Access in 5G Systems. IEEE Commun. Lett. **19**(8), 1462–1465 (2015)
14. Abozariba, R., Naeem, M.K., Patwary, M., Eyedebrahimi, M., Bull, P., Aneiba, A.: NOMA-based resource allocation and mobility enhancement framework for IoT in next generation cellular networks. IEEE Access **7**, 29158–29172 (2019)
15. Wu, Y., Qian, L.P., Mao, H., Yang, X., Zhou, H., Shen, X.: Optimal power allocation and scheduling for non-orthogonal multiple access relay-assisted networks. IEEE Trans. Mob. Comput. **17**(11) 2591–2606 (2018)
16. Liaqat, M., Noordin, K.A., Abdul Latef, T. et al.: Power-domain non orthogonal multiple access (PD-NOMA) in cooperative networks: an overview. Wirel. Netw. **26**, 181–203 (2020)
17. Su, X., Yu, H., Kim, W., et al.: Interference cancellation for non-orthogonal multiple access used in future wireless mobile networks. J. Wirel. Com Netw. **2016**, 231 (2016)

18. Meng, L., Su, X., Zhang, X., Choi, C., Choi, D.: Signal reception for successive interference cancellation in NOMA downlink. In: Proceedings of the 2018 Conference on Research in Adaptive and Convergent Systems (RACS '18). Association for Computing Machinery, New York, NY, USA, pp. 75–79 (2018)
19. Haci, H., Zhu, H., Wang, J.: Performance of non-orthogonal multiple access with a novel asynchronous interference cancellation technique. IEEE Trans. Commun. **65**(3), 1319–1335 (2017)
20. Wei, Z., Guo, J., Ng, D.W.K., Yuan, J.: Fairness Comparison of Uplink NOMA and OMA," IEEE 85th Vehicular Technology Conference (VTC Spring), pp. 1–6. Sydney, NSW (2017)
21. Wang, P., Xiao, J., Ping, L.: Comparison of orthogonal and nonorthogonal approaches to future wireless cellular systems. IEEE Veh. Technol. Mag. **1**(3) 4–11 (2006)
22. Baghani, M., Parsaeefard, S., Derakhshani, M., Saad, W.: Dynamic non-orthogonal multiple access (NOMA) and orthogonal multiple access (OMA) in 5G wireless networks. Loughborough University (2019)
23. Moualeu, J.M., Hamouda, W., Takawira, F.: Relay selection for coded cooperative networks with outdated CSI over Nakagami-m fading channels. IEEE Trans. Wirel. Commun. **13**(5), 2362–2373 (2014)
24. Geng, K., Gao, Q., Fei, L., Xiong, H.: Relay selection in cooperative communication systems over continuous time-varying fading channel. Chin. J. Aeronaut. **30**(1), 391–398 (2017). ISSN 1000–9361
25. Fei, L., Gao, Q., Zhang, J., Xu, Q.: Relay selection with outdated channel state information in cooperative communication systems. IET Commun. **7**(14), 1557–1565 (2013)
26. Liau, Q.Y., Leow, C.Y.: Amplify-and-forward relay selection in cooperative non-orthogonal multiple access. In: IEEE 13th Malaysia International Conference on Communications (MICC), Johor Bahru, pp. 79–83 (2017)
27. Abbasi, O., Ebrahimi, A.: Cooperative NOMA with full-duplex amplify-and-forward relaying. Trans. Emerg. Telecommun. Technol. **29**(7) (2018)
28. Zhong, C., Zhang, Z.: Non-orthogonal multiple access with cooperative full-duplex relaying. IEEE Commun. Lett. **20**(12), 2478–2481 (2016)
29. Do, T.N., da Costa, D.B., Duong, T.Q., An, B.: Improving the performance of cell-edge users in NOMA systems using cooperative relaying. IEEE Trans. Commun. **66**(5), 1883–1901 (2018)
30. Yue, X., Liu, Y., Kang, S., Nallanathan, A., Chen, Y.: Modeling and analysis of two-way relay non-orthogonal multiple access systems. IEEE Trans. Commun. **66**(9), 3784–3796 (2018)
31. Li, X., Liu, M., Deng, C., Mathiopoulos, P.T., Ding, Z., Liu, Y.: Full-duplex cooperative NOMA relaying systems with I/Q imbalance and imperfect SIC. IEEE Wirel. Commun. Lett. **9**(1), 17–20 (2020)
32. Gong, X., Yue, X., Liu, F.: Performance analysis of cooperative NOMA networks with imperfect CSI over Nakagami-m fading channels. Sensors **20**(2), 424 (2020).
33. Cao, H., Fu, L., Dai, H.: Throughput analysis of the two-way relay system with network coding and energy harvesting. In: IEEE International Conference on Communications (ICC), Paris, 2017, pp. 1–6 (2017)
34. Koc, A., Altunbas, I., Yaman, B.: Unified outage performance analysis of two-way/one-way full-duplex/half-duplex fixed-gain AF relay systems. In: 24th International Conference on Telecommunications (ICT), Limassol, pp. 1–5 (2017)
35. Li, E., Wang, X., Wu, Z., Hao, S., Dong, Y.: Outage analysis of decode-and-forward two-way relay selection with different coding and decoding schemes. IEEE Syst. J. **13**(1), 125–136 (2019)

Machine/Deep/Reinforcement Learning in Industries

Integration of Artificial Intelligence for the Analysis and Monitoring of Projects Within Companies

Ouissem Mougari, Sarra Bouzid, and Sihem Saadi(✉)

University of Algiers 1 Ben Youcef Benkhadda, Algiers, Algeria
s.saadi@univ-alger.dz

Abstract. This article focuses on improving the business tracking system by focusing on deadline, costs and resources. The main objective is to use Artificial Intelligence to develop a robust prediction model to facilitate the effective monitoring of projects within companies. The study begins with the collection of information on traditional project monitoring methods, based on the use of predefined and static indicators to assess progress in the three specific aspects of each project. However, these indicators fail to capture the real complexity and dynamics of projects. To obtain more effective monitoring, a study is carried out using two learning models: the first based on the algorithm Support Vector Machine and the second based on the multi-layer perceptron's. An in-depth analysis of the problems of the existing monitoring system is carried out, followed by the collection and preparation of historical data sets from previous projects. Relevant characteristics related to time, costs and resources are carefully selected for analysis. To classify projects according to their progress, the Support Vector Machine and the Multi-layer perceptron's are applied to the data set to obtain a reliable classifier. Finally, a comparative study of the results obtained by the two algorithms is presented. The ultimate goal is to improve project monitoring using Artificial Intelligence-based methods, which should allow more efficient management of time, costs and resources for the whole company.

Keywords: key Performance Indicators · Artificial Intelligence · Support Vector Machine · Multi-layer perceptron's

1 Introduction

In recent years, a multitude of contemporary organizations has embraced the project management model, incorporating dedicated tools and methodologies to mitigate risks that may impact project objectives. This strategic adoption aims to enhance overall performance, meet stakeholder needs and expectations, optimize costs, adhere to tight deadlines, and ultimately achieve the organization's objectives, positioning it favorably in the sector.

However, the conventional methods employed for project monitoring often rely on pre-established and static indicators. These indicators, while useful, fall short in capturing the intricate complexity and dynamic nature inherent in projects. Furthermore, they tend to be formulated in a generic manner, overlooking the specificities and variations unique to each project. This limitation hampers the ability to derive a precise and personalized view of a project's progress and performance. Herein lies the transformative potential of artificial intelligence, encompassing machine learning and deep learning, in offering a fresh perspective.

Advanced techniques within artificial intelligence empower organizations to process vast volumes of data from diverse sources. Through machine learning and deep learning, these methods identify patterns, establish correlations, and extract pertinent information vital for effective project monitoring. The incorporation of artificial intelligence introduces a paradigm shift, enabling a more nuanced and adaptive approach to project management.

Therefore, we will implement an approach centered around the application of support vector machines (SVM) to analyze project data, identify patterns, and predict potential issues related to delays, cost overruns, and resource management. This method significantly contributes to improving the management of deadlines, costs, and resources, thereby enhancing the overall success of projects. In parallel, we will explore the use of multilayer perceptron's (MLP) to classify projects based on their progress and predict the optimal allocation of resources to minimize delays and costs. The effectiveness of these approaches relies heavily on the quality of the data and meticulous preprocessing, especially for the MLP.

The examination of existing research on the application of Artificial Intelligence in project management underscores recent advancements in AI, particularly in the realms of data analysis, planning, and decision-making. This evolving landscape suggests that organizations leveraging these technologies stand to benefit from enhanced insights, greater adaptability, and more informed decision-making processes in the realm of project management.

The review of existing work on the use of Artificial Intelligence in project management highlights recent developments in AI in data analysis, planning and decision-making.

The article by Chen and Lu [2], published in 2020, highlights these advances, while the comprehensive review by Amin and Kamel [1], published in 2019, summarizes the key AI applications by highlighting the benefits, challenges and opportunities. This work highlights the growing importance of AI in improving performance and decision-making in project management, paving the way for future trends and opportunities.

The analysis of Artificial Intelligence methods and techniques adapted to project management explores the use of machine learning, deep learning and Neural Networks to optimize data analysis, planning and decision-making. The work of Roy and Vittal [11] in 2020 as well as the literature review of Dörner and Rostami [3] in 2021 highlight the advantages, limitations and challenges associated with the use of these techniques, underlining the importance of AI to improve project management in an informed manner.

The case study of the application of AI in similar contexts examines the opportunities, challenges and implications of its use in project management. The article by Smith and Kumar [13] in 2018 and the study by Rönnberg and Rydberg [10] in 2020 provide valuable information on the benefits, limitations and lessons learned from the use of AI in these contexts, thus contributing to its advancement and adoption.

The study of researchers and specific work in the field of AI applied to project management identifies key researchers and the most influential work, providing an overview of research trends and advances. The publications of He, Gao and Liu [4] in 2020, Li and Wu [6] in 2019 and Liu and Li [7] in 2020 use bibliometric analysis to highlight the most cited researchers, the most influential journals and the most active areas of research, providing valuable information for professionals and researchers interested in applying AI in project management.

Finally, exploring specific AI approaches to project analysis, monitoring and management highlights the application of techniques such as Neural Networks, genetic algorithms and machine learning to improve project control and management. The studies of Saleh and Okasha [12] in 2019 and Khan and Bao [5] in 2020 examine AI techniques, in particular machine learning and neural networks, to predict project performance and facilitate decision-making, highlighting the benefits and opportunities offered by AI to improve the effectiveness and results of projects.

The primary objective of this article is to address the question of how the application of Artificial Intelligence can enhance the analysis, monitoring, and management of projects within companies, ultimately aiming to elevate performance, efficiency, and strategic decision-making.

Our unique contribution to this endeavor involves the development of robust prediction models utilizing Artificial Intelligence algorithms, specifically employing Support Vector Machines (SVM) and Multi-Layer Perceptron's (MLP). These models are strategically crafted to elevate key aspects of project management, including planning, budgeting, resource management, and the overall advancement of projects. Through the integration of these technological advances, our aim is to streamline processes and fortify project management practices.

To systematically explore and present our contribution, the article is organized into four distinct sections. It begins with an elucidation of the methodology in Sect. 2, providing insights into the approach and techniques employed in the development of our prediction models. The subsequent section, Sect. 3, unveils the results derived from our models, offering a detailed analysis of their performance in various project management dimensions. Following the presentation of results, Sect. 4 engages in a comprehensive discussion, delving into the implications and significance of our findings. This section serves as a critical analysis and interpretation of the observed outcomes, offering insights into the practical applications and potential limitations of the developed models. Lastly, Sect. 5 encapsulates the conclusions drawn from our study, presenting both the key findings and future perspectives. This final section synthesizes the contributions of our research, highlighting the practical implications for companies seeking to optimize their project management through the application of Artificial Intelligence.

2 Methodology

The initial focus of this study revolves around the assessment of the conventional project monitoring methodology implemented at the organizational level, as outlined in reference [9]. This evaluation involves the computation of performance indicators corresponding to pivotal aspects such as deadlines, costs, and resources. Following the calculation of these indicators, their respective values are then subjected to interpretation, providing crucial insights into the overall performance and health of the projects under consideration. This methodical approach allows for a comprehensive understanding of how well projects align with established timelines, adhere to budgetary constraints, and effectively utilize available resources.

2.1 Performance Measurement System

The performance measurement system "deadline", "costs" and "resources" is structured around three main indicators assessed periodically:

Planned Value (PV). An estimate of the expected utility for an agent or intelligent system, taking into account the time, resources and costs associated with planning and executing a sequence of actions in a given environment.

Acquired Value (AV). The actual utility obtained by an agent or intelligent system, taking into account the time, resources used and associated costs when executing a sequence of actions in a given environment.

Real Value of Costs (RV Costs). Actual costs incurred for work performed during a specific period of time.

These three indicators are then used to assess the corresponding periodic indices:

Schedule Performance Index (SPI). Measures the expected completion rate of work of the schedule expressed as the ratio of the acquired value of time to the planned value of time.

SPI = AV Deadline/PV Deadline.

Cost Performance Index (CPI). Measures the performance (productivity) on the cost of mobilized resources expressed as the ratio of the acquired value of costs to the actual value of costs.

CPI = AV Costs/RV Costs.

Resource Performance Index (RPI). Measures the effectiveness or efficiency with which resources are used to achieve a specific objective expressed as the ratio of the acquired value of resources to the planned value of resources.

RPI = AV Resource/PV Resource.

Then, in order to progress towards the second phase which focuses on the integration of Artificial Intelligence which is a field of computer science which focuses on the creation of systems capable of reproducing and simulating human intelligence. A

Table 1. Table of classes with preprocessing.

	Deadline	Cost	Resource
0	Delay	Exceeded	Less resource
1	In time	Respected	Respected
2	Advanced	Not exceeded	More resources

preliminary step is required. This is the pre-processing of initial data, as our database contains qualitative outputs. Table 1 demonstrates the deadline, cost and resource outputs as well as the coding performed:

Each key aspect (time, cost and resources) can be dealt with in three different ways: either behind schedule, with cost overruns and reduced resource utilization; or on schedule, with cost overruns and reduced resource utilization; or ahead of schedule, with reduced cost and increased resource utilization. A coding scheme has been proposed for these categories (0, 1 and 2) respectively for each case encountered.

In the next section, we will introduce the two methods of artificial intelligence that were employed.

2.2 Support Vector Machine (SVM)

As a selected supervised machine learning technique for the study. SVM was chosen due to the presence of three distinct classes corresponding to deadline, cost, and resource variables, thereby creating a multi-class classification challenge. This choice aligns with the specific requirements of the study, where projects are categorized into different classes based on their progress in terms of deadlines, costs, and resource utilization.

The selection of SVM is further justified by the nature of the data, which exhibits a non-linear pattern. SVM is well-suited for handling non-linear relationships in data. To adhere to Vapnik's theory [14], the utilization of a polynomial nucleus became imperative. The formula for this polynomial nucleus is then presented, signifying the mathematical expression employed to capture the complex relationships within the dataset. This choice of SVM, coupled with a polynomial nucleus, reflects a deliberate strategy to accommodate the intricacies of the data and optimize the model's performance in addressing the multi-class classification challenge posed by the project variables.

$$K(x, y) = (\alpha xy + c)^d \quad (1)$$

where

α : is a scaling coefficient.
$\langle xy \rangle$: is the scalar product between the input vectors x and y.
c: a constant.
d: the order of the kernel.

Below is a summary diagram of the Support Vector Machine models associated with each key aspect:

Three models have been created: model 1, model 2, and model 3 for deadlines, costs, and resources. For each model, there are three classes, and one of these classes will be the output of each model, which can specify the project's progress (Fig. 1).

2.3 Multi-layer Perceptron's (MLP)

Artificial Neural Networks, as elucidated in reference [8], are extensively employed in the realm of deep learning due to their capability to effectively model intricate relationships

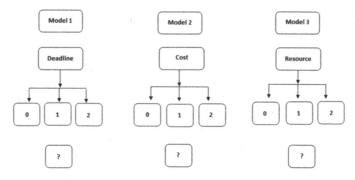

Fig.1. Diagram of Support Vector Machine models.

within data and execute tasks such as classification and prediction. The popularity of Artificial Neural Networks arises from their adaptability and efficiency in handling complex data structures. In our study, the chosen model, Multi-layer Perceptron's, is explicitly tailored to address a multi-label classification problem. Within this context, the model is configured to simultaneously consider and output three predictions for each of the nine available labels. This design reflects a deliberate adaptation to the specific characteristics of the multi-label classification challenge posed by the project variables under consideration (Fig. 2).

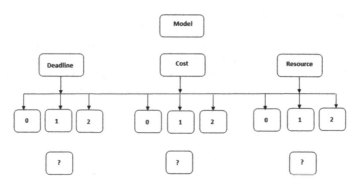

Fig. 2. Diagram of Multi-Layer Perceptron's model

A model has been developed for deadlines, costs, and resources, consisting of nine classes, and three classes will be the output of this model simultaneously, allowing to specify the project's progress.

3 Results

3.1 Performance Indicator Calculation Results

In this section and in the first part, we have introduced the method used for project monitoring within companies. By calculating performance indicators relating to deadlines, costs and resources. These indicators are then evaluated for project progress.

Upon a meticulous examination of performance indicators across crucial facets of the project, a nuanced categorization unveils essential insights into its dynamics. In instances where indicators precisely align with 1, a positive signal is emitted, symbolizing a project that diligently adheres to its predetermined schedule, stays within the allocated budget, and effectively manages resources in accordance with the initial plan. This scenario reflects an optimal state of project management where objectives are being met as intended.

Conversely, when indicators fall below 1, a different narrative unfolds. This situation points towards project challenges, manifesting as delays in the project timeline, exceeding budgetary constraints, and utilizing fewer resources than originally envisaged. Such deviations from the plan signal potential areas of concern that may require immediate attention and strategic interventions to realign the project trajectory.

Furthermore, when indicators exceed 1, a unique set of circumstances comes to light. In such cases, the project is showcasing an accelerated pace, progressing ahead of the predetermined timeline, while still adhering to budgetary constraints. However, the trade-off involves an increased utilization of resources beyond the initial plan, necessitating vigilant monitoring to ensure that resource allocation aligns with the evolving demands of the project.

Following the assessment of system evaluations applied to companies, a comprehensive summary has been compiled, delineating key decisions integral to the project's evaluation. Decision 1 centers on evaluating the project's progress in relation to the established timeline. This decision becomes a pivotal factor in determining whether the project is adhering to its scheduled milestones or encountering delays.

Decision 2, in turn, pertains to the project's advancement with respect to cost considerations. It involves scrutinizing whether the project is staying within the allocated budget or exceeding financial constraints. This decision holds significant weight in assessing the financial health and efficiency of the project. Furthermore, Decision 3 directs attention to the project's progress in terms of resource management. This involves evaluating

Table 2. Project progress table.

Projects	SV	CV	RV	SPI	CPI	RPI
1	−3.23%	−194.50746	42	0.76	0.63	01.05
2	−2.88%	0	341	0.93	1	3.45
3	1.01%	514.697134	10	01.01	02.06	01.01
4	0.00%	34.991162	−149	1	01.06	0.78
5	14.35%	0	123	0.85	1	1.17
6	3.35%	439.454546	386	01.04	1.84	1.7
7	16.94%	−55.994495	−343	1.24	0.9	0.35
8	38.93%	0	412	0.26	1	1.9
9	34.50%	476.485581	452	0.38	3.96	2.26
10	1.71%	−402.26341	0	01.05	0.46	1

whether the utilization of resources aligns with the initial plan or diverges from anticipated levels. Assessing resource management becomes critical in ensuring that the project is adequately supported to meet its objectives.

Preprocessing plays a crucial role as a preliminary step before implementing artificial intelligence methods in decision-making processes. This preparatory phase becomes indispensable for ensuring that the data input into AI models is appropriately formatted, cleaned, and standardized. Within the context of decision-making, the significance of preprocessing is particularly evident, as the quality of the input data directly influences the accuracy and reliability of the AI-driven insights.

Moreover, decisions within this framework are categorized into three distinct classes, providing a structured approach to evaluating project dynamics. In Class 0, the project is characterized by delays, budget overruns, and a minimal utilization of resources. This classification serves as a flag for instances where the project is facing challenges and deviations from the original plan, prompting a need for corrective measures to realign the project trajectory.

Class 1 encapsulates decisions where projects are progressing in a timely manner, staying within budgetary constraints, and actively contributing to overall project advancement. This class represents the ideal scenario, indicating that the project is on track, meeting deadlines, and managing resources efficiently, thus fostering positive project outcomes.

Class 2 denotes decisions where projects are not only ahead of schedule but also within budget constraints, albeit with an additional utilization of resources. This classification suggests an accelerated pace of progress and strategic resource allocation, highlighting opportunities for optimized efficiency and potentially enhanced project outcomes.

Table 3. project's decisions table.

Projects	SPI	CPI	RPI	Decision1	Decision2	Decision3
1	0.76	0.63	01.05	DeLay	Exceeded	More resources
2	0.93	1	3.45	DeLay	Respected	More resources
3	01.01	02.06	01.01	Advanced	Not exceeded	More resources
4	1	01.06	0.78	In time	Not exceeded	Less resources
5	0.85	1	1.17	DeLay	Respected	More resources
6	01.04	1.84	1.7	DeLay	Not Exceeded	More resources
7	1.24	0.9	0.35	Advanced	Exceeded	Less resources
8	0.26	1	1.9	DeLay	Respected	More resources
9	0.38	3.96	2.26	Delay	Not Exceeded	More resources
10	01.05	0.46	1	Advanced	Exceeded	Respected

Table 4. pre-treatment decision table

Projects	SPI	CPI	RPI	Decision1	Decision2	Decision3	Sum	Decision 4
1	0.76	0.63	01.05	0	0	2	2	0
2	0.93	1	3.45	0	1	2	3	1
3	01.01	02.06	01.01	2	2	2	6	2
4	1	01.06	0.78	1	2	0	3	1
5	0.85	1	1.17	0	1	2	3	1
6	01.04	1.84	1.7	2	2	2	6	2
7	1.24	0.9	0.35	2	0	0	2	0
8	0.26	1	1.9	0	1	2	3	1
9	0.38	3.96	2.26	0	2	2	4	1
10	01.05	0.46	1	2	0	1	3	1

To holistically assess the overall progress of the project, we implemented further enhancements by incorporating two additional columns into our database. The first column, aptly labeled "sum", aggregates the values from the three preceding decisions: Decision 1 related to timeline progress, Decision 2 linked to budget considerations, and Decision 3 associated with resource management. This cumulative "sum" column serves as a consolidated measure, providing a holistic perspective on the collective impact of these critical aspects on the project's trajectory.

Simultaneously, the introduction of the second column, designated "Decision 4," is instrumental in encapsulating the comprehensive status of the project's overall progress. This newly introduced column takes into account the synthesized information from the "sum" column, offering a distilled representation of the project's collective performance across the evaluated dimensions. Through "Decision 4," stakeholders gain a consolidated insight into whether the project is currently facing challenges and setbacks, progressing optimally, or showcasing advanced progress with strategic resource allocation.

By incorporating these additional columns, our database now provides a more comprehensive and integrated overview of the project's holistic performance. These refinements not only contribute to a more nuanced understanding of individual decision aspects but also facilitate a streamlined evaluation of the overall project health, empowering stakeholders with the insights necessary for informed decision-making and strategic interventions.

3.2 Support Vector Machine Results

In the subsequent phase, we advanced our project evaluation by employing the "Support Vector Machine" model for classification purposes. This machine learning model enabled us to predict the progress of projects with a remarkable precision level.

Specifically, the model demonstrated an outstanding accuracy of 100% in predicting project progress relative to deadlines, underscoring its reliability in foreseeing whether projects are on schedule, delayed, or ahead of time (Fig. 3).

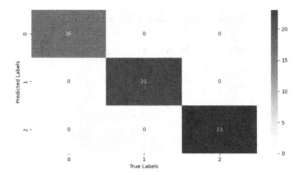

Fig. 3. Deadline Confusion Matrix.

Furthermore, the "Support Vector Machine" model exhibited an impressive precision rate of 92% when predicting project progress in relation to costs. This accuracy level provides valuable insights into the financial aspects of projects, allowing stakeholders to anticipate and address potential budgetary challenges effectively (Fig. 4).

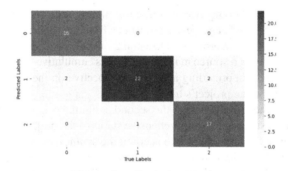

Fig. 4. Cost Confusion Matrix.

Moreover, in the context of resource management, the model achieved a noteworthy precision rate of 100%. This indicates a high degree of accuracy in forecasting how projects are utilizing resources, whether in alignment with the initial plan or requiring adjustments for optimized efficiency (Fig. 5).

Crucially, when assessing the overall status of projects, the "Support Vector Machine" model delivered exceptional results with a 100% accuracy rate. This comprehensive accuracy underscores the model's effectiveness in providing a holistic evaluation of projects, encompassing their timelines, costs, and resource dynamics (Fig. 6).

The following figures present the graphical representation of confusion matrices for delays, costs, and resources for Support Vector Machines (SVM). These diagrams consist of two axes: one for true classes and one for predicted classes. Three classes, 0, 1, and 2, are depicted, with predicted classes displayed in three colors: blue for predicted class 0, purple for predicted class 1, and black for predicted class 2. True classes are also represented in three colors: red for true class 0, orange for true class 1, and green for true class 2 (Fig. 7).

Integration of Artificial Intelligence 77

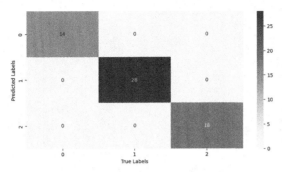

Fig. 5. Resource Confusion Matrix.

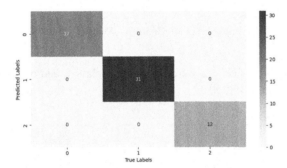

Fig. 6. Project Confusion Matrix.

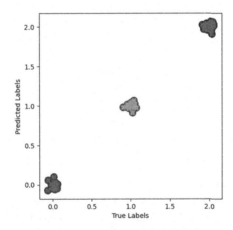

Fig. 7. Scatter plot of the Confusion Matrix Deadline for SVM.

As shown the deadline diagram, it is observed that each true class overlaps with its predicted class. This indicates that all predictions were accurately made (Fig 8).

The cost diagram demonstrates that the majority of projects were predicted accurately. However, there were some prediction errors: two projects of class 0 were classified

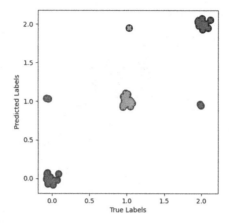

Fig. 8. Scatter plot of the Confusion Matrix Cost for SVM.

as class 1, two projects of class 2 were classified as class 1, and one project of class 1 was classified as class 2 (Fig. 9).

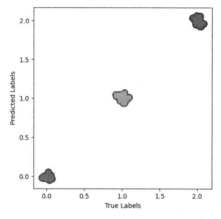

Fig. 9. Scatter plot of the Confusion Matrix Resource for SVM.

Based on the resource diagram, it is observed that each true class overlaps with its predicted class. This indicates that all predictions were accurately made (Fig. 10).

As shown the projects diagram, it is observed that each true class overlaps with its predicted class. This indicates that all predictions were accurately made.

3.3 Multi-Layer Perceptron's Results

In the next phase of our analysis, we implemented the "Multi-layer Perceptron's" model to assess project progress. The model demonstrated an overall accuracy rate of 82%, showcasing its effectiveness in providing insightful predictions.

Integration of Artificial Intelligence 79

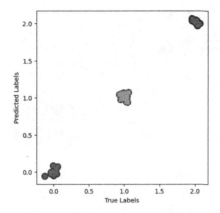

Fig. 10. Scatter plot of the Confusion Matrix Projects for SVM.

Specifically, the model exhibited an accuracy of 56.7% in predicting project progress relative to deadlines, indicating its capability to discern whether projects are adhering to schedules, facing delays, or ahead of time (Fig. 11).

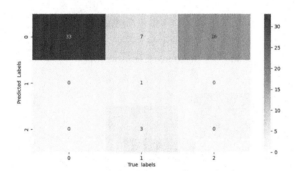

Fig. 11. Deadline Confusion Matrix.

For cost-related predictions, the "Multi-layer Perceptron's" model displayed a remarkable precision of 96.78%. This high accuracy level underscores the model's proficiency in forecasting and evaluating the financial aspects of projects, enabling stakeholders to proactively manage and control costs effectively (Fig. 12).

Similarly, in the realm of resource management, the model achieved a commendable accuracy rate of 95%. This suggests a robust capability to predict and assess how projects are utilizing resources, providing valuable insights for optimizing resource allocation and ensuring efficient project execution (Fig. 13).

Notably, when considering the overall progress of projects, the "Multi-layer Perceptron's" model exhibited a flawless accuracy of 100%. This implies that the model effectively integrates information from various dimensions, offering a comprehensive evaluation of project health encompassing timelines, costs, and resource utilization (Fig. 14).

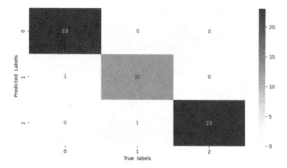

Fig. 12. Cost Confusion Matrix.

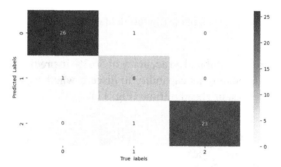

Fig. 13. Resource Confusion Matrix.

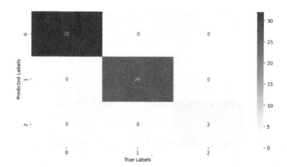

Fig. 14. Project Confusion Matrix.

The following figures present the graphical representation of confusion matrices for delays, costs, and resources for Multi-layer Perceptron's (MLP). These diagrams consist of two axes: one for true classes and one for predicted classes. Three classes, 0, 1, and 2, are depicted, with predicted classes displayed in three colors: blue for predicted class 0, purple for predicted class 1, and black for predicted class 2. True classes are also represented in three colors: red for true class 0, orange for true class 1, and green for true class 2 (Fig. 15).

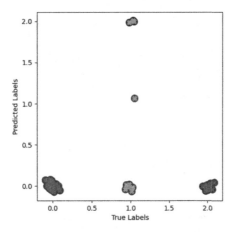

Fig. 15. Scatter plot of the Confusion Matrix Deadline for MLP.

the deadline diagram demonstrates that all class 0 projects were predicted accurately. However, the majority of class 1 and 2 projects were predicted as class 0. Additionally, one class 1 project was correctly classified, and three projects of the same class were classified as class 2 (Fig. 16).

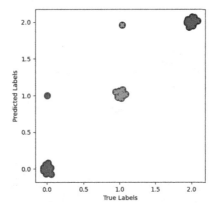

Fig. 16. Scatter plot of the Confusion Matrix Cost for MLP.

Based on this diagram, we observe that the majority of projects have been predicted accurately. However, there were some prediction errors: a class 0 project was classified as class 1, and a class 1 project was classified as class 2 (Fig. 17).

As shown the resource diagram, we observe that the majority of projects have been predicted accurately. However, there were some prediction errors: a class 0 project was classified as class 1, a class 1 project was classified as class 0, and a project from the same class was classified as class 2 (Fig. 18).

The projects diagram demonstrates that each true class overlaps with its predicted class. This indicates that all predictions have been accurately made.

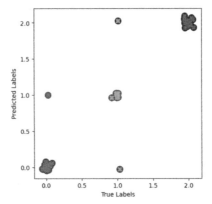

Fig. 17. Scatter plot of the Confusion Matrix Resource for MLP.

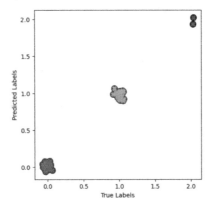

Fig. 18. Scatter plot of the Confusion Matrix Projects for MLP.

4 Discussion

The project monitoring system implemented in enterprises relies on calculating performance indicators related to deadlines, costs, and resources to ascertain a project's status. However, this approach has limitations as it fails to capture the inherent complexities and variations unique to each project. It also requires additional calculations for individual projects, making the process resource-intensive.

In contrast, Support Vector Machine (SVM) and Multi-layer Perceptron's (MLP) present an alternative by offering the capability to simultaneously predict and make decisions for multiple projects without the need for additional calculations. These machine learning methods rely solely on available input data and leverage their ability to learn from provided examples to generalize and predict project statuses.

The accuracy achieved by Support Vector Machine (SVM) in predicting deadlines, costs, and resources can be attributed to the use of a balanced database. This equilibrium in the dataset ensures that the model is trained on a representative mix of instances, preventing biases and enhancing its ability to generalize accurately. In particular, the balanced dataset contributes to SVM's remarkable 100% accuracy in predicting deadlines and resources, underscoring the importance of a well-structured and unbiased training set.

Conversely, the success of Multi-layer Perceptron's (MLP) in achieving accuracy is tied to meticulous adjustments made during the learning process. The fine-tuning of learning rate, batch size, and epochs, coupled with the strategic utilization of hidden layers and the introduction of dropout mechanisms, plays a pivotal role in optimizing MLP's predictive capabilities. These adjustments ensure that the MLP model learns effectively from the provided examples, preventing overfitting and allowing for accurate predictions.

The effectiveness of SVM and MLP varies based on their ranking rates in predicting deadlines, resources, costs, and the overall project status. SVM demonstrates superior efficiency in predicting deadlines and resources, achieving a perfect 100% accuracy for each category. On the other hand, MLP exhibits slightly lower efficiency, with a rate of 56.7% for deadlines and 95% for resources.

In terms of cost prediction, MLP outperforms SVM, attaining a precision rate of 96.78% compared to SVM's 92%. Despite differences in individual aspects, both methods prove highly effective in accurately assessing the overall status of the project, achieving a perfect 100% accuracy.

Upon analysis, it becomes evident that SVM excels in predicting deadlines and resources, while MLP demonstrates greater efficacy in predicting costs. However, both methods exhibit high accuracy and reliability when evaluating the overall status of a project. This nuanced understanding allows stakeholders to leverage the strengths of each method based on specific project requirements, optimizing decision-making processes and resource utilization.

5 Conclusion and Prospects

This article delves into the realm of project monitoring, focusing on critical criteria such as deadlines, costs, and resources. The overarching goal is to develop a predictive model capable of estimating and forecasting the impact of these key factors on the progress of projects.

Drawing on insights from previous research, we devised an approach centered around the application of Support Vector Machines (SVM). The SVM model proved to be robust, enabling the analysis of project data, identification of patterns, and prediction of potential issues related to delays, cost overruns, and resource management. The successful implementation of this model significantly contributed to the enhanced management of deadlines, costs, and resources, thereby bolstering the overall success of projects.

In addition to SVM, we explored the utility of another deep learning method, namely the Multi-layer Perceptron (MLP), which offers substantial advantages in the domain of project management. Leveraging its intricate architecture, our MLP model effectively captured complex interactions between variables, allowing for project classification based on progress and predicting optimal resource allocations to minimize both deadlines and costs. The satisfactory results and efficacy of our MLP model were found to be closely tied to the quality of data and meticulous pre-processing.

To address ongoing challenges and fully harness the potential of Artificial Intelligence (AI) in project analysis, monitoring, and management within companies, several perspectives can be considered. Firstly, there is the prospect of exploring advanced machine learning techniques, incorporating algorithms such as decision trees and genetic algorithms to further enhance prediction accuracy. Additionally, the implementation of a real-time tracking system based on AI presents an opportunity for early anomaly detection, continuous monitoring of key performance indicators, and proactively adapting plans and resources in response to evolving project dynamics.

In summary, the article underscores the significance of predictive models in project management, showcasing the effectiveness of SVM and MLP while also presenting future perspectives for advancing AI applications in project analysis and monitoring.

6 Table of Abbreviations

AI	Artificial Intelligence
SVM	Support Vector Machine
MLP	Multi-layer perceptron's
KPI	key Performance Indicators
PV	Planned Value
AV	Acquired Value
RV COSTS	Real Value of Costs
SPI	Schedule Performance Index
CPI	Cost Performance Index
RPI	Resource Performance Index

References

1. Amin, A.M., Kamel, M.S.: Artificial intelligence in project management: a state-of-the-art review. Procedia Comput. Sci. **160**, 708–715, Location (2019)
2. Chen, J., Lu, M.: Artificial intelligence in project management: recent developments and future trends. J. Constr. Eng. Manag. **146**(1), 04019101, Location (2020)
3. Dörner, L., Rostami, S.: A review on artificial intelligence in project management: a state-of-the-art overview. Comput. Ind. **134**, 103285, Location (2021)
4. He, Z., Gao, P., Liu, X.: Application of artificial intelligence in project management: a bibliometric analysis. IEEE Access **8**, 72057–72068, Location (2020)
5. Khan, S., Bao, Y.: Project performance prediction using artificial intelligence techniques: a systematic literature review. Front. Artif. Intell. **3**, 62, Location (2020)
6. Li, Z., Wu, Q.: Application of artificial intelligence in project management based on bibliometrics and visualization analysis. Future Internet **11**(3), 67, Location (2019)
7. Liu, H., Li, X.: A bibliometric analysis of artificial intelligence in project management. Int. J. Adv. Eng. Res. Sci. **7**(7), 438–448, Location (2020)

8. Michael N.: Neural Networks and Deep Learning, Location (2015)
9. PMBOK (Project Management Body of Knowledge, (Third ed)), Location (2004)
10. Rönnberg, N., Rudberg, M.: Artificial intelligence in project management - a literature review and case study. In Proceedings of the 2020 International Conference on Information Systems, pp. 1–9, Location (2020)
11. Roy, R., Vittal, S.: Artificial intelligence for project management: challenges and opportunities. Int. J. Project Manag. **38**(6), 1–14, Location (2020)
12. Saleh, K., Okasha, M.: Applying artificial intelligence techniques for project control. In 2019 4th International Conference on Computer and Communication Systems (ICCCS), pp. 215–220. IEEE, Rental (2019)
13. Smith, K., Kumar, A.: Artificial intelligence in project management: opportunities, challenges, and implications. In Proceedings of the 2018 International Conference on Artificial Intelligence in Information and Communication, pp. 1–6, Location (2018)
14. Vladimir V.: Statistical Learning Theory, New York: Wiley, Location (1998)

Multi-criteria Inventory Classification with Machine Learning Algorithms in the Manufacturing Industry

Özge Albayrak Ünal[✉] and Burak Erkayman

Department of Industrial Engineering, Engineering Faculty, Ataturk University, 25240 Erzurum, Turkey
{ozgealbayrak,erkayman}@atauni.edu.tr

Abstract. Today, supplying the right quantity of materials and ensuring their availability are key factors in reducing inventory costs and improving the competitiveness of the industry. Companies need to hold large quantities of inventory to meet future demand. Proper inventory control is necessary to easily monitor and manage inventory levels, meet customer needs by controlling inventory levels, and balance items to be procured. To this end, a multi-criteria inventory classification (MCIC) with the integration of an analytic hierarchy process (AHP) and machine learning algorithms is proposed to examine the inventories of a manufacturing industry and create an applicable and effective inventory management system. First, the classes of inventory items are identified by ABC analysis using the AHP method. Then, their performance in inventory classification was evaluated using machine learning (ML) algorithms such as Decision Tree, Naive Bayes, Random Forest, and Support Vector Machine. The results show that the random forest algorithm is more effective than other methods in classifying inventory items.

Keywords: ABC analysis · AHP · Inventory control · Multi-criteria inventory classification · Machine learning

Appendix 1. Full names of the algorithms used in the study

Acronym	Full Name
MCIC	Multi-criteria inventory classification
AHP	Analytic hierarchy process
MCDM	Multi-criteria decision making
FUCOM	Full consistency method
ML	Machine learning
SVM	Support vector machine
CI	Consistency index
CR	Consistency ratio

1 Introduction

Nowadays, in a global business environment, the success of companies depends on effective inventory management. Inventory is an indispensable element to ensure the continuity of business operations of companies and to meet the needs of customers. The main objective of inventory management in companies is to ensure that the necessary materials are available for processing at the right time and the minimum cost. Good inventory management gives companies a competitive advantage by ensuring that operations are more effective and efficient. Optimizing the quantities held in inventory also helps avoid inventory and cost problems and ensures continuity in meeting customer demand. Companies can keep thousands of products in stock. For processes to be carried out punctually and accurately, a certain portion of the products must be kept in stock. Companies that operate with high inventory levels may incur higher warehousing and storage costs, as well as higher service levels. It also leads to a reduction in the value of goods in stock and a shortage of cash. Working with low inventory levels requires more dynamic inventory management and should be supported by effective supply chain planning. It has the advantage of reducing inventory costs, ensuring customer satisfaction by quickly meeting customer needs, etc. In addition, loss of sales, customer dissatisfaction loss of reputation, etc. can also lead to negative situations [1]. For these reasons, companies should establish an inventory policy based on their needs and operational requirements to ensure balanced inventory management and avoid unnecessary inventory. Inventory classification is needed here.

The main purpose of inventory classification is to ensure efficient inventory management by grouping inventories with similar characteristics into the same category. This classification enables stocks to be tracked and managed more effectively and helps support the decision-making process. This classification is referred to in the literature as ABC analysis. The traditional ABC classification technique uses the Pareto principle to divide articles into three classes, A, B, and C, according to their relative importance. Class A is the most valuable class, with 10–20% of the inventory and 60–80% of the total value; values closer to 30% can be obtained with 20% to 25% of Class B items. Class C has a value between 5% and 15% and, with 50–60% of the inventory, is the least important of the classes [2]. However, traditional ABC analysis is largely limited and inadequate for evaluation as it uses only one factor for classification. Therefore, many multi-criteria decision-making (MCDM) methods are used to support ABC analysis in classifying items more accurately by considering multiple criteria. In addition to these methods, machine learning algorithms have also begun to take part in inventory classification in recent years. Although the use of MCDM and ML approaches to inventory management is important, there is still a lack of studies in the literature that include solutions that combine both approaches [3, 4]. In these studies, MCDM methods perform the selection of critical inventory while machine learning increases predictive accuracy in managing the purchase of these products without human intervention [5].

The main purpose of this study is to develop an efficient approach by integrating MCDM and ML methods to analyze, and classify the inventories of a company producing spare parts for the automotive industry and to present powerful methods to support management decisions. This study examined the extent to which the AHP method, one of the MCDM methods, can be used in ABC analysis. As a result of the evaluation of the expert group, the criteria weights are determined by the AHP method, and the inventories are classified. The performance of the Decision Tree, Naive Bayes, Random Forest, and Support Vector Machine algorithms is then evaluated and the accuracy of the classification is examined.

The rest of this article is organized as follows. Section 2 reviews recent developments in the literature on inventory classification, MCDM, and ML methods. Section 3 presents the details of the proposed approaches. Data collection is mentioned in Sect. 4. Section 5 presents the case study and Sect. 6 presents the results.

2 Literature Review

A review of the literature shows that studies are developing new models related to inventory classification and testing the applicability of these models in different sectors. Albayrak Ünal et al. examined the studies in the literature using artificial intelligence methods in inventory classification [6]. Cakir and Canbolat proposed a web-based multi-criteria inventory classification system in an electrical household appliance company based on the AHP method [1]. Liu et al. used an inventory of a Chinese sports equipment manufacturer to perform a multi-criteria ABC analysis with a superiority model-based classification approach. They investigated the ABC classification problem that cannot be compensated between criteria [7]. Yiğit and Esnaf proposed a new three-stage MCIC comprising AHP, the Fuzzy C-Means, and the Revised-Veto stage to demonstrate the applicability of ABC classification principles [8]. Stević et al. Full Consistency Method (FUCOM) integration with ABC analysis is performed for a group of products from Gorenje devices [9].

The use of ML techniques in inventory classification delivers promising results in classification by automating challenging decision-making processes and processing huge amounts of data effortlessly. Some of the studies on this topic are given. Partovi and Anandarajan classify inventory units in a pharmaceutical company using artificial neural networks. This study is also the first to apply machine learning algorithms to MCIC [10]. de Paula Vidal et al. proposed a decision support framework that combines MCDM and ML approaches to assist a railway logistics operator in making maintenance, repair, and operation inventory decisions [3]. Lolli et al. used decision trees and random forest methods to classify inventories with intermittent demand structure of a company that manufactures electrical resistors. The results confirmed the better performance of the proposed methodology [11]. Balaji and Kumar, proposed the MCIC method to classify the inventory of an industry producing rubber parts for the automotive industry. They used the AHP method to estimate the value of the inventory system [12]. Lolli et al. classified multi-criteria inventories with intermittent or non-intermittent demand using SVM with Gaussian kernels and deep neural networks [4]. Kartal et al. have developed a hybrid methodology that integrates MCDM techniques and ML algorithms to analyze the inventory of an automotive company inventory based on multiple criteria. For each method, they determined detailed performance metrics for predicting the algorithms. They concluded that the best classification accuracy was achieved with the support vector machine (SVM) method [13]. Khanorkar and Kane used MCDM and ML models along with ABC analysis to classify inventories. They applied A. Hadi-Vencheh's model for selective classification of inventory in MCDM techniques. Using the K-Means clustering algorithm to classify the items, the results are compared with MCDM techniques [14]. Yu contrasts artificial intelligence-based classification algorithms such as SVM, back-propagation networks, and k-nearest neighbors with traditional multiple discriminant analysis using a dataset of 47 SKUs. The results indicated that, in comparison to other AI-based methods, SVM delivers a more accurate classification [15]. Roy et al. compared the SVM and K-Nearest Neighbor methods with the TOPSIS method for estimating the class of newly added inventory items [16]. Prachuabsupakij has classified the stock-keeping units in the spare parts warehouse of a factory in Thailand according to the ABC classification. Then, it compared the classification accuracy with the Random Forest, Bagging, Adaboost, Dagging, and Decorate methods. The results showed that the Adaboost algorithm is best suited to classify the spare parts of a factory [17].

Based on the studies mentioned above, this study makes various contributions:

- The proposed hybrid approach is considered a real-world application by considering the expert opinions and data of a company operating in the automotive spare parts industry.
- A hybrid method integrating MCDM and ML methods was used to effectively manage and improve inventory management decisions.
- The results were compared using the measures of accuracy, precision, recall, and F-measure and the best method for prediction was determined.

3 Proposed Approach

This research aims to develop a hybrid approach that combines machine learning methods with MCDM methods to perform a multi-criteria inventory analysis for the inventory of a manufacturing company. For this purpose, the AHP method is used to understand which inventories are critical for the company's operational activities and to order and classify the inventories according to their importance. Machine learning methods are then used to assess how accurate the inventory classes are. In addition, this section also presents the definitions of the metrics used to evaluate the performance of the algorithms. The flowchart of the study is shown in Fig. 1.

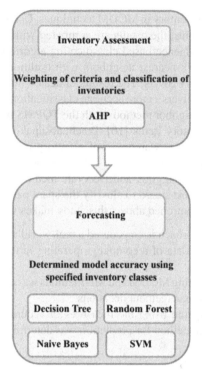

Fig. 1. The flowchart of the study.

3.1 Analytic Hierarchy Process

AHP was developed in 1980 by Thomas L. Saaty to solve and analyze complex problems [18]. AHP is a multi-criteria decision-making technique that shapes the decision-making process in a hierarchical structure and is based on pairwise comparison logic.

It uses the criteria established in the decision-making process to evaluate and prioritize alternatives while considering the decision-maker's subjective opinions. For the pairwise comparisons, Satty's 1–9 scale from Table 1 was used.

Table 1. AHP evaluation scale [18].

Value a_{ij}	Description
1	Criteria i and j are equally important
3	Criteria i is slightly more important than criteria j
5	Criteria i is more important than criteria j
7	Criteria i is most important than criteria j
9	Criteria i is absolutely most important than criteria j
2,4,6,8	Middle values

The AHP process consists of four steps [12]:

1. The problem is broken down into a hierarchical structure that includes a goal, criteria, sub-criteria, and alternatives.
2. The aim, the criteria, the sub-criteria, and the alternatives at the same level are compared pairwise according to their importance for each level of the hierarchical structure.
3. Using Eq. (1), the results of the pairwise comparison of n criteria are given in the n*n comparison matrix A.

$$A = (a_{ij}) \quad (1)$$

where $i, j = 1, 2, 3 \ldots .n$.
The priority weights for the comparison matrix are calculated using Eq. (2).

$$A_w = \lambda_{max} w \quad (2)$$

where A is a comparison matrix with n dimensions. λ_{max} is the biggest eigenvalue of A and w is the eigenvector corresponding to λ_{max}.

4. To evaluate the consistency of the matrix, the consistency index (CI) is calculated using Eq. (3).

$$CI = (\lambda_{max} - n)/(n - 1) \quad (3)$$

The consistency ratio (CR) is calculated with Eq. (4).

$$CR = \frac{CI}{RI} \quad (4)$$

where RI is the random index.

For consistency levels with a CR value of 0.1 or less, the significance level of each element of the pairwise comparison matrix is considered correct. Otherwise, it is best to repeat the process to increase consistency.

3.2 Decision Trees

Decision trees are one of the widely used non-parametric methods for classification or regression problems [19]. There is a predetermined target variable in decision trees. Their organizational structure provides a top-down approach [20]. It consists of roots, branches, and leaves that resemble a tree structure for classifying data. Branches are connected to nodes, provided that each branch is connected to the root above. Each attribute in the data represents a node in the tree after classification. The nodes between this tree structure are classification rules and each leaf is considered as a class. Decision trees have the advantage that they are easy to create and understand. Parameters used in the study: criterion: gini, splitter: best, min_samples_split:2, and min_samples_leaf:1.

3.3 Random Forest

Random Forest is one of the ensemble techniques and represents an approach where each of the classifiers is a decision tree classifier so that the classifiers form a collection of "forests" [17]. A random forest is a design in which the number of criteria randomly chosen for the division of items at each node and the number of trees in the forest are both varied. In the test phase, each random forest provides classification performance. Once a combination is identified, its performance is evaluated by averaging all folds, and the combination with the best performance is selected [11]. See [21] for the steps of the Random Forest method. Parameters used in the study for the Random Forest method: n_estimators = 100, criterion = 'gini', max_depth = None.

3.4 Naive Bayes

The Naive Bayes method is based on Bayes' Theorem and is a supervised classification method that predicts labels based on a probability function. The mathematical expression of Bayes' theorem can be found in Eq. 5.

$$P(X|Y) = \frac{P(X).P(Y|X)}{P(Y)} \quad (5)$$

Here $P(X)$ refers to the probability of event X and, $P(Y)$ refers to the probability of event Y. $P(X|Y)$ shows the probability of event X if event Y occurs, and $P(Y|X)$ shows the probability of event Y if event X occurs.

Although Naïve Bayes is a computationally difficult method, it works quite fast after the dataset is trained. Each attribute is considered independent of the other attributes of the class. Forecasting is done by classifying a new situation by combining the effects of the independent variable on the dependent variables. This classifier calculates the probability of an event occurring based on probability by performing some simple calculations. The classification is then made according to the one with the highest probability value. Naive Bayes is very useful for very large data sets and additional analysis. It is a simple classification method that performs well for complex problems [22]. Parameters used in the study; priors: None, var_smoothing:1e-9.

3.5 Support Vector Machine

Support Vector Machine is a supervised learning technique used for regression and classification problems. Rather than minimizing the mean squared error in the training dataset, SVM applies the notion of structural risk minimization, which allows a restriction on the generalization error of a model [23]. Using the closest data points for each class, the algorithm finds the hyperplane with the maximum distance between them. The data points closest to the hyperplane—known as support vectors—are employed to identify the hyperplane. Ideal for complex but small and medium data sets. In the SVM method generally favors linear, polynomial, radial, and sigmoid kernel functions. In this study, the Radial Base Function was used, which gave better results than the others. The kernel function used is given by Eq. (6).

$$G(X_i, X_j) = exp\left(-\|X_i - X_j\|^2\right) \quad (6)$$

where $\|X_i - X_j\|$ is the Euclidean distance between X_i and X_j.

3.6 Performance Metrics

The metrics mentioned in this section are the basic metrics to better evaluate the performance of the classification model and should be considered together in the evaluation. The calculations of accuracy, precision, recall and F-measure are given in Eqs. (7)–(10).

Accuracy: Refers to how accurate a measurement or estimate is. A high accuracy value indicates that the overall performance of the model is good.

$$Accuracy = \frac{TP + TN}{TP + TN + FP + FN} \quad (7)$$

Precision: For each class, it shows how many of the predictions for that class are correct. A high precision value indicates that the positive predictions of the model are largely accurate.

$$Precision = \frac{TP}{TP + FN} \quad (8)$$

Recall: It shows how many of all real examples belonging to the positive class were correctly predicted. A high recall value indicates how successful the model is without missing true positives.

$$Recall = \frac{TP}{TP + FP} \quad (9)$$

F-Measure: By calculating the harmonic mean of precision and recall, it evaluates the classification model's performance. It is used to select the best model when we have the precision and accuracy values of more than one model. The maximum value of the F1 measure is 1, while the minimum value is 0.

$$F - Measure = \frac{2 * TP}{2 * TP + FP + FN} \quad (10)$$

where TP indicates that the model correctly predicts the positive class, i.e. it represents the number of true positives. FP indicates that the model incorrectly predicts the positive class, i.e. it is the number of false positives. FN indicates that the model incorrectly predicted the negative class, i.e. it represents the number of false negatives. TN indicates that the model correctly predicts the negative class, i.e. it represents the number of true negatives.

4 Data Collecting

The inventory classification multi-criteria issue is implemented using real-world data from a manufacturing company in the automobile spare parts market. Before the model was created, the data underwent pre-processing stages such as the completion of missing data and the removal of erroneous data. The data set consists of 50 inventory items used by the company and is evaluated based on five criteria. A four-person expert group consisting of the company's procurement, planning, and warehousing managers was formed for the evaluation. In selecting and evaluating the criteria, the opinions of the expert group and the comprehensive literature review listed in Table 2 were considered. As a result, five important criteria were determined concerning their suitability for the company's data: Price, Demand, Delivery Time, Storage, and Criticality. The definitions of the criteria used in the study can be found in Fig. 2.

Table 2. List of criteria for the classification of the inventory.

References	Criteria
Flores and Whybark [24]	Average Unit Cost, Annual Dollar Usage
Flores et al. [25]	Price, Lead Time, Criticality, Annual Dollar Usage
Partovi and Anandarajan [10]	Unit Price, Ordering Cost, Demand, Lead Time
Cakir and Canbolat [1]	Price/Cost, Annual Demand, Blockade Effect in Case of Stockout, Availability of The Substitute Material, Lead Time, Common Use
Hadi-Vencheh and Mohamadghasemi [26]	Annual Dollar Usage, Limitation of Warehouse Space, Average Lot Cost, Lead Time
Rezaei [27]	Unit Price, Annual Demand, Stock Ability, Lead Time, Certainty of Supply
Yu [15]	Price, Lead Time, Criticality, Annual Dollar Usage
Soylu and Akyol [28]	Average Unit Cost, Annual Dollar Usage, Lead Time, Critical Factor
Šimunović et al. [29]	Annual Cost Usage, Criticality, Lead Time 1, Lead Time 2
Keshavarz Ghorabaee et al. [30]	Average unit cost, Annual dollar usage, Lead time
Kabir and Akhtar Hasin [31]	Price, Demand, Criticality, Expiration Date, Durability
Kartal et al. [13]	Criticality, Cost, Supply, Demand, Unit size
Kheybari et al. [2]	Average Unit Cost, Critical Factor, Annual Dollar Usage, Lead Time
Eraslan & Ic [32]	Price, Demand, Lead Time, Criticality, And Volume

(*continued*)

Table 2. (*continued*)

References	Criteria
Lolli et al. [4]	Demand, Lead Time, Purchasing, And Holding Costs
Razavi Hajiagha et al. [33]	Stochastic Demand, Unit Price, Current Stock Value, Lead Time, Criticality
Yiğit and Esnaf [8]	Average Unit Cost, Annual Dollar Usage, Criticality Factor, Lead Time
Tavassoli and Farzipoor Saen [34]	Average Unit Cost, Annual Dollar Usage, Critical Factor, Durability, Substitutability, Reparability
de Paula Vidal et al.[3]	Lead Time, Impact of The Shortage, Total Cost of Demand, Frequency of Use
Khanorkar and Kane [14]	Usage Value, Lead Time, And Unit Cost
Kansarn et al. [35]	Unit Cost, Usage, Annual spending, Lead Time, No. of potential suppliers, Product Lifecycle
Octaviany and Ishak [36]	Cost, Lead Time, Demand, Annual Usage, Life Time

Price (C1): A stock item's currency expression. The value of an item is a prevalent control feature for products, and it is preferable to avoid storing high value products in big quantities.

Demand (C2): This is the amount of demand for the stock item over the course of a year. If there is a high demand for the product in which the part is used, the part becomes important.

Lead Time (C3): It is the number of days that pass between the order of a stock item and its delivery to the company. The lead time is important because it indicates inventory levels in the face of unpredictable demand and the time it takes to respond to a crisis.

Storage (C4): It is the criterion that indicates the status of the associated material and how much of it is in the warehouse. It is important that, due to the current limitations on stocking items in the warehouse.

Criticality (C5): It is about how necessary inventory items are for the uninterrupted continuation of production and service processes. If the lack of parts disrupts production and service processes, the part be-comes more critical.

Fig. 2. Definitions of criteria.

5 Case Study

This study was conducted in one of the international car manufacturers in Turkey. First, the AHP method was applied to classify 50 different items in the company's warehouse. Then, machine learning methods were applied to perform the prediction for multi-criteria inventory analysis.

5.1 Determination of Criteria Weights Using the AHP Method

In this section, information on the ABC classification is first provided and shown in Fig. 3. ABC inventory management systems use the widely used Pareto principle to classify inventory items into A, B, and C categories according to their importance.

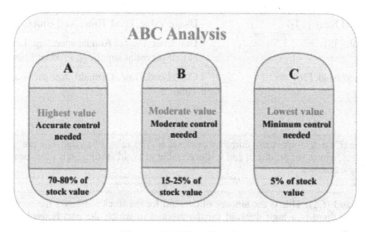

Fig. 3. ABC Classification.

Class A: This is the smallest category, representing the highest quality and most valuable items of inventory and customers owned. These products are designed to make a significant contribution to overall profits without much cost to the seller's resources.

Class B: It includes inventory items that are less critical than class A items and more critical than class C items. In this category, there is the possibility of moving up to class A if sales are good, and the possibility of falling to class C if sales fall.

Class C: The inventories in this class are the largest category that contributes the least to the company's profit. These products provide a steady and stable income to maintain the business continuously.

The steps of the AHP method were started. In this step, five different criteria (price, demand, delivery time, storage, and criticality) were taken as input, and product classes were selected as output. The AHP method was applied using Excel. The experts created a decision matrix with pairwise comparisons of the criteria using the scale given in Table 1. The weights of the criteria were then calculated and listed in Table 3.

Table 3. Criterion weights.

Criteria	C1	C2	C3	C4	C5	Criteria weights (w)
C1	1	0.500	5	7	7	0.365
C2	2	1	3	5	5	0.380
C3	0.200	0.333	1	3	3	0.130
C4	0.143	0.200	0.333	1	0.333	0.047
C5	0.143	0.200	0.333	3	1	0.076
Consistency Ratio: 0.085						

According to Table 2, the demand criterion is the most important criterion with the highest weight (38%) among the criteria. The demand criterion is followed by the criteria price, lead time, criticality, and storage. The consistency ratio of the pairwise comparison matrix, which was created based on the criteria, was specified as 0.085. As this value is below 0.1, it shows that the comparison matrix is consistent.

The steps of the AHP method are repeated when comparing the alternatives for each criterion and the total weighted score for each inventory item is calculated and ranked in descending order. The percentage cumulative weights are then determined and the inventory items are assigned to classes A, B, or C according to the Pareto principle. The most important inventory class is class A, which accounts for the first 20% of inventory items. The next 30% are designated as class B, while the last 50% represent class C.

Inventories are classified into 10 items for class A, 15 items for class B, and 25 items for class C. Table 4 shows the ranking of the items and the final classifications. Class A items should be reviewed at short intervals and more attention should be paid to demand forecasting. In addition, they should be stocked in locations that allow easy movement and handling and in regional warehouses that are closer to the market. Class B products should be reviewed at medium-term intervals and demand forecasting should be carried out carefully. Storage areas should be optimized according to the demand level and rotation speed of these products. Class C items, on the other hand, should be reviewed at longer intervals and the demand forecast should be less sensitive than for other product classes, as this class contains products with low demand. Low stock levels can be by determining for Class C products and they should be stored in more remote locations.

Table 4. Cumulative sums and items classes.

Items Ranking	Weighed Sums	Cumulative Sums	Class
3	0.039	3.86%	A
1	0.038	7.61%	A
12	0.037	11.34%	A
5	0.036	14.95%	A
14	0.035	18.45%	A
⋮	⋮	⋮	⋮
22	0.020	58.70%	B
13	0.020	60.66%	B
21	0.020	62.61%	B
34	0.019	64.48%	B
47	0.019	66.35%	B
⋮	⋮	⋮	⋮
27	0.010	96.19%	C
18	0.010	97.16%	C
45	0.010	98.12%	C
50	0.009	99.06%	C
49	0.009	100.00%	C

5.2 Implementation of Machine Learning Algorithms

Supervised machine learning algorithms are applied to determine the accuracy of the inventory classes. Demand, criticality, price, lead time, and storage criteria are taken as inputs and the inventory classes determined using the AHP method are selected as outputs.

6 Result and Discussion

6.1 Performance Metrics Results

A dataset of 50 inventory items is analyzed in the Python programming language using a Decision Tree, Random Forest, Naive Bayes, and five-fold cross-validation with Support Vector Machine. To create the model, the data set was divided into 80% for training and 20% for testing. Since accuracy alone is not sufficient to evaluate the classification mechanisms, a detailed analysis based on confusion matrices, including precision, recall, and F-measure was performed. The classification accuracy of the algorithms is shown in Fig. 4.

Multi-criteria Inventory Classification with Machine Learning Algorithms

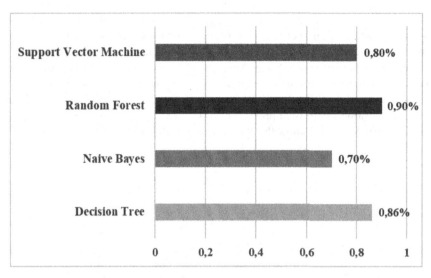

Fig. 4. Accuracy of the algorithms.

As a result of these evaluations, it was determined that the random forest method classifies the inventory items more accurately than the other methods. An accuracy rate of 90% means that the classification model correctly predicted 90% of the total samples. This rate indicates that the overall performance of the model is high. The Decision Tree and SVM methods also performed better than Naïve Bayes.

The performance metrics are shown in Table 5 and Figs. 4, 5 and 6.

Table 5. Performance metrics.

	Class	Precision	Recall	F-Measure
Decision Tree	A	0.67	1.00	0.80
	B	1.00	0.67	0.80
	C	0.88	1.00	0.93
Naive Bayes	A	0.00	0.00	0.00
	B	0.50	1.00	0.67
	C	1.00	0.80	0.89
Random Forest	A	1.00	0.80	0.89
	B	0.50	1.00	0.67
	C	1.00	1.00	1.00
SVM	A	1.00	0.50	0.67
	B	0.67	1.00	0.80
	C	1.00	0.75	0.86

Random Forest predicted class A with high precision and slightly lower recall. This suggests that while the model correctly predicts most instances of class A, it misses some of the actual class. Although the model can detect the entire class B, only half of these predictions are correct. The model correctly detected all of class C and did not make any incorrect predictions. Examining the precision, recall, and F-measure values of Naive Bayes, we find that the model does not predict class A correctly at all. On the other hand, it shows that it both predicts class C correctly and detects most of the real class C. Figures 5, 6, and 7 show the comparison of precision, recall, and F-measure for the algorithms according to classes A, B, and C.

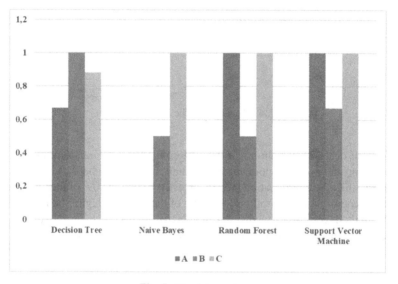

Fig. 5. Precision values.

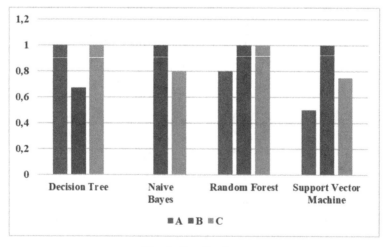

Fig. 6. Recall values.

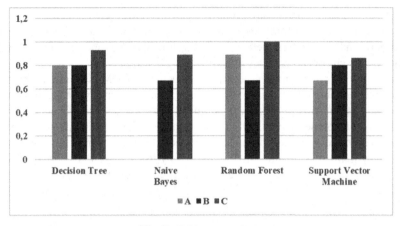

Fig. 7. F-Measure values.

Prediction and training time are a critical component in evaluating the performance, effectiveness, and usability of a machine-learning model. The training time indicates how quickly and efficiently the model can be trained, while the prediction time reflects the model's ability to make quick and effective predictions for new data. The training and prediction times for the algorithms are shown in Table 6.

Table 6. Training, estimation, and total times by algorithms.

Algorithms	Training Time	Prediction Time	Total Time
Decision Tree	0.841	0.909	1.750
Naive Bayes	1.273	0.688	1.961
Random Forest	0.960	0.608	1.568
SVM	0.876	0.978	1.854

Looking at Table 6, the algorithm with the lowest training time is the decision tree with 0.841, while the algorithm with the lowest prediction time is the random forest with 0.608. However, if the total time is considered, it can be seen that the calculation of the random forest method takes less time than the others.

7 Conclusion

Today, many companies need to properly classify and effectively manage their inventory to improve their capabilities such as stock tracking, inventory management, and cost control. In this study, MCIC and machine learning methods are integrated to classify inventories according to their importance. A real-world case study is presented to validate inventory classification using the data set of 50 items from an automobile spare parts manufacturing company. First, the weights for the items were determined by analyzing the criteria and alternatives using the AHP method. Based on the cumulative values, the inventories were categorized into classes A, B, and C. Then the methods Decision Tree,

Naive Bayes, Random Forest, and Support Vector Machine were applied to predict the identified classes with five-fold cross-validation. The accuracy and performance metrics of the algorithms were determined. The results show that the random forest algorithm classifies the inventory items better than the other algorithms.

The integration of machine learning algorithms with MCDM methods has the potential to provide valuable information for the development of inventory management strategies and decision-making processes. In addition, it can make a significant contribution to increasing the competitiveness of companies and will enable companies to manage inventory data more effectively by increasing operational efficiency. In future studies, the inventory items will be effectively classified by using more than one MCDM and different ML methods. In addition, the classification process will be performed with a larger number of criteria and inventory items.

References

1. Cakir, O., Canbolat, M.S.: A web-based decision support system for multi-criteria inventory classification using fuzzy AHP methodology. Exp. Syst. Appl. **35**(3), 1367–1378 (2008)
2. Kheybari, S., Naji, S.A., Rezaie, F.M., Salehpour, R.: ABC classification according to Pareto's principle: a hybrid methodology. Opsearch **56**, 539–562 (2019)
3. de Paula Vidal, G.H., Caiado, R.G.G., Scavarda, L.F., Ivson, P., Garza-Reyes, J.A.: Decision support framework for inventory management combining fuzzy multicriteria methods, genetic algorithm, and artificial neural network. Comput. Indust. Eng. **174**, 108777 (2022)
4. Lolli, F., Balugani, E., Ishizaka, A., Gamberini, R., Rimini, B., Regattieri, A.: Machine learning for multi-criteria inventory classification applied to intermittent demand. Prod. Plan. Control **30**(1), 76–89 (2019)
5. Kourentzes, N.: Intermittent demand forecasts with neural networks. Int. J. Prod. Econ. **143**(1), 198–206 (2013)
6. Albayrak Ünal, Ö., Erkayman, B., Usanmaz, B.: Applications of artificial intelligence in inventory management: a systematic review of the literature. Archiv. Comput. Methods Eng. **30**(4), 2605–2625 (2023)
7. Liu, J., Liao, X., Zhao, W., ZYang, N.: A classification approach based on the outranking model for multiple criteria ABC analysis. Omega **61**, 19–34 (2016)
8. Yiğit, F., Esnaf, Ş: A new Fuzzy C-Means and AHP-based three-phased approach for multiple criteria ABC inventory classification. J. Intell. Manuf. **32**(6), 1517–1528 (2021)
9. Stević, Ž, et al.: An integrated ABC-FUCOM model for product classification. Spect. Eng. Manag. Sci. **1**(1), 83–91 (2023)
10. Partovi, F.Y., Anandarajan, M.: Classifying inventory using an artificial neural network approach. Comput. Ind. Eng. **41**(4), 389–404 (2002)
11. Lolli, F., Ishizaka, A., Gamberini, R., Balugani, E., Rimini, B.: Decision trees for supervised multi-criteria inventory classification. Procedia Manufact. **11**, 1871–1881 (2017)
12. Balaji, K., Kumar, V.S.: Multicriteria inventory ABC classification in an automobile rubber components manufacturing industry. Procedia CIRP **17**, 463–468 (2014)
13. Kartal, H., Oztekin, A., Gunasekaran, A., Cebi, F.: An integrated decision analytic framework of machine learning with multi-criteria decision making for multi-attribute inventory classification. Comput. Ind. Eng. **101**, 599–613 (2016)
14. Khanorkar, Y., Kane, P.: Selective inventory classification using ABC classification, multi-criteria decision-making techniques, and machine learning techniques. Mater. Today: Proc. **72**, 1270–1274 (2023)

15. Yu, M.-C.: Multi-criteria ABC analysis using artificial-intelligence-based classification techniques. Exp. Syst. Appl. **38**(4), 3416–3421 (2011)
16. Roy, A., et al.: Comparative analysis of KNN and SVM in multicriteria inventory classification using TOPSIS. Int. J. Inf. Technol. **15**(7), 3613–3622 (2023)
17. Prachuabsupakij, W.: ABC Classification in spare parts for inventory management using ensemble techniques. In: 2019 IEEE Asia Pacific Conference on Circuits and Systems (APCCAS). IEEE (2019)
18. Saaty, T.L.: The Analytic Hierarchy Process. Education (1980)
19. Bhargava, N., Purohit, R., Sharma, S., Kumar, A.: Prediction of arthritis using classification and regression tree algorithm. In: 2017 2nd International Conference on Communication and Electronics Systems (ICCES), pp. 606–610. IEEE (2017)
20. Kantardzic, M.M.: Data Mining: Concepts, Models, Methods, and Algorithms. Wiley (2020)
21. Ali, M.R., Nipu, S.M.A., Khan, S.A.: A decision support system for classifying supplier selection criteria using machine learning and random forest approach. Decis. Analyt. J. **7**, 100238 (2023)
22. Jackins, V., Vimal, S., Kaliappan, M., Lee, M.Y.: AI-based smart prediction of clinical disease using random forest classifier and Naive Bayes. J. Supercomput. **77**, 5198–5219 (2021)
23. Cervantes, J., Garcia-Lamont, F., Rodríguez-Mazahua, L., Lopez, A.: A comprehensive survey on support vector machine classification: applications, challenges and trends. Neurocomputing **408**, 189–215 (2020)
24. Flores, B.E., Whybark, D.C.: Implementing multiple criteria ABC analysis. J. Oper. Manag. **7**(1–2), 79–85 (1987)
25. Flores, B.E., Olson, D.L., Dorai, V.: Management of multicriteria inventory classification. Math. Comput. Model. **16**(12), 71–82 (1992)
26. Hadi-Vencheh, A., Mohamadghasemi, A.: A fuzzy AHP-DEA approach for multiple criteria ABC inventory classification. Expert Syst. Appl. **38**(4), 3346–3352 (2011)
27. Rezaei, J.: A fuzzy model for multi-criteria inventory classification. In: Analysis of Manufacturing Systems, pp. 167–172 (2007)
28. Soylu, B., Akyol, B.: Multi-criteria inventory classification with reference items. Comput. Ind. Eng. **69**, 12–20 (2014)
29. Šimunović, K., Šimunović, G., Šarić, T.: Application of artificial neural networks to multiple criteria inventory classification. Strojarstvo: časopis za teoriju i praksu u strojarstvu, **51**(4), 313–321 (2009)
30. Keshavarz Ghorabaee, M., Zavadskas, E.K., Olfat, L., Turskis, Z.: Multi-criteria inventory classification using a new method of evaluation based on distance from average solution (EDAS). Informatica **26**(3), 435–451 (2015)
31. Kabir, G., Akhtar Hasin, M.A.: Multi-criteria inventory classification through integration of fuzzy analytic hierarchy process and artificial neural network. Int. J. Ind. Syst. Eng. **14**(1), 74–103 (2013)
32. Eraslan, E., Ic, Y.T.: An improved decision support system for ABC inventory classification. Evol. Syst. **11**, 683–696 (2020)
33. Razavi Hajiagha, S.H., Daneshvar, M., Antucheviciene, J.: A hybrid fuzzy-stochastic multi-criteria ABC inventory classification using possibilistic chance-constrained programming. Soft. Comput. **25**(2), 1065–1083 (2021)
34. Tavassoli, M., Farzipoor Saen, R.: A stochastic data envelopment analysis approach for multi-criteria ABC inventory classification. J. Ind. Prod. Eng. **39**(6), 415–429 (2022)
35. Kansarn, P., Sumrit, D., Vanichchinchai: Applied Multiple Criteria Inventory Classification for General Spare Parts: A case Study in Cement Industry in Thailand (2023)
36. Octaviany, C.D., Ishak, D.P.: Analysis of Multi Criteria Classification and Inventory Planning on Flexible Packaging Industry (2022)

Analyzing Results of Business Process Automation with Machine Learning Methods

Elif Yigit(✉) and Seda Özmutlu

Department of Industrial Engineering, Uludag University, Bursa, Turkey
Elif.yigit1623@gmail.com

Abstract. Production and management systems and enterprise resource management systems are constantly creating data. Businesses use integrated systems to monitor all processes such as sales, planning, production and logistics systems. During the follow-up and use of these systems, office employees perform routine, repetitive and non-value-added transactions. Manual processes such as invoice entries, sales data entry, and order transfer significantly reduce employee satisfaction and cause some personal errors. Due to the developing requirements, the concept of Robotic Process Automation (RPA), which can operate like humans in many programs and customer systems, have emerged in recent years. There are softwares that work as a white collar employee in enterprises to perform RPA-defined, non-interpretation-based, rule-based and standard tasks. In this study, we study the task of uploading invoices to the customer system, that is one of the standard and routine transactions in Logistics Processes. These tasks are automated with RPA. Software robots repeat the processes and purify the process from non-value-added transactions. However, software robots receive some errors in the processes. Data mining methods are used in this study, in order to examine the software outputs and RPA errors. Reports on the results of RPA were analyzed with various machine learning methods using the WEKA software. As a result of the study, the J48 algorithm with an F-value of 75% gave the best result in the estimation of RPA outputs. In future studies, analyses will be made to examine and eliminate the root causes of the errors.

Keywords: Robotic Process Automation · Machine Learning · WEKA · Digital Transformation

1 Introduction

Most of the daily work in businesses consists of routine tasks such as, invoice entries, receipt control, order entries, uploading of loading documents to the relevant systems, reporting, etc. In addition, most of the office workers' time is spent setting up meetings and ensuring that the work is done efficiently and without error. Since these tasks are standard, repetitive, and non-value-added, they reduce employee motivation and loyalty. This environment creates a good application ground for automatic processes. Automation of standard and repetitive tasks prevents manual errors in processes and affects employee satisfaction.

Companies play a proactive role in adopting new technologies due to competitive market conditions. Different technologies have been developed in recent years for automation activities in business processes. These technologies are generally labeled as software robots. Software robots are systems that use machine learning and computer perception to perform any task performed on computer systems. Software robots do not have any hardware components like physical robots but are only found in computer systems. Robotic Process Automation (RPA), on the other hand, is a technology that uses software robots that can communicate with each of the digital systems and imitate the actions of a person to carry out a business process. Robots are used for rule-based and well-structured repetitive tasks. RPA uses Optical Character Recognition (OCR) technology to detect and extract data from a document as a preliminary step in processing a document. Optical Character Recognition (OCR) is a technology for extracting data from image-based documents. Once the text is extracted from the images, these texts can be entered into any platform such as computers, web pages, and ERP systems as editable text [13, 15]. Figure 1 shows the general structure of process automation.

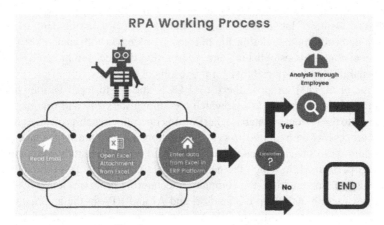

Fig. 1. Process Automation [22].

Although one of the goals of business process automation is to carry out automation with the least error as possible; RPA robots can nonetheless make mistakes just as humans. Errors, on the other hand, should be as low as possible in order to increase the efficiency of the process and to get a return on the investment of the software made in the RPA robot. Otherwise, the benefit of process automation will be less than expected.

In this study, the automation of the processes in the logistics department of a large textile enterprise is considered. The reason for choosing the logistics operation is that there are a lot of routine transactions and paperwork in the logistics department, which are mostly routine and do not involve critical decisions. It is expected that by integrating the RPA process, significant time and efficiency gains will be observed in the process.

In this study, we aim to examine the outputs and errors of the RPA process. Afterwards, prediction of the output of RPA will be performed with various machine learning classification algorithms. After output prediction, the root causes of errors can be studied, and activities related to the reduction of errors can be started.

2 Literature Research

2.1 Digitalization and RPA

Since digitization of business processes is a relatively new area of research, the literature available on the topic is limited. In this section, we share several papers on the topic. Salmen [14] investigated the time savings to be obtained from the processes of Small and Medium-Sized Enterprises (SMEs) using Artificial Intelligence and RPA applications. While mentioning that companies have increased their needs and investments in digitalization after the Corona epidemic, they also questioned the necessity of this investment in SMEs. In the study, the cases where SME order entry processes carried out manually and with RPA, were evaluated. While they expect RPA to save at least 50% of the process duration, the actual time savings compared to traditional tools resulted as 64.55%.

Filzmoser and Koeszegi [4] investigated the requirements and stages in the digitalization of processes using robotics and AI. There are some roles that human and robotic automations play in the integration of these technologies into business processes. There are four important roles for managing and sustaining this collaborative work. These roles are (i) division of labor-distribution of duties and roles, assignment, (ii) process integration-determining the relationship of roles in processes with each other, (iii) delegation of decision rights-establishing and determining the decision hierarchy structure, (iv) accountability and responsibility for the results of activities.

Rohaime et al. [13], propose the use of OCR and AI to input various templates such as invoices, receipts and forms, which are among the daily and routine works of businesses, into the system or database. Performing operations such as invoice entries and document processing with robotic process automation, not with the help of an employee, is beneficial both to save time and reduce manual errors. In this study, automation of a process was performed, in which the necessary details are taken from the invoice and this information is entered into the company's accounting program. It uses OCR in the process of extracting data from the invoice, and AI and RPA in the automation of the process. In the study, two different RPA bots were compared and it was seen that the accuracy rates of both were 100%. The bots compared are TagUI - PyTesseract and AA-IQ Bot, respectively. The bot that traded faster than these two tools was the AA IQ Bot. The open source structure of the IQ Bot has provided cost-efficiency and flexibility in its application.

Syed et al. [16] present a literature review that identifies the themes and challenges of studies on RPA in their article. In this study, 125 sources were examined. The processes that bots will automate are manually integrated into the system. Changes in business processes need to be defined to bots. Therefore, due to the bots' adherence to outdated rules, they run the risk of producing inaccurate results and their performance decreasing over time. There is a need to monitor the work of bots, identify changes in business rules, and develop proactive approaches in response.

Ling et al. [10] studied the automatic processing of business documents. They propose a model by incorporating automatic document processing into the system with human-computer interaction. OCR was used effectively together with RPA technology in the study. First, computer vision technology is used through OCR to decode the document in PDF or image format and extract relevant information from the documents.

Document numbers, titles and sources are determined. As a second step, text features are extracted from documents using natural language processing technology and text segmentation. The study includes 9653 documents. 91% classification accuracy is achieved with the Bayesian method. The model proposed in this study significantly reduced the document processing employees by transforming the document processing process via an AI algorithm through RPA. They have increased the efficiency to over 200% and the error rate to a controllable level of 0.05%.

Terminio and Gilabert [17] discuss the effects of robotic automation and artificial intelligence (RAAI) in the work environment. It is emphasized that RAAI will constantly confront employees with rapidly advancing technological developments. Traditional education might not be sufficient to adapt to transition. Research shows that involving individuals in this transition process produces positive results and reduces social costs. All parties, especially the relevant staff, should be a part of the process. For effective individual participation, it is necessary to determine the right strategy to motivate change and prevent resistance. The paper emphasizes the importance of adopting a comprehensive and inclusive approach to address the challenges in the evolving business environment.

Agostinelli et al. [1] present SmartRPA technology in their study as an innovative contribution that improves standard RPA applications. RPA process flow diagrams are applied for jobs with a specified standard. After the process is automated, any changes to the process must be defined to software robots. SmartRPA technology is a technology that directly observes the user's behavior on the interface and is developed automatically. In order to reduce the disadvantages of standard RPA applications, SmartRPA offers an approach that interprets user interface logs and selects the most appropriate process among the multiple processes it examines. SmartRPA is an important first step towards intelligent and fully automated software robots. The main weakness of the approach is that real-world user interfaces are more complex than the interface logs considered and recorded in this study. Normally, many different processes are executed simultaneously in the user interface logs, making the selection of possible variations difficult. In future studies, it is aimed to increase the depth of the algorithm by including more heterogeneous user interface logs.

Gruzauskas and Pacesaite [6] discuss the opportunities created in the business world with the rise of Industry 4.0 and the importance of AI-based systems that change supply chains. AI, automation and sustainability are critical to keep pace with rapidly changing dynamics in business. While the research emphasizes the complexity of using artificial intelligence-based systems to solve the transportation route planning problem, it also presents how these technologies are applied in the transportation and logistics industry and the design of a transportation coordination expert system that can provide companies with a competitive advantage.

2.2 Machine Learning Algorithms

Machine learning is a branch of AI designed to ensure that computer systems have the capacity to learn from data based on previous experience, in order to perform a specific task, solve a problem, or gain a skill. By extracting patterns and relationships from data sets, algorithms can make predictions against unknown future data and optimize results [8].

The main goal of machine learning is to improve the ability of algorithms to understand patterns obtained from data and predict future events by using the generalizing capabilities of algorithms. This is used to analyze, model, and make sense of information in complex and large data sets. Furthermore, machine learning includes various techniques such as classification, regression, clustering, and reinforcement learning, which makes it possible to be used effectively in different application areas. Therefore, machine learning plays an important role at the intersection of disciplines such as computer science, statistics, mathematics, and data science [2].

The main categories of machine learning approaches are unsupervised, supervised and reinforcement algorithms as seen in Fig. 2.

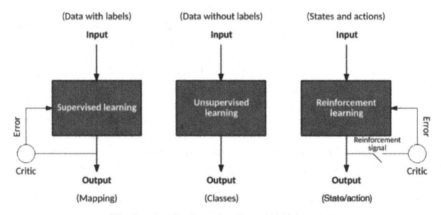

Fig. 2. Machine Learning Core Algorithms [1].

Supervised machine learning is a machine learning paradigm, in which an algorithm learns from a specific input dataset and gains the ability to make predictions on new data. The dataset on which the algorithm is trained has an output label or target value for each sample datum. The algorithm learns a general pattern using these input-output pairs and can then use that model to make predictions or generate responses for new data. Many algorithms such as linear regression, logistic regression, decision support systems, decision trees, random forest are included in this class of learning algorithms [2].

Unsupervised machine learning uses machine learning algorithms to analyze and aggregate unsupervised data clusters, commonly related to clustering, density estimation, and representative learning tasks.

Reinforcement learning is a learning paradigm based on learning from experience with the aim of improving an agent's ability to move within a given environment. This type of learning seeks to learn how an agent interacts with its environment to choose the best course of action based on the feedback it receives from those interactions [2].

Classification results include a variety of metrics that are used to evaluate the performance of a machine learning model. These measurements are used to evaluate the accuracy, sensitivity, recall, and specificity of the model. In a binary classification scenario, four basic values stand out:

- **True Positive (TP):** The number of cases that the model correctly predicted as positive.
- **True Negative (TN):** The number of cases that the model correctly predicted as negative.
- **False Positive (FP):** The number of cases that the model predicted as positive but were actually negative. Refers to a false alarm condition.
- **False Negative (FN):** The number of cases that the model predicted as negative but were actually positive [19].

Along with these values, a number of other performance measures can also be obtained. Key metrics include:

- **Precision:** The percentage of true positives, against what is predicted as positive

$$Precision = \frac{TP}{TP + FP} \quad (1)$$

- **Recall:** The percentage of the true positives that have been detected.

$$Recall = \frac{TP}{TP + FN} \quad (2)$$

- **Accuracy:** The proportion of total states that were correctly predicted.

$$Accuracy = \frac{TP + TN}{TP + TN + FP + FN} \quad (3)$$

- **F1 Score:** The harmonic mean of the Precision and recall values.

$$F1 = 2 \cdot \frac{Precision \text{x} Recall}{Precision + Recall} \quad (4)$$

These metrics are used to evaluate the performance of a classification model and to understand how effective the model is. Depending on the characteristics of the model and its application area, which the choice of the critical metric may vary [11].

2.3 J48 Algorithm

In this study, we use the J48 algorithm to predict the output of RPA in logistics tasks, since it provided the best results for the task. In the WEKA data mining tool, J48 is an open source Java implementation of the C4.5 algorithm [7], and an extension of ID3. This algorithm generates the rules, from which particular identity of that data is generated. It calculates the output value of a new sample based on various attributes of available data. Different attributes are represented by different internal nodes of a decision tree, the possible values that these attributes can have in the observed samples are denoted by the branches between the nodes, and the terminal nodes give the final output [3]. The objective is to progressively develop a decision tree to achieve a balance between flexibility and accuracy [9].

3 Methodology

3.1 Process Automation with RPA

In this study, the business process of "Uploading Payment Documents to the Customer System", which is one of the business processes of the Logistics unit in a textile company, was automated with RPA. The reason for choosing this business process is that there are many repetitive work steps involving no human decision. The work on improving logistics business processes with RPA started in February 2023.

Automation of computer tasks that are rule-based and not interpretation-based is done with RPA. Performing the repetitive tasks with no added value for office workers, RPA bots increases the efficiency of the process and reduces the error rate. Ling et al. [10] measured the productivity as high as 200% and the error rate as 0.05% as a result of document processing with RPA.

There are some main steps for the implementation of the RPA business process. These steps are referred to as the Robotic Process Automation lifecycle (see Fig. 3). After the logistics process was identified as the application area in accordance with the process in Fig. 3, the process was analyzed before automating the process with RPA.

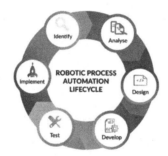

Fig. 3. Robotic Process Automation Life Cycle [23].

Business Process Definition and Analysis. In order to test the suitability of the process to RPA, the office employees' rework of the process and the process inputs, process stages and outputs of the process were observed. Observations are usually made with business analysis teams, who do not own the business but have a good command of the process flows. Business analysts evaluate the suitability of the examined process for RPA automation. While making this evaluation, the value to be gained by the automation with RPA of the process is calculated. In line with these calculations, the time spent on the process is kept with the help of time studies and the return on investment (ROI) is calculated for automation. If the cost of the investment to be made (software and hardware investment, workload and project cost) is less than the profit to be obtained, the automation of the process is decided. However, ROI is not the only factor influencing this decision. Some non-value-added jobs can be automated because they significantly affect employee satisfaction, even if they are not high-yielding. In line with these studies on the suitability of the process for automation, the next stage is passed.

Creation of Process Flow Diagrams and Ideal Case Design. Initially, flow diagrams of the current situation are drawn. After standard, repetitive, non-value-added jobs are eliminated, new flow diagrams are created, in which the ideal/desired process design is presented. This stage is concluded with a group work of project stakeholders, process owners and analyst teams.

In this case study, comprehensive process flow diagrams were created for the current and ideal processes. The ideal process is designed in three stages: In the first stage, the payment documents received by e-mail are saved in public folders by the end users. These saved documents are added to the work list in the public folders according to the order in which they are uploaded to the system. At the second stage, the order of loading information and time of payment documents is combined into a single format. In the third stage, the file, which is turned into a single pdf format, is uploaded to the system. A section from the process flow diagrams is given in Fig. 4.

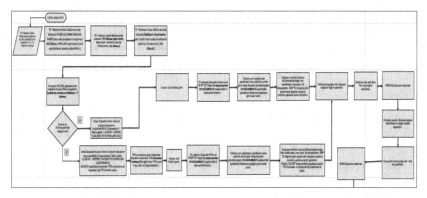

Fig. 4. A section of the process flow diagram of the current logistics invoice process.

Process Automation with RPA-Software, Testing and Go-Live Phase-Implementation. After the ideal process design is made, bot development studies are initiated with the participation of IT Analysts and software teams from the project stakeholders. The automation of the process was implemented with RPA bots developed by Robusta, a Turkish software company (https://robusta.ai/tr/). Flows of the software bot (see Fig. 5) were created according to the ideal situation modeled by the business units. After the automation development is completed, the testing phase begins. After the successful completion of the process, software and testing phase, the process was taken live. Of course, continuous development requires measuring the performance of the process, and making improvements.

3.2 RPA Performance Analysis

RPA might end a process successfully or with errors. RPA consistently documents the work done with the list of the tasks it performs, reports with data on many parameters. A sample report output is shown in Fig. 6.

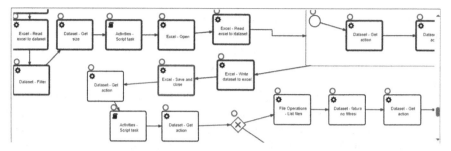

Fig. 5. RPA Bot Process Diagram.

Fig. 6. RPA output reports.

The process went live on April 1, 2023 and data up to August 31, 2023 were considered, since this was the time where the research began. There are a total of 854 data points. 169 data points from before Go-Live were excluded, resulting in 685 data points. In the report, the parameter under the heading "EndedDate" refers to the end time of the process. Input values were obtained from the end date information. The input and output parameters and levels obtained from the RPA robot are as follows:

Input Values;

- Month (April 2023 – May 2023 – June 2023 – July 2023 – August 2023),
- Days of the month (1–31),
- Day (Monday – …….. – Sunday),
- Time (01:00 -…… - 24:00),
- Shift (1- 2 - 3)

Output (Result) Value;

- End status (SUCCESS, FAILED) (Fig. 7)

When systems are Go-Live, errors are expected. Descriptive analyses and graphs were used to examine the relationship between the process parameters and the end status. Figure 8 shows the breakdown of end status by month. In April, the first month after the bot went live, more errors are apparent. April data can be excluded in future studies, however is currently kept, due to scarcity of data.

The ratio of the end status to the days of the week is shown in Fig. 9. Day input does not seem to have a clear effect on the process. Errors seem to fall towards the second half of the week. Mondays, Wednesdays and Thursdays might require further analyses, since percentage of errors is the highest.

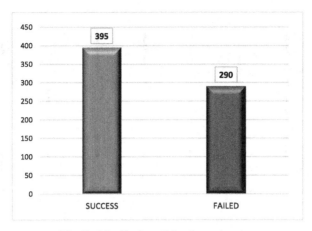

Fig. 7. Distribution of data by end status.

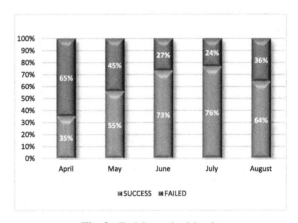

Fig. 8. End Status by Month.

The effect of shifts on the end status is in Fig. 10. More errors were made in the 1st shift, i.e. the regular office hours. One reason could be that concurrently programming the process and working the process with the robot causes more errors. When end users log in to the common shared folder, together with the robot, errors are received in file access. The work hierarchy or task order of robots and office staff should be orchestrated and regulated.

Graphs for evaluating the process output according to the hours of the day and the days of the month are given in Figs. 11, and 12, respectively. During some hours and days, the error rate seems to be increasing. Reasons and root causes should be investigated.

3.3 RPA Output Prediction with Machine Learning Algorithms

The aim of this study is to predict the output or result of RPA in a logistics invoice process. In order to predict outputs of the RPA process, several machine learning algorithms

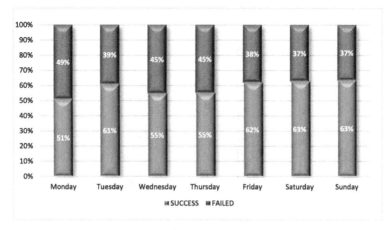

Fig. 9. End Status by Day of the Week.

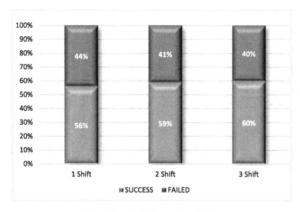

Fig. 10. End Status by Shift.

were experimented within the WEKA environment (https://www.cs.waikato.ac.nz/ml/weka/). WEKA is a Java-based open-source software package that stands for "Waikato Environment For Knowledge Analysis".

This study used the following algorithms to estimate the end status in RPA:

- **J48 (C4.5):** J48 is the implementation of the C4.5 decision tree algorithm in WEKA. It creates a decision tree by analyzing the structure of the data set. The decision tree uses a tree-like graph and acts as a decision support system [21]
- **RandomForest:** Random forest is a method of classification by combining many decision trees [8].
- **SMO (Sequential Minimal Optimization):** SMO is a training algorithm for support vector machines (SVM). SVM classifies by creating a hyperplane between two classes.
- **k-Nearest Neighbor (KNN):** The k-Nearest Neighbor algorithm uses the labels of its nearest neighbor k to label a data point.

Analyzing Results of Business Process Automation 115

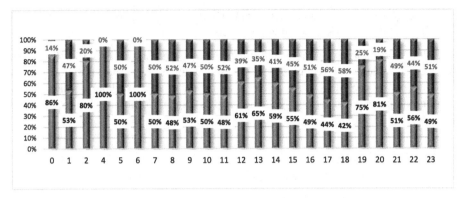

Fig. 11. End status by hour.

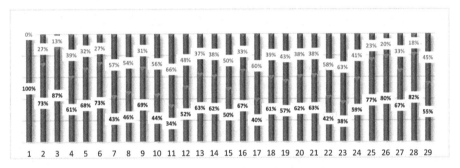

Fig. 12. End Status by Days of the Month.

- **NaiveBayes:** The Naive Bayes classifier works on the basis of Bayes' theorem. Its implementation in WEKA classifies between traits using the assumption of independence.
- **Logistic:** Logistic regression is a method used to estimate the probability of a dependent variable. The Logistic classifier in Weka implements the logistic regression model.
- **MultilayerPerceptron:** This is a multilayer neural network (MLP) classifier. Artificial neural networks are used in a variety of problems because of their ability to model complex relationships.
- **Kstar:** Kstar is an instance-based classifier, that is the class of a test instance is based upon the class of those training instances similar to it, as determined by some similarity function.
- **Random Tree:** Random Tree is one of the most common learning methods like random trees. This algorithm is designed to perform classification or regression tasks using decision trees. Random Tree aims to achieve more powerful and generalizable models by adding randomness as you build each tree [5].
- **Decision Table:** Creates a decision table to classify the samples in the data set. The decision table contains predictions of various classes based on specific combinations

of values of input attributes (attributes). This classifier is particularly suitable for working with categorical (nominal) data sets [20].

SMOTE was applied to remove the imbalance in the data set. SMOTE (Synthetic Minority Over-sampling Technique) produces synthetic samples of the minority class by oversampling each data point. SMOTE selects the closest neighboring sample to a minority sample and creates new synthetic samples using the difference between the two samples [18].

The data distribution before and after 50% SMOTE is applied, is as in Table 1.

Table 1. Data distribution before and after SMOTE.

SMOTE	FAILED	SUCCESS	TOTAL
YES	435	395	830
NO	290	395	685

10-fold cross-validation is applied as the testing method. The dataset is divided into 10 equal parts, with 9 groups of data being used as training data, while the remaining one group of data is being used as test data. This test is repeated in such a way that all combinations are test data in order. The average of the results obtained as a result of each test gives the final result. The cross-validation working logic is shown in the image in Fig. 13 [12].

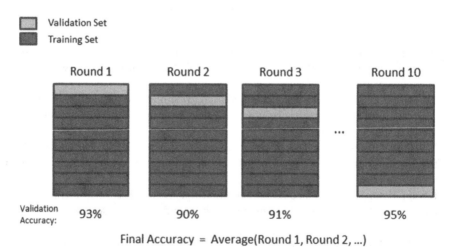

Fig. 13. Cross Validation [9].

In the study, 10 different classification algorithms were used to predict results or end status of RPA. Some algorithms were used several times with various parameters. The results of the classification algorithms are given in Table 2. After 50% SMOTE was applied, the data size increased to 830. The number and percentages of the correctly classified instances as a result of different classification algorithms are available in Table 2. It is seen that the highest estimation percentage among all models is in the J48 algorithm, with the confidence factor parameter 0.75. The J48 algorithm parameters were tested with different values, but the default values provided the best results, except the confidence factor. The higher the confidence factor, the lower the pruning value. Various confidence factor levels were investigated in the study (0,25, 0,50, 0,75, 0.80 and 0,85) for the confidence factor, but the best results were obtained with 0.75. The J48 algorithm correctly predicted 298 of the 395 SUCCESS results as in the confusion matrix results. The algorithm correctly predicted 323 of 435 FAILED results. In total, 621 out of 830 data points were predicted correctly. J48 was chosen among the other algorithms because it has the highest accuracy on the estimation of End Status in the entire dataset.

Table 2. Classification Results

SMOTE	Test mode	Classifier	Quantity — Correctly Classified Instances	Percent — Correctly Classified Instances	Total Number of Instances
50%	10-fold cross-validation	Naive Bayes	556	67%	830
50%	10-fold cross-validation	Logistic	586	71%	830
50%	10-fold cross-validation	MultilayerPerceptron	606	73%	830
50%	10-fold cross-validation	SMO	577	70%	830
50%	10-fold cross-validation	KNN	598	72%	830
50%	10-fold cross-validation	Kstar	607	73%	830
50%	10-fold cross-validation	Decision Table	562	68%	830
50%	10-fold cross-validation	J48-CF:0.75	621	**75%**	830
50%	10-fold cross-validation	RandomForest	615	74%	830
50%	10-fold cross-validation	RandomTree	564	68%	830

Precision, recall and F1 score values for classification models were also examined. The results are given in Table 3. According to the results, the highest estimation was obtained in the J48 algorithm.

Table 3. Classification Results with Performance Measures of Classification.

Classifier	%	Detailed Accuracy By Class								
		SUCCESS			FAILED			Weighted Avg.		
		Precision	Recall	F-Measure	Precision	Recall	F-Measure	Precision	Recall	F-Measure
Naive Bayes	67%	0,669	0,605	0,636	0,670	0,729	0,698	0,670	0,670	0,668
Logistic	71%	0,711	0,643	0,676	0,702	0,731	0,410	0,706	0,706	0,705
MultilayerPerceptron	73%	0,713	0,724	0,719	0,746	0,736	0,741	0,730	0,730	0,730
SMO	70%	0,692	0,648	0,669	0,698	0,738	0,717	0,695	0,695	0,694
KNN	72%	0,702	0,716	0,709	0,738	0,724	0,731	0,721	0,720	0,721
Kstar	73%	0,739	0,673	0,705	0,726	0,784	0,754	0,732	0,731	0,73
Decision Table	68%	0,691	0,582	0,632	0,668	0,763	0,712	0,679	0,677	0,674
J48-CF:0.75	**75%**	**0,727**	**0,754**	**0,740**	**0,769**	**0,743**	**0,756**	**0,749**	**0,748**	**0,748**
RandomForest	74%	0,746	0,691	0,717	0,737	0,786	0,761	0,741	0,741	0,740
RandomTree	68%	0,668	0,661	0,659	0,690	0,706	0,698	0,679	0,680	0,679

4 Conclusion

In this study, the invoice related tasks in the logistics department of a large textile company were digitized with RPA. The current manual process and the desired RPA bot processes were mapped initially, after which Go-live and testing phases were performed. The operation of RPA produces errors, and the errors should be minimized for efficiency of the bot within the concept of continuous improvement. For this reason, errors of the RPA were examined. First, the process factors which might affect RPA, such as shift, hour, day, were determined, as well as output factors. The single output factor was End Status. Next, considering the input factors, 10 classification algorithms were used to predict End Status. The highest estimation of End Status was reached with the J48 algorithm with confidence factor of 0.75.

To reduce RPA errors, error definitions must be analyzed correctly. The parameters that cause errors should be investigated and work should be done to improve these parameters. Future studies include expanding the analyses with more data and a wider array of classification algorithms. Using association rules might be an option, in order to identify the situations causing an error as a process result, followed by taking actions in the relevant situations. The process stage at which the errors occur can be investigated, so that it would be possible to work on the process step that causes the error. The root causes of the errors in the system are then to be determined and preventive actions can be performed in the system accordingly.

Errors that occur increase human intervention in the bot and reduce reduce return on investment. The findings in these studies pave the way to developing Smarter RPAs, ie improving the capabilities of SmartRPAs. Understanding the behavior of RPA bots extends towards enhancing their working principles and algorithms. RPA bots are active in different jobs. It might be possible to create dashboards for all jobs that are scheduled in the RPA bots' work plan. In order to minimize human intervention in future studies, potential problems can be predicted on the model trained with the data obtained from RPA reports, and warning systems can be installed on the dashboards to indicate the types of errors, and time and the stages at which they occur.

Moreover, this study provides a basis for establishing software error prevention or warning systems. It is a starting point for the development of simulating systems where outcome predictions can be estimated and processes are improved, especially in businesses where investment costs are high.

References

1. Agostinelli, S., Lupia, M., Marrella, A., Mecella, M.: Reactive synthesis of software robots in RPA from user interface logs. Comput. Industry **142**, 103721 (2022)
2. Akman, G., Yorur, B., Boyaci, A.I., Chiu, M.C.: Assessing innovation capabilities of manufacturing companies by combination of unsupervised and supervised machine learning approaches. Appl. Soft Comput. **147**, 110735 (2023)
3. Pachauri, A.: Comparative study of J48, Naive Bayes and one-R classification technique for credit card fraud detection using WEKA. Adv. Comput. Sci. Technol. **10**(6), 1731–1743 (2017)
4. Filzmoser, M., Koeszegi, S.T.: Integrating robotics and artificial intelligence in business processes—an organization theoretical analysis. Front. Artific. Intell. Appl. **311**, 177–184 (2018)
5. Flores, K.R., de Carvalho, L.V.F.M., Reading, B.J., Fahrenholz, A., Ferket, P.R., Grimes, J.L.: Machine learning and data mining methodology to predict nominal and numeric performance body weight values using Large White male turkey datasets. J. Appl. Poultry Res. **32**(4), 100366 (2023)
6. Gruzauskas, V., Pacesaite, K.: Expert system for the freight coordination based on artificial intelligence. J. Manag. **37**(2) (2021)
7. Ihya, R., Namir, A., Elfilali, S., Guerss, F.Z., Ait Daoud, M.: Modeling the Acceptance of the E-Orientation Systems by Using the Predictions Algorithms, pp. 27–31 (2020)
8. Josso, P., Hall, A., Williams, C., Le Bas, T., Lusty, P., Murton, B.: Application of random-forest machine learning algorithm for mineral predictive mapping of Fe-Mn crusts in the World Ocean. Ore Geol. Rev. **162**, 105671 (2023)
9. Kaur, G., Chhabra, A.: Improved J48 classification algorithm for the prediction of diabetes. Int. J. Comput. Appl. **98**(22), 13–17 (2014)
10. Ling, X., Gao, M., Wang, D.: Intelligent document processing based on RPA and machine learning. In: Proceedings Chinese Automation Congress, pp. 1349–1353 (2020)
11. Liu, W.X., Cai, J., Wang, Y., Chen, Q.C., Zeng, J.Q.: Fine-grained flow classification using deep learning for software defined data center networks. J. Netw. Comput. Appl. **168** (2020)
12. Ofer, D.: Machine Learning for Protein Function, Ph.D. Dissertation, The Hebrew University of Jerusalem (2015)
13. Rohaime, N.A., Abdul Razak, N.I., Thamrin, N.M., Shyan, C.W.: Integrated invoicing solution: a robotic process automation with AI and OCR approach. In: 2022 IEEE 20th Student Conference on Research and Development, pp. 30–33 (2022)

14. Salmen, A.: Employing RPA and AI to automize order entry process with individual and small-sized structures: a SME business case study. Acta Acad. Karviniensia **22**(2), 78–96 (2022)
15. Shidaganti, G., Salil, S., Anand, P., Jadhav, V.: Robotic process automation with AI and OCR to improve business process. In: Proceedings of the 2nd International Conference on Electronics and Sustainable Communication Systems, ICESC, pp. 1612–1618 (2021)
16. Syed, R., et al.: Robotic process automation: contemporary themes and challenges. Comput. Industry **115**, 103162 (2020)
17. Terminio, R., Rimbau Gilabert, E.: The digitalization of the working environment: the advent of robotics, automation and artificial intelligence (RAAI) from the employees perspective a scoping review. Front. Artif. Intell. Appl. **311**, 166–176 (2018)
18. Turlapati, V.P.K., Prusty, M.R.: Outlier-SMOTE: a refined oversampling technique for improved detection of COVID-19. Intell.-Based Med. **3–4**, 100023 (2020)
19. Wang, W., Jian, S., Tan, Y., Wu, Q., Huang, C.: Representation learning-based network intrusion detection system by capturing explicit and implicit feature interactions. Comput. Secur. **112**, 102537 (2022)
20. Workie Demsash, A.: Using best performance machine learning algorithm to predict child death before celebrating their fifth birthday. Informat. Med. Unlocked **40**, 101298 (2023)
21. Yadav, A.K., Chandel, S.S.: Solar energy potential assessment of western Himalayan Indian state of Himachal Pradesh using J48 algorithm of WEKA in ANN based prediction model. Renew. Energy **75**, 675–693 (2015)
22. https://krify.co/tag/types-of-rpa/
23. https://motivitylabs.com/robotic-process-automation-development-life-cycle/

Predictive Analysis of Surface Defects in Engineering Structures Using Machine Learning Technologies

Roman Mysiuk[1](✉), Iryna Mysiuk[1], Volodymyr Yuzevych[2], Roman Shuvar[1], Svyatoslav Tsyuh[3], and Nataliia Pavlenchyk[4]

[1] Ivan Franko National University of Lviv, 1 Universytetska Street, Lviv 79000, Ukraine
mysyukr@ukr.net
[2] Karpenko Physico-Mechanical Institute of the NAS of Ukraine, 5 Naukova Street, Lviv 79060, Ukraine
[3] Lviv University of Business and Law, 99 Kulparkivska Street, Lviv 79021, Ukraine
[4] Ivan Boberskyi Lviv State University of Physical Culture, 11 Kostiushka Street, Lviv 79000, Ukraine

Abstract. Bridges, pipelines and roads can be engineering objects. As a result, the basis of the material for each of them is different, as well as certain parameters of their strength. The main defects considered in the work are corrosion and cracks. Based on the statistical analysis of these parameters and machine learning classifiers, it is possible to predict the defect or not. Due to the insufficient amount of known data, the formation of the model requires more complex dependencies. The described methodology can be used to complement the diagnosis of changes in the state of surface defects of engineering structures. The advanced model includes the set parameters of the components of the main mixture, the presence of protection of materials and selected main ones that affect the rate of change in the state of damage to objects. The most accurate prediction 96% for the collected data was found in the Decision Tree Classifier. The accuracy of the forecasting of the collected parameters makes it possible to assess the criticality of the system's condition and the planning of restoration works in the studied areas. Thus, in such monitoring or diagnostic systems, it is possible to predict and track a year of defects in structures. The obtained results can be useful for improving the durability and checking the condition of the structures under study.

Keywords: predictive analysis · machine learning · classification · statistical modeling · data analytics · surface defect

1 Introduction

Monitoring systems usually monitor the current state. Engineering objects are usually the most necessary connections that ensure the vital activity of communities. Predicting the change in the state of macro defects allows you to prevent system failures or disasters. Collecting a universal set of parameters for assessing the possibility of the change in the

© The Author(s), under exclusive license to Springer Nature Switzerland AG 2025
A. Mirzazadeh et al. (Eds.): ODSIE 2023, CCIS 2204, pp. 121–132, 2025.
https://doi.org/10.1007/978-3-031-81455-6_8

state of a defect is not an easy task. Statistical analysis shows the dependence of values on certain criteria. Such results can be applied in the field of defect detection. Since many aspects can affect the change in the state of cracks or corrosion in materials, it is worth focusing on the main well-known parameters.

A large number of things can be predicted today. Predicting the change in the state can be used to assess the criticality of some repair work and the duration of operation. Moreover, with machine learning, such work involves training a model on a certain set of data and testing it on real ones. The next stage after the forecast is the verification of the correctness of the macro defect detection in the material using real devices or means.

In works [1, 2], the main parameters for conducting regression analysis of data are the depth of the defect and temperature, and an overview of other causes of damage (for example: the degree of corrosion and other physical indicators).

In order to form a model for further analysis at the initial stage, it is worth considering known sets of parameters that are important in the context of the analysis. This will make it possible to supplement the model with a larger number of parameters and specify them during the formation of the model of changes in the states of defects on the surfaces.

In the process of analyzing the existing approaches, it is worth highlighting the regression and classification classes of problems in the process of intellectual data analysis. The difference between these approaches mostly lies in the formation of the research goal. Regression analysis is used to predict values, and classification is used to assign each object one of the predefined classes based on its defined parameters.

The topic of data research on parameters and state changes is not sufficiently well researched, given the complexity of forming data sets and impact parameters. Among the practical consequences of predicting changes in the state of defects is the possibility of automatic monitoring and assessment of the resource of elements of engineering structures. Forecasting changes in the state of defects will allow, on the basis of historical (statistical) data, to improve known methods and means of acoustic, vibration and surface diagnostics in the context of preliminary analysis of dependencies between parameters. In addition, the work can highlight some unique parts of the approach and methodology, which are based on the use of existing data in the study of predicting the change in defect states. As a result, it is proposed to conduct data analysis based on logistic regression, support vector method (SVM), decision trees, random forest and others. The mentioned methods belong to the class of machine learning methods, which allow them to be used to support decision-making regarding the resource of an engineering structure based on statistical data. An appropriate brief review of the research objectives allows us to assert the perspective of the topic and the relevance of the research direction.

The main differences between the classification methods can be the speed of learning and the ability to interpret. Some methods, such as logistic regression, are known for their simplicity and interpretability, which makes them popular for use in various industries. While other methods, such as neural networks, can detect high accuracy, but require more data and computing resources for training. Decision tree and random forest methods are mostly more efficient in the class of problems for working with large data sets because they work efficiently with numerical and categorical features. The advantage of the corresponding approaches is that they do not require complex pre-processing of the data. The choice of the most optimal method can be reduced to the search for the minimum error during training. This means that the method with the smallest deviations based on the validation data set will be the most effective in terms of accuracy.

2 Literature Review and Problem Statement

In work [3], an overview of the causes of pipe damage and factors that affect the durability of their operation is considered. Such examples as causes of pipeline failures, dents combined with cracks and examples of monitoring systems make it possible to assess the current state and methods of their detection. When detecting and localizing defects as described in [4], it is possible to assess the state of damage using the wave method. This paper describes in sufficient detail how data analytics and the defect detection process are carried out. The results of the work make it possible to distinguish the types of defects, the main places where they appear. A larger number of important parameters such as shape of pipe, pipe thickness, inner or outer surface, existing welds, joints and previous repair are presented in [5]. However, this set of parameters can be expanded with some parameters from work [6], such as: age, weather, protection, pressure, hydrogeology, soil hazards. These parameters should be taken into account when forming a statistical model. Features of operation and parameters that affect the strength of the metal and the monitoring system are described in work [7, 8]. However, this is the effect of external temperatures and the aging of pipes over time. Methods of durability assessment are described in [9]. These parameters and factors relate mainly to underground pipelines, which should be taken into account when creating statistical tables. In bridge structures, the main elements are reinforced parts in concrete. Some features of operation and assessment of aging over the years, taking into account seasonal temperatures, are described in [10]. For the formation of decision-making systems that can predict the change of the state of cracks, an important role is played by investment attractiveness [11, 12], and risk when implementing technologies and managing business structures [13–16].

3 The Purpose and Objectives of the Research

The main goal of the work is the implementation of analysis predictions based on statistical models formed from parameters that affect the change in the state of macro defects in engineering objects.

The main tasks in the work are the following points:

– formation of statistical models and basic parameters that influence the change in the state of macro defects in materials. The appropriate data set is important for the formation of dependencies and determination of importance in the context of further analysis.
– use machine learning classifiers and conduct training based on statistical analytics. Comparison of the accuracy of each of the selected machine learning methods (Random Forest, DecisionTree, GaussianNB, K-Neighbors, SVC), which allows for predictive analytics.

– describe the forecasting process based on changing state of surface defect. Analysis of the classification result based on a test set of data that can be integrated into the system for monitoring changes in defect states and assessing the resource of an engineering object (pipeline).

4 Materials and Methods of Research

Statistical modeling is used in the work, for this, at the first stage, all data are collected in a comma separated values (CSV) file. This file is processed by a program written in the Python programming language with connected libraries matplotlib, seaborn for graphs, sklearn - for classifiers and working with data sets. Among the classifiers used: RandomForestClassifier, DecisionTreeClassifier, GaussianNB, KNeighborsClassifier, SVC. All these machine learning algorithms are popular, but each of them has its own characteristics:

- RandomForestClassifier uses multiple decision trees to improve generalization and reduce overfitting and works well with large datasets.
- DecisionTreeClassifier is easier to interpret and useful for explaining the model decision.
- GaussianNB features are considered to be conditionally independent, often used for text classifications, but not always realistic results.
- KNeighborsClassifier uses the k-nearest neighbors' algorithm with given distances, predictions can be slow for large datasets.
- SVC (Support Vector Classifier) works well for large volumes of data but is sensitive to changing parameters.

The confusion matrix shows the dependence and correlation between parameters, how similar these values are. The paper shows the dependence between the parameters of the statistical model.

Regarding the size of the data set used, the parameters considered during the analysis, the number of values is equal to 131553. Label coding was used among the pre-processing steps applied to the data before modeling. Each of the classifiers can be used for this type of tasks and data, their main difference is in accuracy, which is considered in the work. The analysis of each of the above-mentioned methods allows you to preform the advantages and disadvantages of each approach.

5 Results

5.1 Statistical Models and the Influence of the Main Parameters

An important part is the search for data that will make it possible to conduct certain studies of the effectiveness and the ability to evaluate the change in the state of the defect. Usually, historical data is collected for years, especially since the process of aging of materials and the appearance of defects is quite long-term. The statistical model is a set of parameters that should be taken into account in the values to highlight the importance for assessing the possibility of the change in the state of macro defects in materials. The following Table 1 gives an important parameter from open sources on the latest research. The main parameters of the statistical model with possible values make it possible to combine for different materials. Therefore, for engineering objects such as bridges and pipelines, some similar parameters can be combined.

Another important parameter for the evaluation of the entire engineering object is the load and length of the investigated element. However, due to the insufficient amount of information about these cases, it can be considered separately.

It is known from works [5, 6] that cracks can appear at the joints in the seams, which affect the spread of damage. Pipes or parts of the bridge that are in contact with the ground are more exposed to moisture and the risk of defects is greater. In Fig. 1 different pipe connections are shown and the connection points most vulnerable to defects are highlighted in red.

Another important parameter from [5, 6] is the object's operating time. The number of years affects the deterioration of the defect and allows you to see the trends of their change based on this data. It is also possible to draw a conclusion about the most vulnerable places in connections, which are: on the border of the joint, bend, branching and switching modes. Obviously, the layer of material from which these pipes are made will be thicker. At the same time, the pressure in such areas will be much higher and will be influenced by other substances. That is, the wear of such structural elements occurs faster than in a continuous straight pipe.

5.2 Learning Process and Classification of Models

Simpler trends can be seen by constructing graphs of parameter dependencies, but classification tasks require a large amount of data under various combinations and conditions. A large amount of data can guarantee a better forecast, so the selected dataset contains more than 100,000 records for the years 2011 to 2021 [17]. In addition, it is possible to evaluate the importance of parameters based on regression methods of machine learning using similar algorithms as for classification. The underlying data for training and testing [17] allows better evaluation and statistical inference about attributes. This data is useful for calculating and finding the main dependencies between them and establishing certain relationships between them. Data sets from the parameters contained in the list of mechanical damage can be applied similarly as in Table 1 with some deletions in unnecessary columns such as city names, company names, etc.

Often the data are text and tape types that need to be converted into numerical values for processing in algorithms using, for example, Label Encoder. This stage of data transformation is important considering that most of the data is text. Many data contain missing values that must be filled in for correct calculations. Often for this, approaches with the selection of the most frequent values are used.

The learning process consists in training the model based on statistical data with the results of defects or no defects. Accordingly, training takes place on all data to improve forecast accuracy. Then there is a stage of testing the model based on several characteristics. After training, the accuracy_score, precision_score, recall_score and f1_score values are formed, which show the training results.

Table 1. Important parameters for the statistical model of evaluating the change in the state of macro defects in engineering objects based on [17] and [18].

Parameter	Description	Value
Material name	The name of the main material, which is the main component of the engineering structure	FITTING_MATERIAL_TEXT
Material type	A mixture of concrete with other nanoparticles or the name of the metal	MODEL_NUMBER
Dimensions	The diameter of the pipe or the thickness of the bridge beams	FIRST_PIPE_NO MINAL_SIZE_TE XT, SECOND_PIPE_ NOMINAL_SIZE _TEXT
Presence joint	Boolean value of the presence or absence of connections between parts	yes - present, no - absent
Connection type	Changing the position of the structure (availability of an angle of inclination, bending)	bend or horizontal
Substance type	Type of leaking substance, if any	for pipes: water, oil or gas
Age	Duration of operation of the structure	YEAR_INSTALLED
Underground	Boolean value about underground or above ground use (pipes only)	LEAK_LOCATION_A_TEXT
Protection	Presence or absence of additional protection of the material coating	LEAK_LOCATIO N_A_TEXT
Moisture	The presence of moisture from the environment	LEAK_LOCATION_B_TEXT
Wall thickness	Pipe wall thickness or bridge deck thickness	In mm, specified according to MODEL_NUMBER and
Repair count	Number of completed repairs	For example, 30
Defect location	Indication of the place of examination or the presence of defects in the past	LEAK_OCCURRED_TEXT
Defect type	Indicated which macrodefects are being investigated	LEAK_CAUSE_TEXT
Risk level	The level of risk that can be with the specified parameters. Can be calculated based on previous values	High, medium or low
Defect	Boolean value of the presence or absence of defect based on the previous values	yes - present, no - absent

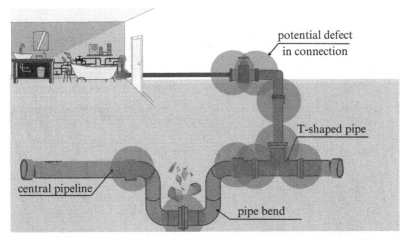

Fig. 1. Locations of potential defects based on different types of pipe connections.

When processing the data, many values among the parameters are omitted and this can affect the training results. Therefore, the most frequent combinations were used to fill in the omission of important values (see Fig. 2).

The most optimal way to check the accuracy of the determination is to divide the data into two parts in the proportion of, for example, training and testing data 80 by 20.

The Decision Tree classifier was the most accurate 96% after training, followed by the Random Forest classifier 93%. The obtained accuracy results can be explained by the features of the internal structure of the algorithms. The algorithms chosen will have different results for different data sets. It is not necessary that the selected algorithm will also display data under other conditions. The Random Forest classifier is a partial implementation of a decision tree, so it showed a similar but worse result. Given the imbalance of the datasets, other algorithms may have shown lower accuracy when used for this task.

Additionally, similar data analysis techniques can be applied to bridge structures using data from reference [18]. This involves examining the data related to bridge conditions and employing various analytical methods to assess factors such as structural integrity, load-bearing capacity, and the presence of defects. Reference [19] specifically delves into the analysis of bridge data, focusing on techniques aimed at detecting and mitigating defects like cracks. These techniques often involve sophisticated data processing and interpretation methods to accurately identify and characterize structural flaws.

The importance of parameters and their correlation can be seen in the graphical representation. The comparative confusion table with all used values has the form of a heat map. Saturation of colors signals a greater correlation of values (see Fig. 3).

Values with a higher value are more important and have a greater impact on the overall result. A value with a negative value is the result of the correlation of confusion matrix. And can be considered as having less influence on the relationship with other parameters. The parameters that will affect the result are the main ones when obtaining such a statistical model. Therefore, the following parameters were selected for the study: year,

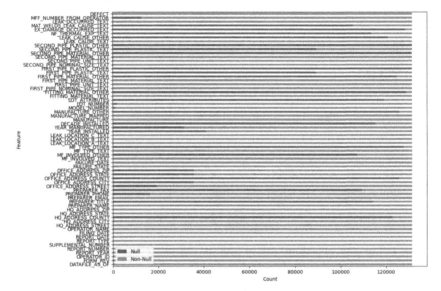

Fig. 2. Missing and filled values from reports [17].

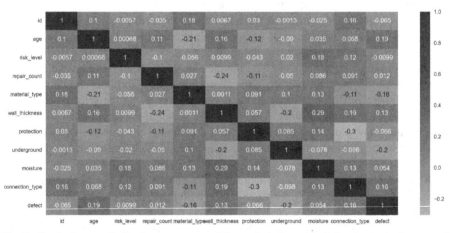

Fig. 3. Confusion matrix for the parameters of the statistical model of the change in the state of macro defects.

level of risk of damage to the object, number of repairs carried out during operation in each element, type of material, thickness of the pipe walls, presence of surface protection of the object, determination of the type of operation (underground or above-ground objects), humidity value in the soil near the object, the type of connection and the definition of the defect.

After training, you can run tests on different values of this model and check the accuracy. Some combinations from the set were used as a test and the forecast result coincides.

5.3 Forecasting the Change in the State of Defects for Bridges and Pipelines

The example of pipelines shows the general scheme of forecasting based on statistical data on the change in the state of macro defects in materials. The first stage is the classification of pipelines according to the content of the substance flowing in the pipes. The second stage is the selection of features for each of the selected classes. At the third stage, statistical models are formed, and at the fourth stage, forecasting is performed. Prediction of results whether there is a defect or not on the last one (see Fig. 4). In addition to analyzing and summarizing models based on the principle of analysis of connections on pipelines, it is possible to pay attention and emphasize the number of years of operation of pipelines.

The early stages allow you to form the desired data sets based on certain patterns or shapes. Thus, in the process of data transmission and preprocessing, a set of data is obtained. Already at the stage of using machine learning classifiers, you can choose the most effective one, taking into account accuracy. The result signals a possible defect or not based on the transmitted similar input parameters.

Such a structural scheme can be extended using input data from the real world. For example, using sensors or a camera [8] can be transmitted in a processed form to a prognostic application, for example regarding risks or current humidity. The next stage of the work may be the integration of the use of computer vision and predictive modeling for the purpose of testing work in real time mode. That is, such an analysis showed the possibilities and selection of the optimal classifier based on the data set and makes it possible to implement a more complex analysis for solving problems in this field.

Furthermore, fuzzy logic presents an alternative approach to traditional machine learning methods in analyzing structural data. Fuzzy logic allows for dealing with uncertainty and imprecision in data by using variables and fuzzy sets. In [20, 21] showcase examples of how fuzzy logic is applied in structural analysis, demonstrating its effectiveness in handling complex data related to engineering structures. This approach can provide valuable insights into the assessment and management of structural integrity, particularly when dealing with incomplete or uncertain information.

A scheme for bridges is built in a similar way, since similar characteristics can be found for both engineering objects. The result of this approach is the possibility of entering previously formed values and receiving an instant response about the change in the state of the defect.

Models based on statistical data are important, because using them you can proceed to the next stages of working with data, for example, such as forecasting. In addition to the forecast in the time domain, there is an opportunity to classify objects into several groups, which is used in the work. The implementation of classifiers determines the optimality of the solution for each case separately. As shown in [22–24], the results may differ for different data sets.

With the use of different classifiers, the accuracy metric was obtained, which was used to evaluate the algorithm for the selected data set and application area. The results of the work can be used in the future to assess the risks and validity of the system, taking into account information about changes in the state of defects. Although in this context there are different computational approaches from reliability theory to neural networks, it can be argued that the result should be information about the defect.

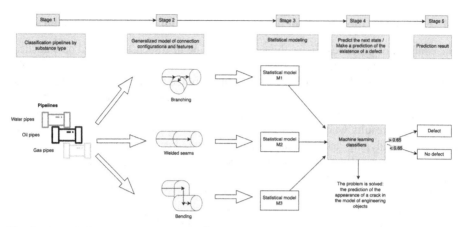

Fig. 4. General diagram of the stages of forecasting the change in the state of defects in pipelines based on connection type.

6 Discussion

In statistics, the more data there is, the more accurate the analysis for the researched topic can be. According to the expansion of the parameters of the statistical model, it is possible to better characterize the change in the state of the defect in specific situations. Among the limitations of this approach are the limited amount of data and the balance of the data. Determining which of the parameters may be more or less important in the context of the data set. This question can be solved taking into account the application of various methods of machine learning.

When processing the data, many values among the parameters are omitted and this can affect the training results. Therefore, the most frequent combinations were used to fill in the omission of important values.

Since there are many methods and approaches to solve this problem, only a few RandomForestClassifier, DecisionTreeClassifier, GaussianNB, KNeighborsClassifier, Support Vector Classifier are selected. Applying them on the same data set with possible defects showed that the decision tree is the most accurate. Better efficiency and accuracy can be in different algorithms depending on the data set and the studied objects [21, 23]. The steps shown in Fig. 4 show a conceptual diagram of the data processing and forecasting process.

7 Conclusion

The paper shows the possibility of performing predictive analytics based on machine learning methods, which can help in evaluating changes in the state of defects on the surfaces of engineering structures. To achieve this goal, pre-processing of data, statistical analysis and formed conditions affecting the change in the state of defects of engineering structures have been translated. On the basis of known data, the values of important parameters that affect the change in the state of defects are formed. Such groups of parameters are combined into statistical models. Machine learning classifiers were used to predict the presence of defects based on this data. The most effective classifier in terms of accuracy was the Decision Tree 96%. The forecasting process and its features for bridges and pipelines are also described.

The main researched part of this work has been completed, namely, a statistical model has been formed, which shows the dependencies between parameters and makes it possible to make predictions based on historical data.

The practical aspects of the research results are the use of intelligent data analysis in monitoring systems. A refined approach to the prediction of changes in defect states can benefit engineering facilities in terms of cost savings, maintenance efficiency and improved safety. Among the limitations and problems encountered during the research process, a small amount of data related to more complex model structures can be highlighted. As a result, the next stage can be proposed for potential future studies using more complex analytical processes. This demonstrates a reflexive approach and will open up opportunities for further research in the field of predictive analysis of surface defect states.

References

1. Omar, I., et al.: Comparative analysis of machine learning models for predicting crack propagation under coupled load and temperature. Appl. Sci. **13**(12), 7212 (2023)
2. Bolzon, G., Gabetta, G., Nykyforchyn, H. (eds.): Degradation Assessment and Failure Prevention of Pipeline Systems. Lecture Notes in Civil Engineering, vol. 102, pp. 203216. Springer, Cham (2021)
3. Vishnuvardhan, S., Ramachandra Murthy, A., Choudhary, A.: A review on pipeline failures, defects in pipelines and their assessment and fatigue life prediction methods. Int. J. Pressure Vessels Piping **201**, 104853, (2023). https://doi.org/10.1016/j.ijpvp.2022.104853, ISSN 0308-0161
4. Ghavamian, A., et al.: Detection, localisation and assessment of defects in pipes using guided wave techniques: a review. Sensors **18**, 4470 (2018). https://doi.org/10.3390/s18124470
5. Wang, K., et al.: Failure risk prediction model for girth welds in high-strength steel pipeline based on historical data and artificial neural network. Processes **11**, 2273 (2023). https://doi.org/10.3390/pr11082273
6. Barton, N.A., et al.: Improving pipe failure predictions: factors affecting pipe failure in drinking water networks. Water Res. **164**, 114926 (2019). https://doi.org/10.1016/j.watres.2019.114926. ISSN 0043-1354
7. Mysiuk, R.V., et al.: Determination of conditions for loss of bearing capacity of underground ammonia pipelines based on the monitoring data and flexible search algorithms. Archiv. Mater. Sci. Eng. **115**(1), 13–20 (2022). https://doi.org/10.5604/01.3001.0016.0671
8. Mysiuk, R., Yuzevych, V., Mysiuk, I., Tyrkalo, Y., Pavlenchyk, A., Dalyk, V.: Detection of surface defects inside concrete pipelines using trained model on JetRacer kit. In: 2023 IEEE 13th International Conference on Electronics and Information Technologies (ELIT), Lviv, pp. 21–24, (2023). https://doi.org/10.1109/ELIT61488.2023.10310691
9. Yuzevych, L., Skrynkovskyy, R., Koman, B.: Development of information support of quality management of underground pipelines. EUREKA: Phys. Eng. **4**, 49–60 (2017). https://doi.org/10.21303/2461-4262.2017.00392

10. Dzhala, R., et al.: Simulation of corrosion fracture of nano-concrete at the interface with reinforcement taking into account temperature change. In: 4th International Workshop on Modern Machine Learning Technologies and Data Science, MoMLeT&DS 2022, CEUR Workshop Proceedings 3312, Leiden, Lviv, The Netherlands–Ukraine, 25–26 November 2022, pp. 123–133 (2022). https://ceurws.org/Vol-3312/paper10.pdf
11. Skrynkovskyi, R.M.: Methodical approaches to economic estimation of investment attractiveness of machine-building enterprises for portfolio investors. Actual Prob. Econ. **118**(4), 177–186 (2011)
12. Skrynkovskyi, R.: Investment attractiveness evaluation technique for machine-building enterprises. Actual Prob. Econ. **7**(85), 228–240 (2008)
13. Popova, N., et al.: Marketing aspects of innovative development of business organizations in the sphere of production, trade, transport, and logistics in VUCA conditions. Stud. Appl. Econ. **38**, 4 (2021). https://doi.org/10.25115/eea.v38i4.3962
14. Popova, N., Kataiev, A., Skrynkovskyy, R., Nevertii, A.: Development of trust marketing in the digital society. Econ. Annals-XXI **176**(3–4), 13–25 (2019). https://doi.org/10.21003/ea.v176-02
15. Mysiuk, R., et al.: Detection of structure changes in lightweight concrete with nanoparticles using computer vision methods in the construction industry. In: Proceedings of Eighth International Congress on Information and Communication Technology (ICICT 2023). LNNS, vol. 694, Springer, Singapore (2023). https://doi.org/10.1007/978-981-99-3091-3_27
16. Kniaz, S., Brych, V., Heorhiadi, N., Tyrkalo, Y., Luchko, H., Skrynkovskyy, R.: Data processing technology in choosing the optimal management decision system. In: 2023 13th International Conference on Advanced Computer Information Technologies (ACIT), Wrocław, Poland, pp. 372–375 (2023). https://doi.org/10.1109/ACIT58437.2023.10275581
17. Mechanical Fitting Failure Data from Gas Distribution Operators. U.S. Department of Transportation. Pipeline and Hazardous Materials Safety Administration. https://www.phmsa.dot.gov/data-and-statistics/pipeline/mechanical-fitting-failure-data-gas-distribution-operators. Accessed 28 Mar 2024
18. 2023 - Download NBI ASCII files - National Bridge Inventory - Bridge Inspection - Safety Inspection - Bridges & Structures - Federal Highway Administration. Federal Highway Administration. https://www.fhwa.dot.gov/bridge/nbi/ascii2023.cfm. Accessed 28 Mar 2024
19. Chordia, A., et al.: Surface crack detection using data mining and feature engineering techniques. In: 2021 IEEE 4th International Conference on Computing, Power and Communication Technologies (GUCON). Kuala Lumpur, pp. 1–7 (2021). https://doi.org/10.1109/GUCON50781.2021.9574002
20. Jain, R., Sharma, R.S.: Predicting severity of cracks in concrete using fuzzy logic. In: 2018 International Conference on Recent Innovations in Electrical, Electronics & Communication Engineering (ICRIEECE), Bhubaneswar, p. 2976 (2018)
21. Victoria Biezma, M., Agudo, D., Barron, G.: A fuzzy logic method: predicting pipeline external corrosion rate. Int. J. Pressure Vessels Piping **163**, 55–62 (2018). ISSN 0308-0161
22. Han, B.: Comparison of different machine learning algorithms in classification. J. Phys.: Conf. Ser. **2037**(1), 012064 (2021). https://doi.org/10.1088/1742-6596/2037/1/012064
23. Mysiuk, I., Mysiuk, R., Shuvar, R., Yuzevych, V., Hudyma, V., Vizniak, Y.: Category classification of content from Instagram business pages. In: 2023 13th International Conference on Advanced Computer Information Technologies (ACIT), Wrocław, pp. 570–573 (2023). https://doi.org/10.1109/ACIT58437.2023.10275458
24. Sheth, V., Tripathi, U., Sharma, A.: A comparative analysis of machine learning algorithms for classification purpose. Procedia Comput. Sci. **215**, 422–431 (2022). https://doi.org/10.1016/j.procs.2022.12.044. ISSN 1877-0509

Enhancing Autonomous Industrial Navigation: Deep Reinforcement Learning for Obstacle Avoidance in Challenging Environments

Richa Mohta[1], Vaishnavi Desai[2], Tanvi Khurana[3], M. R. Rahul[4], and Shital Chiddarwar[4](✉)

[1] Carnegie Mellon University, Pittsburgh, USA
[2] Exxon Mobil, Bangalore, India
[3] BlackRock, Mumbai, India
[4] Visvesvaraya National Institute of Technology, Nagpur, India
shitalsc@mec.vnit.ac.in

Abstract. Industrial settings require autonomous navigation to automate dangerous, time-consuming, or difficult tasks. Avoiding obstacles is crucial to autonomous robot safety and effectiveness in complex and changing industrial settings. Because of uneven ground, poor grip, and limited sensor data, rough terrain is difficult to navigate. This paper shows how deep reinforcement learning helps industrial robots avoid obstacles. The goal is to create a solid framework that lets a robotic agent autonomously travel from a starting point to an endpoint, avoiding obstacles. The suggested method uses reinforcement learning algorithms and deep neural networks without affecting each other. The computer agent learns how to plan its path and avoid obstacles. The study involves building simulation environments with MATLAB Simulink and planning experiments to test the framework. The manuscript also discusses hardware limitations, real-time operations, and deep reinforcement learning model challenges. This manuscript advances industrial robotics by providing a practical obstacle avoidance solution using twin-delayed deep deterministic policy gradient (TD3) and Deep Q-Network (DQN) deep reinforcement learning algorithms, which could revolutionize manufacturing and other industrial tasks.

Keywords: Obstacle Avoidance · Deep Reinforcement Learning · twin-delayed deep deterministic policy gradient (TD3) · Deep Q-Network (DQN) · robot motion planning · Artificial Intelligence

1 Introduction and Literature Review

Deep reinforcement learning is being used to improve the autonomous navigation of industrial robots, enabling them to navigate through complex and dynamic environments. The method integrates deep neural networks and reinforcement learning algorithms, allowing the agent to learn optimal path planning and obstacle avoidance. The research, which addresses hardware limitations, real-time operations, and model complexity, shows that the method can navigate quickly and safely through industrial obstacles, demonstrating its potential for real-world industrial automation.

Off-road vehicles, ranging from drones to unmanned ground vehicles, have become increasingly prevalent in various applications, including agriculture, mining, search and rescue, and surveillance [1–3]. Equipped with cutting-edge navigation systems, these vehicles are built to conquer challenging environments with robust and adaptive capabilities. [4]. Developing obstacle avoidance systems for off-road vehicles necessitates overcoming challenges through innovative approaches and advanced technologies for safe and efficient navigation. Traditional planning approaches, heavily reliant on predefined maps or rule-based algorithms [5, 6], may not be suitable for navigating uncertain and complex environments. They often come with several disadvantages, including limited adaptability and inability to handle unforeseen situations. This has led to a growing interest in leveraging machine learning techniques, such as deep reinforcement learning (DRL) [5, 6], to enable autonomous off-road navigation. Deep reinforcement learning is a subfield of machine learning that combines deep neural networks with reinforcement learning algorithms, allowing an agent to learn optimal actions based on feedback from the environment [7]. This approach offers the potential to develop motion planning systems that can adapt and optimize their actions in real time based on changing conditions and learned experiences. Furthermore, the availability of affordable and powerful hardware, such as Raspberry Pi - a popular single-board computer, and Simulink - a powerful modeling and simulation tool, has opened new possibilities for implementing navigation systems. Raspberry Pi offers a cost-effective and versatile platform for onboard computation, while Simulink provides a comprehensive environment for designing, simulating, and implementing control systems. Combining Raspberry Pi, Simulink models, and deep reinforcement learning can enable the development of efficient, adaptable, and autonomous navigation systems.

Reinforcement learning (RL) is a framework that facilitates the learning of optimal decision-making through interaction with an environment [8]. Within this framework, an agent learns to take actions in the environment to maximize its cumulative rewards over time. The agent explores the environment, receives feedback in the form of rewards or penalties, and updates its decision-making policy based on this feedback. The environment represents the external world where the agent operates, and it is typically defined by its states, actions, and rewards. The agent's ultimate goal is to learn an optimal policy that enables it to make the best decisions to achieve its objectives in the environment [9]. Deep reinforcement learning (DRL) forms a subset of RL by seamlessly integrating deep neural networks with reinforcement learning algorithms [10]. This integration empowers DRL to tackle high-dimensional and complex state and action spaces through the utilization of deep neural networks as function approximators. DRL further enables the agent to learn continuous and parameterized policies, allowing it to directly process raw sensory inputs like images or sounds from the environment. This capability to learn intricate representations of the environment unlocks the potential for sophisticated decision-making based on raw sensory data [10]. Deep reinforcement learning (DRL) encompasses two primary categories: value-based [11] and policy-based [12] methods. Value-based algorithms indirectly derive the agent's policy through iterative updates to the value function, which estimates the expected return from a given state and action. Q-learning, a popular value-based algorithm, leverages deep neural networks in conjunction with the Q-learning update rule to learn optimal Q-values [13, 14]. Conversely, policy-based algorithms directly establish a policy network and optimize its parameters to maximize the expected return [15, 16]. Examples of policy-based algorithms include Proximal Policy Optimization (PPO) [17], Asynchronous Advantage Actor-Critic (A3C) [18], and Deep Deterministic Policy Gradient (DDPG) [19–21]. This paper explores the use of Deep Q-Network and Twin Delayed Deep Deterministic Policy Gradient (TD3) algorithms in a framework for obstacle avoidance in autonomous systems operating in complex environments. The framework trains agents to learn optimal path planning and avoidance policies, enabling them to navigate from a starting point to a fixed endpoint while avoiding obstacles. The effectiveness of the framework is evaluated through simulation environments and comprehensive experiments.

2 The Proposed Framework

Establishing a robust framework for autonomous navigation using deep neural networks and reinforcement learning involves several key components. The proposed framework consists of 1. Environment and robot modeling that includes model of mobile robot, arena, interaction between robot and arena and sensors 2. Deep Neural Network (DNN) Architecture that includes design of a neural network architecture for learning path planning and obstacle avoidance, essentially, recurrent neural networks (RNNs) for handling spatial and temporal information. It covers Continuous v/s Discrete Action Space Setup, Reward Functions, Observation, Action, and Environment Reset Function. 3. Training Process which includes Training of the deep neural network using the defined reinforcement learning algorithm and Utilization of carefully planned experiments to generate diverse training scenarios and improve the model's generalization. 4. Optimal Path Planning that Leverages the trained neural network for optimal path planning from the initial point to the predefined endpoint and Incorporates algorithms that enable dynamic path adjustments based on real-time perception. 5. Obstacle Avoidance this Integrates obstacle avoidance strategies into the neural network's decision-making process and ensures that the agent can adeptly navigate around obstacles by learning from the reinforcement learning framework. 6. Simulation and Experimentation conducted using various environment and scenarios to assess the effectiveness of the trained model and measurement of the performance metrics such as success rate, navigation time and collision avoidance capabilities. 7. Performance evaluation of the model in terms of robustness, adaptability to different environment and generalization to unseen scenarios. Each step of the framework is discussed in details in the subsequent sections.

2.1 Environment and Robot Modelling

Physical modeling involves simulating and modeling systems that include tangible physical components. It adopts a network-based approach, where Simscape blocks represent real-world elements like wheels and motors. This technique is especially valuable for comprehending the physical manipulation or utilization of an object by a user. In the context of this manuscript, the environment comprising the arena and the mobile robot was developed as a physical model, allowing for an accurate representation and analysis of their interactions and behaviors.

Fig. 1. Six Wheel All Terrain robot considered for this work. (https://robu.in/product/black-6wd-search-rescue-platform-smart-car-chassis-damping-off-road-climbing-wifi-car/)

Mobile Robot. In the model, the mobile robot (Fig. 1) is represented as a six-wheeled rover using Simscape Multibody blocks. These blocks have the ability to simulate bodies, joints, constraints, force elements, and sensors. The modeling process begins with the inclusion of the World Frame block, which serves as the global reference frame for the model. This frame is inertial and remains stationary throughout the simulation. The main chassis of the mobile robot is generated as a rectangular Geometry Solid block. Additional components such as the base, fenders, and links for the wheels are also created using Solid blocks, each customized to the desired shape. These components are then connected to the chassis block using WeldJoint blocks, replicating the behavior of a welded joint. Rigid Transform blocks are also employed to define the orientation and location of the geometries accurately. To simulate the movement of the wheels in a lifelike manner, the wheel model is imported as a CAD file and converted into a Simulink block. Joint blocks, mimicking a Revolute joint, establish the connection between the wheels and the chassis body. To establish the connection between the base link and the World frame block, a Rigid Transform block is used. This block aids in setting the coordinate frame of the mobile robot, which in this particular model aligns with the World Coordinate Frame. Additionally, mechanism configuration and solver configuration blocks are connected to the model to set parameters such as gravity and solving parameters, respectively. Figure 2 illustrates the Simulink model representing the mobile robot. Overall, this modeling approach creates a functional representation of the mobile robot, enabling accurate simulation and analysis of its movements and behaviors within the given environment.

Arena. To generate the arena, the initial step involves creating the 'ground,' which serves as the surface upon which the rover operates. This is achieved by generating a rectangular geometry solid block, which is then connected to the World Frame. Once the ground is established, obstacles are introduced in the form of cubes and cuboids of various sizes. These obstacles are created using solid geometry blocks and connected to the ground using Rigid Transform blocks, which define their precise location and orientation relative to the ground. Furthermore, the arena's boundaries are modeled as walls, employing a similar approach. By following this process, the generation of the arena is successfully completed. Figure 3 showcases the Simulink model representing the arena, as described in the previous paragraph.

Interaction Between the Robot and Arena. To enable interaction and movement between the robot and the arena, a 6-DOF Joint Block is introduced, establishing a connection between them. This joint block allows the robot to freely translate and rotate

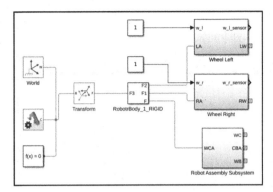

Fig. 2. Simulink Model of Robot.

in any direction relative to the World Frame, enabling it to traverse the ground. Additionally, the initial position of the robot is set using x and y coordinates with respect to the ground. Random values are assigned to x and y coordinates so that with each reset, the robot starts at a different arbitrary location on the arena while the final position remains fixed. In order to facilitate movement across the ground, it is essential to simulate the normal force and friction of the wheels when in contact with the ground. These forces can be effectively modeled using the Spatial Contact Force block. This block establishes connections between the wheels of the rover and the ground plate, as well as between other components of the chassis and individual obstacles. These connections ensure a physically accurate model, where the robot moves on the ground without sinking or lifting, collides with obstacles without passing through them, and halts at the arena's walls without falling off. To initiate movement, the user is required to provide input in the form of angular velocity. This input is applied to the Revolute Joint blocks that connect the wheels to the body of the robot, determining the motion of the wheels. By incorporating these components and connections, the Simulink model achieves a realistic representation of the interaction and movement between the robot and the arena, facilitating dynamic and accurate simulations.

Sensors. The Transform Sensor block from Simscape Multibody is utilized to measure the relative spatial relationship between the follower frame (robot) and the base frame (arena) to which it is connected. This block is configured to measure various parameters in the frame of the robot, including the instantaneous position (x and y), linear velocity (vx), angular velocity (wz), and orientation angle. By employing this block, the model accurately captures and monitors these essential metrics. Additionally, the Spatial Contact Force block, besides setting the normal and frictional forces, serves as a proximity sensor. Through the "Separation Distance Sensing" functionality, this block determines the shortest distance between the robot's body and all the obstacles present in the arena. When this distance reaches zero, it indicates a collision between the robot and an obstacle. In Figs. 3 and 4, the Simulink model visually represents the robot-arena system, encompassing the various components and connections discussed above. This model enables a comprehensive understanding and analysis of the interaction and behavior between the robot and the arena, facilitating simulations and evaluations of their performance and dynamics.

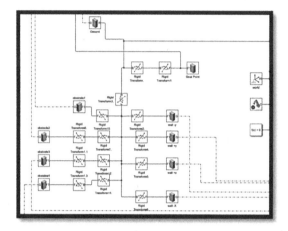

Fig. 3. Simulink Model of Arena.

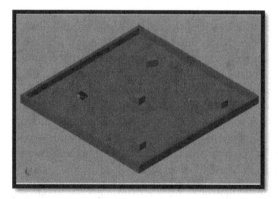

Fig. 4. Robot Arena System.

Deep Reinforcement Learning. Deep reinforcement learning (DRL) combines reinforcement learning's dynamic, trial-and-error approach to maximize outcomes with the power of deep learning for tackling complex tasks with high-dimensional state spaces. Unlike traditional reinforcement learning, DRL does not require extensive prior knowledge. Instead, it leverages deep learning's ability to extract meaningful representations from large datasets, allowing machines to learn complex relationships and develop effective strategies without explicit instructions. This approach fosters a more natural and goal-oriented learning process, mimicking how humans learn through experience. Instead of being programmed with specific actions, the learning agent explores the environment and discovers the actions that lead to the most desired outcome, known as the reward. This continuous exploration and feedback loop allows DRL to handle increasingly intricate problems with minimal prior knowledge (Fig. 5).

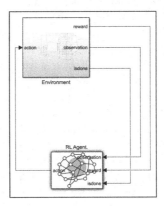

Fig. 5. Illustration of reinforcement learning.

Two main components make up Reinforcement Learning: The Agent solves uncertainty-based decision-making problems. This computer program must make crucial decisions to navigate the problem space. Agent decision-making is heavily influenced by its environment. The Agent can take actions at each environmental state. The Agent must choose the best action from this set. This choice affects the environment, which may change. The Agent interacts with everything in the Environment, which represents the problem. All external factors that affect the Agent's behavior and goals are included. Agents receive feedback on their actions from the environment through observations, states, and rewards. The environment usually has a clear task and may reward the Agent directly. This reward shows how well the last action worked. This reward is determined by the reward function. DRL is closed-loop. Environmental model of arena and robot. Observation is the sensor outputs and action are the wheel input angular velocity.

Continuous v/s Discrete Action Space Setup. The generation of an environment and agent requires the definition of several observations and actions. Limits need to be set on the actions to keep the motion of the rover in check- action space. The definition of this action space sets apart the continuous and discrete model setup. The continuous one i.e., a continuous actions space allows the input angular velocities of both the wheels to lie in a certain range without any relational constraint between them, i.e., both the wheels function as separate entities. The number of possible decisions, thus, can be infinite, providing a motion without jerk. In the discrete setup i.e., a discrete action space, actions have been defined specifically. Angular velocities to both wheels cannot function independently. Only four actions are permitted: both wheels have the same positive angular velocities (moving forward), both wheels have the same negative angular velocities (moving backward), one wheel has a certain angular velocity while that of the other wheel is zero (left turn and right turn, depending on which wheel's motion is arrested). Another difference between the setups is the value of initial velocity, Continuous setup requires higher initial velocity to induce linear motion and hinder circular motion. This is because spinning about its own axis is a possibility and hence a potential issue. The discrete setup can operate on lower/no initial velocity since spinning is curbed anyway.

Reward Functions. In reinforcement learning, the reward function guides an agent's learning. It converts each state (or state-action pair) the agent encounters into a numerical value indicating its desirability. By selecting one action from a set of options, this function measures the agent's immediate or future rewards. Deep Reinforcement Learning (DRL) allows the agent to learn through interactive trial-and-error in its environment, where an algorithm maximizes reward function. DRL is divided into Positive and Negative Reinforcement, like human learning. Positive reinforcement associates a behavior with positive outcomes to increase its frequency. Positive reinforcement helps DRL models optimize task performance and adopt consistent and sustainable behavior patterns. Negative reinforcement: Removing or avoiding negative consequences encourages a behavior. However, its main purpose is to maintain a minimum performance standard rather than maximize model potential. Negative reinforcement prevents the model from doing harm, but it doesn't encourage positive ones. Designing effective reward functions requires understanding these reinforcement mechanisms. This report's model uses positive reinforcement learning to keep rewards positive. This supports the robot's goal of reaching an endpoint while minimizing travel distance and avoiding obstacles. The Reward functions designed are as follows:

1. High positive reward for being in a defined vicinity around the goal point (based on the separation distance data from the Transform Sensor block)
2. Positive reward based on Euclidean Distance from the goal point - lesser the distance, greater the reward. This also includes a constraint which ascertains that the reward can only be received if the linear velocity of the rover is above a threshold value (based on the separation distance data from the Transform Sensor block)
3. Negative reward for spinning - higher the value of angular velocity of the body about z axis, higher the penalty (based on the wz data from the Transform Sensor block)
4. Negative reward for collision with an obstacle - if the body of the robot collides with or is within a threshold distance from any of the obstacles, a penalty ensues. This reward is stopped when there is no motion in the wheels of the rover, so as to prevent accumulation of negative values when the robot is stuck.
5. Negative reward for collision with the boundary walls of the arena - if the body of the robot collides with or is within a threshold distance from any of the obstacles or any of the walls, a penalty ensues. This reward is stopped when there is no motion in the wheels of the rover, to prevent accumulation of negative values when the robot is stuck.
6. Positive reward for maintaining a certain orientation - an angle 'theta', which is the angle the robot needs to move in to reach the goal in a straight line, is calculated. This is compared with the orientation angle of the robot at every instant. If this angle lies within a certain range (± 5 deg) from 'theta', then a reward has ensued.
7. Positive reward for higher linear velocity - if the linear velocity of the rover lies in a certain range, the reward increases as the velocity increases. Beyond the upper and lower limit, the value of this reward is zero.

The sum of all these rewards defines the total reward and is represented as the total reward function. The gain values are set so that the overall function value is positive in most cases. This is done to ensure that maximization of higher positive reward is carried out. The function varies slightly from continuous to discrete, in terms of gain values as well the individual reward functions themselves. In discrete setup, there is no need for the negative reward for spinning or the positive reward for higher linear velocity, so they are omitted.

Observation. A state represents the agent's immediate and specific situation within the environment. It encapsulates the current place, time, and configuration of the agent in relation to significant elements like obstacles and goals. This state can be either the current situation faced by the agent or a potential future state envisioned by the environment as feedback. However, the agent does not possess access to the entire state of the environment. Instead, it can only perceive a subset of the state information, known as the observation. This observation constitutes the agent's sole source of information for making decisions and determining its next course of action. The observations for this model are the parameters that will most significantly affect the action in a positive manner, they are as follows:

1. Distance of the rover from obstacles - Dist (separation distance data from the Transform Sensor block)
2. Linear velocity of the rover (vx data from the Transform Sensor block)
3. Orientation of the rover (angle data from the Transform Sensor block)
4. Angular velocity of the rover with respect to Z direction (wz from the Transform Sensor block)
5. Angular velocity of the wheels on the right side of the rover - wr (at a previous time step, from the action)
6. Angular velocity of the wheels on the left side of the rover - wl (at a previous time step, from the action)

The observations also differ from continuous to discrete setup. The discrete one does not require wz as an observation. Furthermore, for deployment without sensor, the

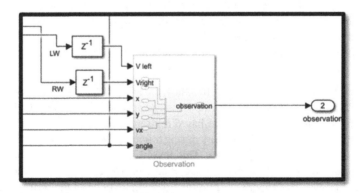

Fig. 6. Simulink representation of the Observation.

observations were reduced to only two – wr and wl. The Simulink model representation of the observation is as shown in Fig. 6.

Action. At each state, the environment generates a set of possible actions - action space. The agent chooses one amongst these, this is known as the action. It is basically the movement made by the agent in the environment based on the observations. The model is defined by two actions: 1. Angular velocity to the wheels on the right side of the robot – wr and 2. Angular velocity to the wheels on the left side of the robot - wl. These actions control the movement of the rover in the arena. They are the same for discrete and continuous setups; although their natures are widely different as described earlier. The Simulink model representation of the action function is as shown in Fig. 7.

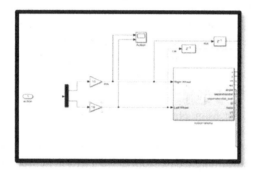

Fig. 7. Simulink representation of the Action function.

Environment Reset Function. The reset or the IsDone function in the DRL system is a flag to terminate episode simulation. This signal is used to end an episode if the agent reaches its assigned goal or goes irrecoverably far from its goal.

The IsDone Function Consists of the Following.

1. When the robot is out of bound i.e., it crosses the X and Y limits of the arena (X and Y coordinate data from the Transfrom Sensor block).
2. When the robot is stuck at one or more obstacles for a defined period (using delay timer and triggered subsystem).
3. When the robot has reached its goal point (separation distance data from the Transform Sensor block).

Algorithms. Model-free, online, and off-policy, the Deep Q-Network (DQN) algorithm is a powerful reinforcement learning method. Instead of an actor network, DQN estimates rewards using a critic network. A value-based reinforcement learning agent. The DQN agent constantly updates its critic network during training. The agent uses epsilon-greedy exploration to properly explore actions. This involves selecting a random action with probability ϵ or the action with the highest predicted value using the value function with probability 1-ϵ. The agent's current understanding suggests this greedy action is best. The DQN agent uses a circular experience buffer to store and use past experiences. This buffer samples mini-batches of experiences for critic network updates to help the agent learn from its past interactions. Remember that the DQN algorithm is limited to discrete action spaces, making it ideal for this discrete setup. Twin-Delayed Deep Deterministic Policy Gradient (TD3) is a model-free, online, off-policy reinforcement learning algorithm for continuous action spaces. TD3 uses an actor network and a critic network to learn a policy and estimate state values, unlike DQN. This makes it an actor-critic reinforcement learning agent that seeks the best policy to maximize cumulative long-term reward. At each training step, the TD3 agent updates the actor and critic networks. A circular experience buffer stores past experiences for efficient learning. The agent updates the actor and critic networks with mini-batches of randomly sampled experiences from this buffer. TD3's stochastic noise model perturbs the policy-selected action at each training step, which is unique. This encourages exploration and prevents local optima trapping. The continuous setup uses the TD3 algorithm for training because

Table 1. Training parameters used for DQN.

Parameter	Description	Value/Range used for simulation
Learning Rate	Estimates the step size during gradient descent	0.001
Discount Factor	Balances immediate and future rewards	0.9
Exploration – Exploitation	Decays over time while balancing exploration and exploitation	Started with 1.0 and decayed to 0.1
Replay buffer size	Size of buffer storing past experience for training	1 million transitions
Mini batch size	Number of experiences sampled from replay buffer for each training iteration	256
Target network update frequency	Frequency at which target network parameters are updated	0.1 update per learning iteration
Huber loss parameter	Determines the threshold point where the loss function transitions from quadratic (like MSE) to linear (like MAE)	1.0
Gradient Clipping	Prevents exploding gradient during backpropagation	9
Weight initialization	Method for initializing neural network weights	He technique
Learning rate schedule	Adjusts the learning rate over time	Adaptive learning rate, ADAM
Number of training episodes	Total number of episodes agent experiences during learning	1000

it can handle continuous action spaces. In contrast, DQN is limited to discrete action spaces and used in the discrete setup.

Training. With the environment created and the algorithm chosen, training was conducted. The following training parameters (Table 1 and Table 2) were selected after many trials:

Table 2. Training parameters used for TD3.

Parameter	Description	Values used for training
Learning rate (Actor and Critic)	Determines the step size during gradient descent for both the actor and critic networks	0.001
Discount factor	Discount factor for future rewards	0.9
Target Actor Update Interval	Frequency at which the target actor network is updated	1
Target Critic Update Interval	Frequency at which the target critic network is updated	1
Discount Factor (γ)	Discount factor for future rewards	0.9
Target Noise	Amount of noise added to the target action	0.1 5
Noise Clip	Clip range for the noise added to the action	0.5
Policy Update Frequency	Frequency of updating of network	2
Replay Buffer Size	Size of the replay buffer for storing experiences	1 million transitions
Batch Size	Number of transitions sampled from the replay buffer for each update	128
Policy Update Steps	Number of policy updates to perform per train step	1

3 Setup for Deployment

The Trained Agent is converted to a *function block* with input as observations and output as the actions. This is done using the 'Generate Greedy Policy' option of the Agent in the Simulink model. This function block performs inference on trained policy. It gets stored as a Policy Mask. For deployment, a voltage and frequency output are to be sent to the Raspberry Pi controller of the robot. The output of the *DRL function block* is the action (angular velocities). These velocities are converted to potential difference via Pulse Width Modulation (PWM) since that is the governing parameter of velocity regulation in the motor driver. Since the deployment was done without sensors, the reference speeds are set using velocity profiles- *Trapezoidal Velocity Profile Generator*. These values are also used as observations by the *DRL function blocks*. The angular velocities given by the profile generator and those given by the DRL function block are compared. A P-I-D controller is used to minimize the error. The output of the P-I-D controller, multiplied by gain values, gives the voltage that needs to be provided to the motor. This voltage is compared with the voltage across the batteries and converted to a PWM Duty cycle. A signal which gives a value between 0 to 1 is used to vary the frequency of the PWM signal. A value of 1 indicates the motor running at maximum speed and a value of 0 indicates a stopping condition. This represents frequency pulses sent to the designated PWM pins on the RPi board. Simulink support package for Raspberry Pi Hardware provides a *GPIO PWM* block and the *GPIO Write* block, with the help of which we can send these input signals to the RPi and hence, to the motor driver. For the motion of the motors in a set direction, *GPIO Write* blocks have been used to send signal values to the desired pins.

3.1 Deployment

The deployment of a deep reinforcement learning (DRL) agent on hardware using MATLAB involved several steps and tools. First, necessary packages like Reinforcement Learning Toolbox, Instrument Control Toolbox, and Data Acquisition Toolbox were installed. The RL agent was trained using algorithms like TD3 and DQN within MATLAB's Reinforcement Learning Toolbox. A MATLAB function block was created to encapsulate the trained RL agent's behaviour. Simulink was then utilized to design a system integrating hardware, communication protocols, and the RL agent's MATLAB code. The Simulink model was converted to C/C++ using MATLAB's Embedded Coder and deployed on Raspberry Pi hardware. Observing the hardware system's behaviour post-deployment allowed for further refinement of the RL agent's performance through training with hardware data (Fig. 8).

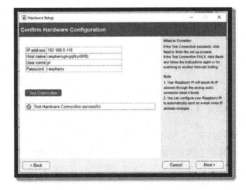

Fig. 8. Image of connection of Rpi

3.2 Testing

The performance of the proposed framework was tested on the considered robot in a variety of simulated environments. These environments included different numbers and locations of obstacles, as well as varying start and goal positions for the robot. The positions of the obstacles were identified using binary occupancy map as shown in the Fig. 9. The parameters give in Tables 1 and 2 were used for running simulations. The results obtained after simulation using both the algorithms are presented in the next section.

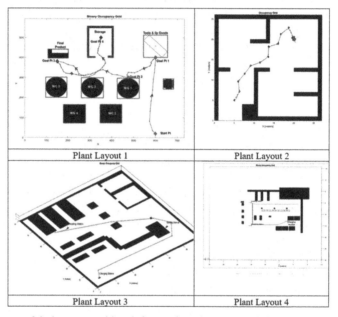

Fig. 9. Layouts of industry considered for testing the proposed framework (Four different arrangements of obstacles were considered).

4 Results

4.1 Training Statistics for TD3

The TD3 algorithm navigates complex obstacle courses well, but its performance decreases as the environment becomes denser with more obstacles (Table 3). As challenges increase, success and reward decrease. Despite these complex scenarios, the TD3 algorithm continues to learn well (Table 4). Interestingly, as the environment becomes more complex, the TD3 algorithm prioritizes using proven strategies over trying new ones. The action distribution shows that the agent relies more on forward movement in simpler layouts and turns and stops more in complex environments. Table 5 shows

Table 3. General Performance of the framework when trained with TD3 algorithm.

Parameter	Layout 1	Layout 2	Layout 3	Layout 4
Average Episode Reward	100	90	80	70
Success Rate	95%	90%	85%	75%
Average Collision Rate	0.2	0.3	0.5	0.8
Average Path Length	20	25	30	35
Average Execution Time	10 s	12 s	15 s	18 s

Table 4. Learning Progress of the framework when trained with TD3 algorithm.

Parameter	Layout 1	Layout 2	Layout 3	Layout 4
Average Loss (Actor & Critic)	0.05	0.07	0.1	0.15
Average Policy Entropy	0.8	0.7	0.6	0.5
Target Q-value Improvement	0.02	0.015	0.01	0.005
Exploration vs. Exploitation Ratio	0.7:0.3	0.6:0.4	0.5:0.5	0.4:0.6

Table 5. Hardware Utilization for the framework when trained with TD3 algorithm.

Parameter	Layout 1	Layout 2	Layout 3	Layout 4
Average CPU/GPU Usage	70%	80%	90%	100%
Average Memory Usage	400 MB	500 MB	600 MB	700 MB
Training Speed (Iterations/Second)	100	80	60	40

Table 6. Environment Specific Metric when trained with TD3 algorithm.

Parameter	Layout 1	Layout 2	Layout 3	Layout 4
Number of Obstacles Encountered	5	8	12	15
Average Distance to Obstacles	1.5 m	1.2 m	1 m	0.8 m
Time Spent Avoiding Obstacles	5 s	7 s	10 s	12 s

that computational demands rise with environment complexity. The TD3 algorithm efficiently adjusts resource use to meet these demands. Complex environments require more precise navigation because the robot encounters obstacles more often and operates closer to them. Reward breakdown shows a focus shift (Table 6). Distance reward dominates in

Table 7. Other Parameters for the framework when trained with TD3 algorithm.

Parameter	Layout 1	Layout 2	Layout 3	Layout 4
Reward Breakdown	Distance reward: 80%, Collision penalty: 20%	Distance reward: 70%, Collision penalty: 30%	Distance reward: 60%, Collision penalty: 40%	Distance reward: 50%, Collision penalty: 50%
Action Distribution	Forward: 80%, Turn left/right: 10%, Stop: 10%	Forward: 70%, Turn left/right: 15%, Stop: 15%	Forward: 60%, Turn left/right: 20%, Stop: 20%	Forward: 50%, Turn left/right: 25%, Stop: 25%

simpler environments, but collision penalty increases in complex environments, emphasizing the importance of safe navigation. The TD3 algorithm negotiates obstacles and succeeds in complex environments despite a slight decline (Table 7).

4.2 Training Statistics for DQN

Table 8. General Performance of the framework when trained with DQN algorithm.

Parameter	Layout 1	Layout 2	Layout 3	Layout 4
Average Episode Reward	80	70	60	50
Success Rate	85%	75%	65%	55%
Average Collision Rate	0.4	0.5	0.7	0.9
Average Path Length	25	30	35	40
Average Execution Time	12 s	15 s	18 s	21 s

Table 9. Learning Progress of the framework when trained with DQN algorithm.

Parameter	Layout 1	Layout 2	Layout 3	Layout 4
Average Loss	0.1	0.12	0.15	0.2
Exploration Rate	0.8	0.7	0.6	0.5
Q-value Improvement	0.015	0.01	0.005	0.002

DQN performs moderately in simple obstacle layouts like layout 1, but its performance degrades as environment complexity increases (Table 8). This decline is especially noticeable in layouts 2, 3, and 4's complex obstacle configurations. DQN achieves goals less efficiently than simpler layouts, as shown by its lower average episode reward (Table 9). As complexity increases, DQN's average collision rate rises, demonstrating its inability to avoid obstacles, especially in difficult environments. Complex layouts have longer average execution times (Table 10), demonstrating DQN's inefficiency compared

Table 10. Hardware and Resource Utilization for the framework when trained with DQN algorithm.

Parameter	Layout 1	Layout 2	Layout 3	Layout 4
Average CPU/GPU Usage	80%	90%	100%	110%
Average Memory Usage	500 MB	600 MB	700 MB	800 MB
Training Speed (Iterations/Second)	80	60	40	20

Table 11. Environment Specific Metric for the framework when trained with DQN algorithm.

Parameter	Layout 1	Layout 2	Layout 3	Layout 4
Number of Obstacles Encountered	5	8	12	15
Average Distance to Obstacles	1.4 m	1.1 m	0.9 m	0.7 m
Time Spent Avoiding Obstacles	7 s	10 s	13 s	15 s

to simpler environments. With similar resources, DQN learns less efficiently, as shown by its higher average loss and slower Q-value improvement in complex layouts (Table 9). DQN prefers forward movement in simpler layouts, but its lack of exploration and adaptation in complex environments leads to inefficient strategies. DQN also operates closer to obstacles in complex layouts and spends more time avoiding them, suggesting poor navigation strategies compared to simpler environments. DQN's reward breakdown shifts from distance reward in simpler layouts to collision penalty in complex layouts, reflecting its safety struggles in challenging situations. The performance limitations of

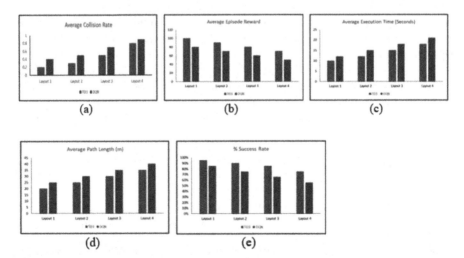

Fig. 10. Comparison of performance of TD3 algorithm-based framework and DQN based framework when implemented to 4 layouts.

DQN limit its applicability in real-world industrial environments with complex obstacle configurations (Table 11). Figure 10 shows that DQN has lower average episode rewards, success rates, and obstacle distances than TD3, indicating lower goal achievement and collision avoidance. DQN has a higher average loss and slower Q-value improvement than TD3, suggesting less efficient learning. DQN uses more CPU/GPU and memory than TD3, indicating more computational demand for similar performance. DQN spends more time and operates closer to obstacles than TD3, suggesting less effective obstacle avoidance.

5 Discussion

The framework proposed in this paper trains robots to autonomously navigate from starting points to predefined endpoints while avoiding obstacles using deep neural networks and reinforcement learning algorithms. The details the framework's environment and robot modeling, deep neural network architecture, training, optimal path planning, obstacle avoidance strategies, simulation and experimentation, and performance evaluation methods are also provided. The paper compares TD3 and DQN reinforcement learning algorithms for obstacle avoidance. The comparison shows that TD3 outperforms DQN in navigation efficiency, safety, and effectiveness, especially in complex environments, through extensive simulations. The results emphasize TD3's robustness, reliability, and resource efficiency, making it a more practical and promising industrial robot navigation option.

6 Conclusions

The work presented in this paper emphasizes obstacle avoidance for self-governing systems in complex and ever-changing environments. The deep reinforcement learning used to create a framework that allows an agent to travel from an initial position to a predetermined destination while avoiding obstacles. The robot/agent can learn and navigate to the goal while avoiding obstacles using MATLAB Simulink simulation environments and experimental methods. The study shows that deep reinforcement learning for obstacle avoidance is practical and effective, providing valuable knowledge for self-directed navigation systems. DRL paths can be compared to probability or RRT* paths in the future. Advanced training methods, such as TD3 on a trained DQN, can improve results further.

References

1. Shahmoradi, J., Talebi, E., Roghanchi, P., Hassanalian, M.: A comprehensive review of applications of drone technology in the mining industry. Drones **4**, 34 (2020). https://doi.org/10.3390/drones4030034
2. Bonadies, S., Gadsden, S.A.: An overview of autonomous crop row navigation strategies for unmanned ground vehicles. Eng. Agricult. Environ. Food **12**, 24–31 (2019). https://doi.org/10.1016/j.eaef.2018.09.001

3. Ni, J., Hu, J., Xiang, C.: A review for design and dynamics control of unmanned ground vehicle. Proc. Inst. Mech. Eng. Part D: J. Automob. Eng. **235**, 1084–1100 (2021). https://doi.org/10.1177/0954407020912097
4. Wang, H., Yuan, S., Guo, M., Chan, C.-Y., Li, X., Lan, W.: Tactical driving decisions of unmanned ground vehicles in complex highway environments: a deep reinforcement learning approach. Proc. Inst. Mech. Eng. Part D: J. Automob. Eng. **235**, 1113–1127 (2021). https://doi.org/10.1177/0954407019898009
5. Li, Y., Zhang, S., Ye, F., Jiang, T., Li, Y.: A UAV path planning method based on deep reinforcement learning. In: 2020 IEEE USNC-CNC-URSI North American Radio Science Meeting (Joint with AP-S Symposium), pp. 93–94. IEEE, Montreal (2020). https://doi.org/10.23919/USNC/URSI49741.2020.9321625
6. Honda, K., Yonetani, R., Nishimura, M., Kozuno, T.: When to Replan? An Adaptive Replanning Strategy for Autonomous Navigation using Deep Reinforcement Learning. (2023). https://doi.org/10.48550/ARXIV.2304.12046
7. Xue, X., Li, Z., Zhang, D., Yan, Y.: A deep reinforcement learning method for mobile robot collision avoidance based on double DQN. In: 2019 IEEE 28th International Symposium on Industrial Electronics (ISIE), pp. 2131–2136. IEEE, Vancouver (2019). https://doi.org/10.1109/ISIE.2019.8781522
8. Wang, X., et al.: Deep reinforcement learning: a survey. IEEE Trans. Neural Netw. Learn. Syst. **35**(4), 5064–5078 (2022). https://doi.org/10.1109/TNNLS.2022.3207346
9. Kiran, B.R., et al.: Deep reinforcement learning for autonomous driving: a survey. IEEE Trans. Intell. Transp. Syst. **23**, 4909–4926 (2022). https://doi.org/10.1109/TITS.2021.3054625
10. Perez-Dattari, R., Celemin, C., Ruiz-del-Solar, J., Kober, J.: Continuous control for high-dimensional state spaces: an interactive learning approach. In: 2019 International Conference on Robotics and Automation (ICRA), pp. 7611–7617. IEEE, Montreal (2019). https://doi.org/10.1109/ICRA.2019.8793675
11. da Silva, A.R., Rezeck, P.A.F., Luz, G.S., Alves, T.R., Macharet, D.G., Chaimowicz, L.: A DRL approach for object transportation in complex environments. In: 2022 Latin American Robotics Symposium (LARS), 2022 Brazilian Symposium on Robotics (SBR), and 2022 Workshop on Robotics in Education (WRE), pp. 1–6. IEEE, São Bernardo do Campo (2022). https://doi.org/10.1109/LARS/SBR/WRE56824.2022.9995823
12. Wang, Y., Yao, F., Cui, L., Chai, S.: An obstacle avoidance method using asynchronous policy-based deep reinforcement learning with discrete action. In: 2022 34th Chinese Control and Decision Conference (CCDC), pp. 6235–6241. IEEE, Hefei (2022). https://doi.org/10.1109/CCDC55256.2022.10033562
13. Low, E.S., Ong, P., Cheah, K.C.: Solving the optimal path planning of a mobile robot using improved Q-learning. Robot. Auton. Syst. **115**, 143–161 (2019). https://doi.org/10.1016/j.robot.2019.02.013
14. Sichkar, V.N.: Reinforcement learning algorithms in global path planning for mobile robot. In: 2019 International Conference on Industrial Engineering, Applications and Manufacturing (ICIEAM), pp. 1–5. IEEE, Sochi (2019). https://doi.org/10.1109/ICIEAM.2019.8742915
15. Carlucho, I., De Paula, M., Acosta, G.G.: Double Q-PID algorithm for mobile robot control. Expert Syst. Appl. **137**, 292–307 (2019). https://doi.org/10.1016/j.eswa.2019.06.066
16. Jin, Z., Wu, J., Liu, A., Zhang, W.-A., Yu, L.: Policy-based deep reinforcement learning for visual servoing control of mobile robots with visibility constraints. IEEE Trans. Industr. Electron. **69**, 1898–1908 (2022). https://doi.org/10.1109/TIE.2021.3057005
17. Schulman, J., Wolski, F., Dhariwal, P., Radford, A., Klimov, O.: Proximal Policy Optimization Algorithms. (2017). https://doi.org/10.48550/ARXIV.1707.06347
18. Babaeizadeh, M., Frosio, I., Tyree, S., Clemons, J., Kautz, J.: Reinforcement Learning through Asynchronous Advantage Actor-Critic on a GPU (2016). https://doi.org/10.48550/ARXIV.1611.06256

19. Jesus, J.C., Bottega, J.A., Cuadros, M.A.S.L., Gamarra, D.F.T.: Deep deterministic policy gradient for navigation of mobile robots in simulated environments. In: 2019 19th International Conference on Advanced Robotics (ICAR), pp. 362–367. IEEE, Belo Horizonte (2019). https://doi.org/10.1109/ICAR46387.2019.8981638
20. Jiang, H., Wan, K.-W., Wang, H., Jiang, X.: A dueling twin delayed DDPG' architecture for mobile robot navigation. In: 2022 17th International Conference on Control, Automation, Robotics and Vision (ICARCV), pp. 193–197. IEEE, Singapore (2022). https://doi.org/10.1109/ICARCV57592.2022.10004320
21. Prasuna, R.G., Potturu, S.R.: Deep reinforcement learning in mobile robotics – a concise review. Multimed. Tools Appl. (2024). https://doi.org/10.1007/s11042-024-18152-9

Fast and Accurate Right-Hand Detection Based on YOLOv8 from the Egocentric Vision Dataset

Van-Dinh Do[1], Trung-Minh Bui[2], and Van-Hung Le[2(✉)]

[1] Electrical Department, Sao Do University, Chi Linh, Vietnam
[2] Tan Trao University, Tuyen Quang, Vietnam
van-hung.le@mica.edu.vn

Abstract. To build systems for device control, entertainment, and human-machine interaction using hand gestures, the hand data area needs to be detected, hand posture needs to be estimated, and hand gestures need to be recognized. Hands need to be done quickly and accurately. Therefore, each step in the system construction model needs to be performed quickly and accurately. The advent of deep learning (DL) has brought very impressive results in solving computer vision problems, especially CNN for object detection (OD) and recognition problems. From there, DL is a good approach to quickly and accurately solve the hand detection (HDe) problem. In this study, we fine-tuned the right-HDe model using the YOLOv8 (YOLOv8n, YOLOv8m, YOLOv8l, and YOLOv8x) on the HOI4D dataset, which is the latest DL network of the YOLO family for OD. Right-HDe results on YOLOv8x are the best ($P = 99.2\%$, $R = 98.81\%$, $mAP@50 = 99.4\%$). The HDe results must be carried out for each hand action of the HOI4D dataset. At the same time, the computation time on the GPU and CPU of YOLOv8 is (YOLOv8n = 197 fps, YOLOv8m = 194 fps, YOLOv8l = 192 fps, YOLOv8x = 188 fps) and (YOLOv8n = 1.73 fps, YOLOv8m = 1.64 fps, YOLOv8l = 1.62 fps, YOLOv8x = 1.55 fps), respectively.

Keywords: Fast and Accurate Right-Hand detection · YOLOv8 and its variants · HOI4D · Egocentric Vision · Deep Learning

1 Introduction

To build applications to control robotic arms [1, 2] to perform complex operations such as human hands, or applications to help visually impaired people [3, 4] grasping objects, the data to train the model must be the data collected from the bearing sensors (Egocentric Vision camera - EVC). To build these applications, hand pose (HP) needs to be estimated and gestures recognized as in the study of Le et al. [5]. Detecting the hand on the image though is a pre-processing step to perform the next steps such as 2D, and 3D hand skeleton estimation, and hand action recognition. This step is very important because the HDe results directly affect the results of HP estimation and activity recognition. It is recommended to detect the data area of the hand on the image that needs to be researched and tested to choose the best method for building an overall model of the application using the pose and activity of the hand. At the same time, to meet the calculation execution time close to the real-time of the control system, the HDe step needs to be performed quickly.

The dataset obtained from the EVC is the dataset obtained from the camera mounted on the head or chest of a person, it is similar to the reality when building the above applications. Hand data obtained from EVC is fraught with many challenges such as fingers being obscured by the direction of the view-point, fast hand movement speed making the data blurry and noisy, and hands being obscured by objects when performing grasping. These issues are clearly shown in [6].

In DL models for OD, YOLO (You Only Look Once) version 1 (v1) [7] is a one-stage CNN network and is considered the method with the fastest computation time and reasonable accuracy in OD [8] in RGB images. With the strong development of computer hardware, the YOLO network is continuously evolving with the development of hardware [9] with versions like YOLOv1 [7] in 2015, YOLOv2/9000 [10] in 2016, YOLOv3 [11] in 2018, YOLOv4 [12] in 2020, YOLOR [13] and YOLOx [14] in 2021, YOLOv5 [15], YOLOv6 [16] and YOLOv7 [17] in 2022, YOLOv8 [18] and YOLO-NAS [19] in 2023. With recent development and continuous improvement, YOLOv8 [18] was introduced and published in May 2023. YOLOv8 [18] is considered to have high accuracy and the fastest calculation time for the OD problem on the COCO dataset [20].

To enrich learning data for DL models, the HOI4D dataset has been built and published by Liu et al. [21]. HOI4D [21] is a benchmark dataset obtained from the EVC RGB-D, which is for evaluating models for HP estimation and hand action recognition in 3D space. This dataset provides color data, depth images, point clouds, and native data types. At the same time, this dataset is also a new dataset published in CVPR 2022 and is still not widely used by the research community. HOI4D was collected on two hands (right and left hand). In particular, the data on right-hand is much more because this dataset is collected on Asian (Chinese) people, so most people use their right-hand to perform actions [22].

In this study, we perform normalization of the HOI4D dataset for HDe evaluation on the RGB images. We also performed fine-tuned the right-HDe model of the HOI4D dataset with YOLOv8 and variations. The HDe results in the image are evaluated and analyzed for each hand action. Fine-tuning the HDe model based on data from four cameras and the fine-tuned model of each camera.

The main contributions of the paper are as follows:

- We perform fine-tuning the right-HDe model of the data of four cameras and the data of each camera of the HOI4D dataset on the YOLOv8 and variations (YOLOv8n, YOLOv8m, YOLOv8l, and YOLOv8x).
- We normalize the HOI4D dataset for HDe evaluation. From there, select good results to apply to the 3D HP estimation model and hand gesture recognition.
- We evaluate the right-HDe results on the fine-tuned model of the HOI4D dataset of the YOLOv8 and variations by the precision, recall, and mAP@50 measurements according to each hand action of the right hand. The computation times of the YOLOv8 variants are also compared on both GPU and CPU.

The remaining structure of the paper is presented as follows: Sect. 2 presents the related research of HDe using DL. The architecture, new improvement of YOLOv8, and details of the fine-tuning process of the right- HDe model are presented in Sect. 3. Section 4 presents the experiment and h HDe results of YOLOv8 on the HOI4D dataset. Finally, conclusions and future research are presented in Sect. 5.

2 Related Works

Today, with the strong development of the 4.0 industrial revolution, it has brought many achievements to humans such as artificial intelligence, blockchain, etc. Previously, to solve problems of artificial intelligence and blockchain, machine learning and traditional algorithms [23–25] were often used. With the strong development of computer hardware and the advent of DL, solving artificial intelligence and blockchain problems has become easier, especially being able to perform calculations on GPUs.

OD is a very important problem of computer vision. HDe is a pre-processing step and is important in performing research such as tracking, estimating, classifying, and recognizing hand activities [26, 27]. Currently, to solve the problem of detecting hands in images, CNNs are often used and there have been many impressive results on this problem. One-stage CNNs (YOLO family [28], SSD family [29]) have lower accuracy than two-stage CNNs (RCNN [30], Fast RCNN [31], Faster RCNN [32], RFCN [33], Mask RCNN [34]), but the computation time of one-stage CNNs is many times faster than two-stage CNNs [28, 35]. Regarding one-stage CNNs, there have been many studies on HDe using YOLOv2 and YOLOv4 networks. In Chen et al. [36]'s research, YOLOv2 was improved to MS-YOLOv2 with an increase in HDe accuracy of more than 2% compared to YOLOv2. Van et al. [37] conducted a study comparing HDe results on the Epic Kitchens-100 dataset based on using YOLOv3 [11] and VGG16 [38]. The results show that VGG16 has higher accuracy. Gao et al. [39] have improved the SSD network for detecting and locating hands in images to prepare for human-robot interaction. By HDe in the egocentric dataset categories, Bandini et al. [40] conducted a survey and analysis of hand egocentric vision datasets, in which the authors also systematized related problems such as HDe, activity recognition, and egocentric vision estimation, so, follow, etc. Nguyen et al. [5] have conducted a comprehensive study comparing the detection and classification of hand activities from hand sensor data with versions of the YOLO family such as YOLOv4, YOLOv5, and YOLOv7. The hand egocentric vision dataset is FPHAB [6]. The results show that YOLOv7 has the best accuracy results and the fastest running time. In addition, many other studies on HDe in images based on DL are also presented [41–46].

3 Fine-Tuning YOLOv8 for Right-Hand Detection

3.1 Background

YOLO (You Only Look Once) [7] is a typical CNN and is used a lot for OD problems. YOLO has a simple architecture as its name suggests, as presented in Fig. 1, including a backbone network and extra layers.

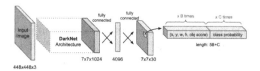

Fig. 1. The general architecture of YOLO [7].

Fig. 2. Time to announce versions of YOLO.

The development and improvement process of YOLO is shown in Fig. 2. In Fig. 2, versions of YOLO were announced as YOLOv1 [7] in 2015, YOLOv2/9000 [10] in 2016, YOLOv3 [11] in 2018, YOLOv4 [12] in 2020, YOLOR [13] and YOLOx [14] in 2021, YOLOv5 [15], YOLOv6 [16] and YOLOv7 [17] in 2022, YOLOv8 [18] and YOLO-NAS [19] in 2023.

YOLO architecture includes: the backbone network is a Conv. network to extract the image features from input images. To analyze features and detect objects, YOLO used the Extra layers, which are the final layers. The backbone commonly used in the YOLO architecture is Darknet (DK). YOLO architectures can be calibrated to different shapes of the input image in later layers of the Extra layers.

$S \times S$ cells are created by dividing the input image into a grid of cells, also known as cells. This image division is not real but divides the output into a matrix A of size $S \times S$. If in the image, $(i, j)^{th}$ cell contains the center of the object, then the corresponding output will be in $A[i, j]$. The number of classes is used to determine the size of the output. To determine the bounding box (BB) for an object in an image, the anchor boxes (ABs) are used for estimating the BB in YOLO. The BB that surrounds the object relatively precisely will be determined based on the previously constructed ABs. After that, the predicted BB of the object is determined by refining the AB based on the regression BB algorithm. In a YOLO model include:

- An AB is created to identify an object in the training image. When there are multiple ABs containing the object, YOLO will choose the AB that the highest valued of IoU relative to a GT BB of the object, as shown in Fig. 3.
- In the input image, each object is assigned to a cell of the feature vector and contains the feature map.

From $Cell_i$, three ABs are defined with blue borders. The GT of the object BB intersects the three ABs. However, the thickest blue border is the AB of the object, its IoU value is the highest when compared to the GT BB.

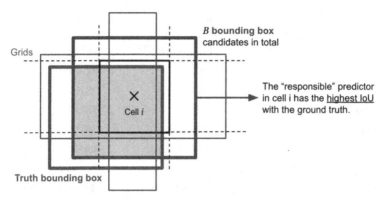

Fig. 3. Defining how to define an AB for an object [10].

When predicting the BB of an object, YOLO predicts multiple BBs for the same object. To reduce the BB number of the prediction, the non-maximum suppression is used to filter out duplicate BBs. Steps of non-max suppression consists of two steps:

Step 1: The first is to use a threshold to filter out BBs based on the probability of containing the object less than a threshold, usually 0.5

Step 2: For BBs with intersection probability greater than a given threshold, a BB with the highest *IoU* containing the object is selected, i.e. the *IoU* index intersecting the GT BB of the object is the highest. Then, calculate the *IoU* noise index for the remaining BBs.

3.2 YOLOv8

There are now many improved versions of YOLO for OD proposed, YOLO versions are shown in Table 1.

YOLOv8 proposed in 2023 is an improvement of YOLOv5. The architecture of YOLOv8 is not much different from previous versions, the architecture of YOLOv8 contains two main components: Backbone and Head. The components of YOLOv8 are illustrated in Fig. 4.

YOLOv8 was developed on the foundation of YOLOv5 by Ultralytics. YOLOv8 has several architectural changes and improvements to improve performance compared to YOLOv5. New enhancements and combinations performed on YOLOv8 compared to YOLOv5 are shown by Ultralytics:

First, YOLOv8 and YOLOv5 have the same backbone, both networks use CSPDarknet53 to select features from input images. $C2f$ is used to replace $C3$ to improve OD accuracy by combining high-level features with contextual information. The first 6×6 Conv. in the body is changed to the 3×3 Conv.. In $C2f$, the remaining connections have the bottleneck is composed of two 3×3 transitions, as shown in Fig. 5, the number of features is shown in "f", the expansion rate and CBS is a block composed of a Conv., a BatchNorm and a SiLU in "e". The output from the last bottleneck is used in $C3$.

Second, an anchor-free model is used to handle the object tasks in YOLOv8, classification, and regression independently. The overall accuracy of the model is improved

Table 1. Timeline of YOLO versions.

YOLO series	Year	Framework	Backbone
YOLOv1 [7]	2015	DK	DK-19
YOLOv2 [10]	2016	DK	DK-19
YOLOv9000 [10]	2016	DK	DK-19
YOLOv3 [11]	2018	DK	DK-53
PP-YOLO [47]	2020	ResNet50-vd	ResNet50-vd
YOLOv4 [12]	2020	DK	CSP-DK-53
YOLOv5 [48]	2020	PyTorch (PT)	Modified CSPv7
YOLOS [49]	2021	PT	Transformer block
PP-YOLOV2 [50]	2021	PT	ResNet50-vd-dcn
YOLOv6 [16]	2022	PT	EfficientRep
YOLOv7 [51]	2022	PT	RepConvN
YOLOv8 [18]	2023	PT	YOLOv8

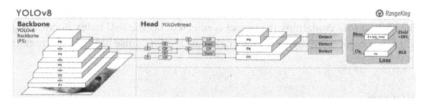

Fig. 4. YOLOv8 architecture [18].

from each branch. In the output layer, the feature score is used as a sigmoid function. From there, the probability of the BB containing the object is represented. A softmax function is presented for the class probability, representing the probability that the object belongs to each possible class.

Third, YOLOv8 has removed two Conv.s (10th and 14th) of the YOLOv5 config. YOLOv8-Seg is added to the YOLOv8 model for semantic segmentation. To maintain the results of OD and object semantic segmentation compared to state-of-the-art methods and fast speed, YOLOv8 used the following loss functions: the loss functions (LFs) for BB loss, binary cross-entropy for classification loss. The BB loss and binary cross-entropy are used to to increase small OD results.

Fourth, the bottleneck of YOLOv8 and YOLOv5 is the same, during the 1st Conv., the size of kernel was changed from 1×1 to 3×3. Figure 6 shows the change to the ResNet block.

Currently, YOLOv8 has five variants for fine-tuning the OD model: YOLOv8n stands for nano YOLOv8, YOLOv8s stands for small YOLOv8, YOLOv8m stands for medium YOLOv8, YOLOv8l stands for large YOLOv8, and YOLOv8x stands for extra large

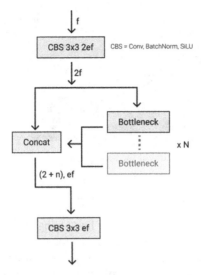

Fig. 5. Architectural illustration of C2f module in YOLOv8.

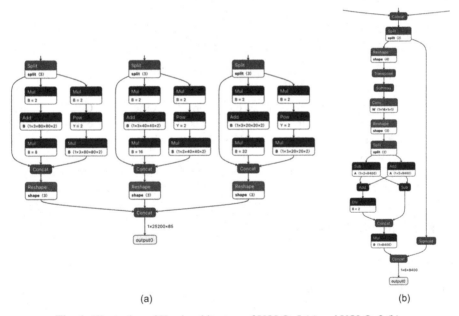

Fig. 6. Illustration of Head architecture of YOLOv5 (a) and YOLOv8 (b).

YOLOv8. Based on the size of the YOLOv8 model, the accuracy will increase with the model size, and the calculation time will also increase with the model size.

Fig. 7. The GT BB of the hand from the egocentric image.

3.3 Right-Hand Detection Model Fine-Tuned

YOLOv8 has been announced and allows the fine-tuning of the HDe models on custom datasets. The source code of YOLOv8 [52] is used for testing to fine-tune the right-HDe model. To synchronize data according to the input data format of the YOLO model of the right-HDe model on the RGB image of the HOI4D dataset, the ground truth (GT) BB is standardized, as illustrated in Fig. 7. The BB format of YOLOv8 is similar to the COCO 2017 dataset [20], as shown in Formula (1).

$$a = \frac{\frac{x\_max + x\_min}{2}}{w\_b}; \quad b = \frac{\frac{y\_max + y\_min}{2}}{h\_b}$$
$$c = \frac{w\_b}{w\_im}; \quad d = \frac{h\_b}{h\_im} \quad (1)$$

where (x_max, y_max) is the coordinates of the top right corner of the GT BB, (x_max, y_max) is the coordinates of the bottom left corner of the GT BB, w_b is the width of the GT BB, h_b is the height of the GT BB, w_im is the width of the image, h_im is the height of the image.

The BB structure in the format of the COCO 2017 dataset [20] used for training and evaluation has the form (l, a, b, c, d), where l is the label of the object.

4 Experimental Results

4.1 Data Collection

The HOI4D dataset [20] is a large-scale 4D radial benchmark dataset. HOI4D dataset contains 2.4 million RGB-D frames. This dataset uses four cameras to collect data from indoor environments.

In this study, we divide HOI4D data by each data collection camera. The hand actions in the HOI4D dataset consist of 16 hand activities (grasping ToyCar, grasping Mug, grasping Laptop, grasping StorageFurniture, grasping Bottle, grasping Safe, grasping Bowl, grasping Bucket, grasping Scissors, grasping Pliers, grasping Kettle, grasping Knife, grasping TrashCan, grasping Lamp, grasping Stapler, grasping Chair) and are shown in Fig. 8.

The GT hand data of the HOI4D dataset includes a 2D hand BB, 3D HP annotation, and hand action label. In study, we only use RGB images and 2D BB annotation of hand. 2D hand BB annotation standardized according to the BB format of the COCO 2017 dataset [20], presented in Sect. 3.3. We only use right-hand data for the fine-tuning HDe model and evaluation model from 603,332 frames. We divide the data of HOI4D database by left hand and right hand activities, by camera label of data collection and 16 hand object grasping activities. We use a ratio (7:3), with 70% data for the training model and 30% data for the testing model, in which the data is divided randomly. After that, the 2D HP annotation is also obtained corresponding to the RGB image that has been divided into the training and testing data.

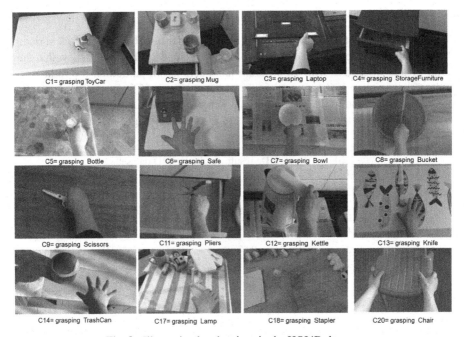

Fig. 8. Illustrating hand actions in the HOI4D dataset.

4.2 Evaluation Metrics

To evaluate the performance of right-HDe, we utilize metrics of evaluation are *IoU* (Intersection over Union), Precision (*P*) as Formula (2), Recall (*R*) as Formula (2), and *mAP* (mean Average Precision) as Formula (3).

$$P = \frac{TP}{TP+FP}; R\frac{TP}{TP+FN}; \qquad (2)$$

$$mAP = \frac{1}{|Classes|} \sum_{c \in Classes} \frac{TP(c)}{TP(c)+FP(c)} \qquad (3)$$

where *TP* is the true predicted hand area, *FP* is the false predicted hand area, *TN* is a false predicted hand area (the label is the true hand area but the predictive model is not hand region), and *FN* is the no-predicted hand annotation BB. The IoU threshold for determining the correct predicted hand region is 0.5. The size of the image as input to the training and testing process is normalized to 1280 × 1280 pixels.

In study, to perform model training and model evaluation, we experiment on a server with the following GPU configuration: RTX 2080 Ti, 12 GB, installed the CUDA 11.2/cuDNN 8.1.0 libraries and CPU: 12th Gen Intel i9-12900K (24) @ 6.500 GHz. The experimental source code is built in Python programming language ($\geq$3.9 version) with the support of some libraries of AI programming and computer vision.

4.3 Results and Discussions

The results of the right-HDe with the fine-tuned model on YOLOv8 and its variants, when evaluated on the HOI4D dataset, are shown in Table 2. The results show that they are all greater than 95% with 16 different hand actions. In particular, the right- HDe results are not reduced when on an action with a sequence of frames containing many hand shapes, and motion phases, and in many frames the hand data is obscured due to the direction of the view-point (only the palm area visible, fingers not visible) or hand

Table 2. Right- HDe results on the HOI4D dataset.

Methods	Hand action	Measurement		
		P(%)	R (%)	mAP50 (%)
	C1	99.5	99	99.2
	C2	99.7	99.5	99.5
	C3	99.5	99.3	99.5
	C4	95.3	92.7	97
	C5	99.5	99.1	99.5
	C6	98.9	98.7	99.3
	C7	99.3	98.5	99.4
	C8	99.3	98.8	99.4
YOLOv8n	C9	99.4	98.4	99.5
	C11	99.1	98.8	99.4
	C12	99.3	98.8	99.5
	C13	99.6	99.2	99.5
	C14	99.5	99.2	99.4
	C17	98.7	98.6	99.3
	C18	98.3	98.4	99.2
	C20	99.2	99	99.3
	Average	**99.00**	**98.5**	**99.24**

(*continued*)

Table 2. (*continued*)

Methods	Hand action	Measurement		
		P(%)	R (%)	mAP50 (%)
YOLOv8m	C1	99.5	99	99.3
	C2	99.8	99.6	99.5
	C3	99.5	99.3	99.5
	C4	94.6	92.6	97.5
	C5	99.6	99.2	99.5
	C6	98.9	98.8	99.3
	C7	99.4	98.3	99.4
	C8	99.5	98.7	99.4
	C9	99.5	98.2	99.5
	C11	99.1	99.1	99.5
	C12	99.5	98.9	99.5
	C13	99.6	99.2	99.5
	C14	99.3	99.1	99.4
	C17	99	98.6	99.4
	C18	98.3	98.3	99.3
	C20	99.4	99.1	99.3
	Average	**99.03**	**98.5**	**99.3**
YOLOv8l	C1	99.7	99.1	99.5
	C2	99.8	99.6	99.5
	C3	99.5	99.3	99.5
	C4	95.3	93.2	97.8
	C5	99.5	99.3	99.5
	C6	99	98.8	99.4
	C7	99.5	98.3	99.4
	C8	99.5	98.8	99.4
	C9	99.5	98.3	99.5
	C11	99.1	99.3	99.5
	C12	99.5	98.9	99.5
	C13	99.6	99.2	99.5
	C14	99.3	99.1	99.4
	C17	99	98.7	99.4
	C18	98.3	98.4	99.3
	C20	99.3	99.2	99.3
	Average	**99.08**	**98.59**	**99.34**

(*continued*)

Table 2. (*continued*)

Methods	Hand action	Measurement		
		P(%)	R (%)	mAP50 (%)
	C1	99.5	99.4	99.5
	C2	99.8	99.6	99.5
	C3	99.6	99.3	99.5
	C4	96	94.4	98.4
	C5	99.7	99.6	99.5
	C6	99.2	98.9	99.4
	C7	99.5	98.6	99.5
	C8	99.5	99	99.4
YOLOv8x	C9	99.3	98.8	99.5
	C11	99.5	99.4	99.5
	C12	99.4	99.3	99.5
	C13	99.7	99.3	99.5
	C14	99.5	99.3	99.5
	C17	99	98.8	99.5
	C18	98.6	98.1	99.4
	C20	99.4	99.1	99.3
	Average	**99.2**	**98.81**	**99.4**

data obscured by other objects. The results of detecting and classifying hands-on action "C4" have the lowest results, because this action is performed quickly, in a complex scene and the hand has a skin color close to the color of the storage furniture, so the image area of the cabinet mistakenly detected and misclassified with the hand. At the same time, the right- HDe and classification results also show that R is lower than P in all hand activities and variations of YOLOv8. That means there are many cases where the right-hand data area is predicted to be another object or the left-hand. The average result of YOLOv8x is the best *(P = 99.2%, R = 98.81%, mAP50 = 99.4%)*. This is a good result that can be applied to the trained model to detect the right-hand in the image obtained from the egocentric vision, which is a pre-processing step to perform the next steps such as 2D and 3D HP estimation, and hand action recognition.

The execution time of HDe on the image of YOLOv8 with its variants is also shown in Table 3. The calculation speed on the GPU of YOLOv8 is greater than 180 fps and on the CPU is 1.5 fps. This is the result consistent with a pre-processing step. The processing time results also show that YOLOv8x (Extra Large) has the highest processing time and YOLOv8n (Nano) has the lowest processing time. This is consistent with the network architecture of the YOLOv8 variants. Tab. 3 also shows that calculation time on GPU is more than 100 times faster than calculation on CPU.

Figure 9 shows the results of right-HDe with chair-holding activity on the HOI4D database using YOLOv8x. The right hand is detected and bounded by a red BB with a label above and a confidence score of the prediction. The value of the confidence score is usually from 0 to 1, in Fig. 9 this value is 0.9, which is a very high value for right-hand prediction. The results show that the right-HDe is very accurate, even though the left-hand is included in the image. However, the detection results are very accurate.

Table 3. The processing time for right-HDe on the HOI4D dataset when performing calculations on GPU and CPU.

Methods	Type	YOLOv8x (fps)	YOLOv8m (fps)	YOLOv8l (fps)	YOLOv8n (fps)
Processing time	GPU	188	194	192	197
Processing time	CPU	1.55	1.64	1.62	1.73

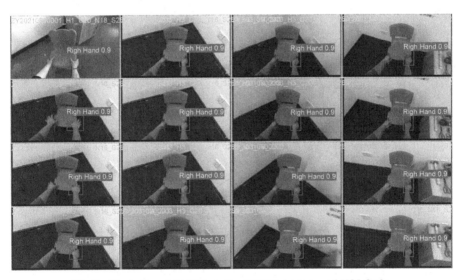

Fig. 9. Illustrating right-HDe in the HOI4D dataset used YOLOv8x.

Figure 10 shows the confusion matrix when detecting the right-hand with labeled ("C1" and "C2") of the HOI4D dataset when using YOLOv8x. The above is the confusion matrix of right-hand prediction with label "C1", below is the confusion matrix of right-hand prediction with label "C2". This result has very high accuracy, only a small number of hand data areas are mistakenly classified as background.

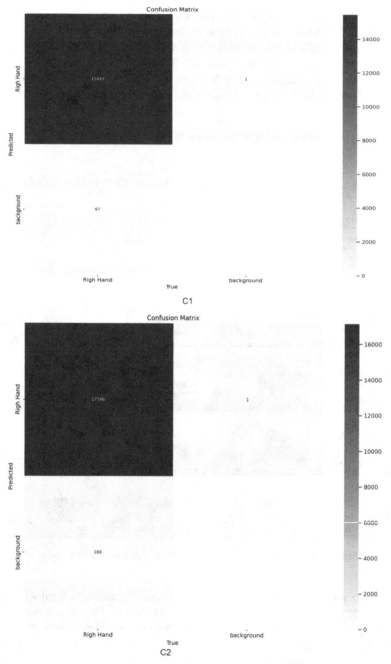

Fig. 10. Illustration of the confusion matrix when detecting the right-hand with labeled ("C1" and "C2") of the HOI4D dataset when using YOLOv8x.

5 Conclusions and Future Works

In the model of building applications for human-machine interaction, entertainment, control, and assisting the blind using hand gestures, detecting hands-on images is an important pre-processing step in building applications that simulate hand movements and recognize HPs. The result of HDe is important to carry out the next steps such as HP estimation, and hand activity recognition. In study, we have performed fine-tuning the right-HDe model on the HOI4D dataset with YOLOv8n, YOLOv8m, YOLOv8l, and YOLOv8x. YOLOv8x model's right-HDe result is the highest ($P = 99.2\%$), and YOLOv8n has the lowest result ($P = 99\%$). The processing of YOLOv8 is also shown when performing calculations on GPU and CPU. Shortly, we will conduct a study comparing the right-HDe results of the HOI4D dataset on some DL networks for typical OD such as SSD VGG16 [53], Mediapipe [54], YOLO-Nas [19], MobileNet V3 small [55], and MobileNet V3 large [55]. We will also use the right-HDe results to limit the hand data area on the image to perform the next steps such as automatic estimation and HP recognition on the HOI4D dataset.

Acknowledgment. This research is supported by Tan Trao University, Vietnam.

References

1. Arshad, J., Qaisar, A., Rehman, A.U., et al.: Intelligent control of robotic arm using brain computer interface and artificial intelligence. Appl. Sci. (Switzerland) **12**(21), 10813 (2022). https://doi.org/10.3390/app122110813. ISSN: 20763417
2. Franceschetti, A., Tosello, E., Castaman, N., Ghidoni, S.: Robotic Arm Control and Task Training Through Deep Reinforcement Learning. LNNS, vol. 412, pp. 532–550 (2022). https://doi.org/10.1007/978-3-030-95892-341. ISSN: 23673389. eprint: 2005.02632
3. Mandikal, P., Grauman, K.: DexVIP: learning dexterous grasping with human hand pose priors from video. In: CoRL 2021 Conference, 2022, pp. 1–11. http://arxiv.org/abs/2202.00164. eprint: 2202.00164
4. Ganesan, J., Azar, A.T., Alsenan, S., Kamal, N.A., Qureshi, B., Hassanien, A.E.: Deep learning reader for visually impaired. Electronics (Switzerland) **11**(20), 1–22 (2022). https://doi.org/10.3390/electronics11203335. ISSN: 20799292
5. Nguyen, H.C., Nguyen, T.H., Scherer, R., Le, V.H.: YOLO series for human hand action detection and classification from egocentric videos. Sensors (Basel, Switzerland) **23**(6), 1–24 (2023). https://doi.org/10.3390/s23063255. ISSN:14248220
6. Garcia-Hernando, G., Yuan, S., Baek, S., Kim, T.-K.: First-person hand action benchmark with RGB-d videos and 3d hand pose annotations. In: Proceedings of Computer Vision and Pattern Recognition (CVPR) (2018)
7. Redmon, J., Divvala, S., Girshick, R., Farhadi, A.: You only look once: Unified, real-time object detection. In: Proceedings of the IEEE Computer Society Conference on Computer Vision and Pattern Recognition, vol. 2016, pp. 779–788 (2016). https://doi.org/10.1109/CVPR.2016.91. ISBN: 9781467388504. eprint:1506.02640
8. Srivastava, S., Divekar, A.V., Anilkumar, C., Naik, I., Kulkarni, V., Pattabiraman, V.: Comparative analysis of deep learning image detection algorithms. J. Big Data **8**(1) (2021). https://doi.org/10.1186/s40537-021-00434-w. ISSN: 21961115

9. Hussain, M.: YOLO-v1 to YOLO-v8, the rise of YOLO and its complementary nature toward digital manufacturing and industrial defect detection. Machines **11**(7), 677 (2023). https://doi.org/10.3390/machines11070677
10. Redmon, J., Farhadi, A.: YOLO9000: better, faster, stronger. In: Proceedings of the 30th IEEE Conference on Computer Vision and Pattern Recognition, CVPR 2017, vol. 2017, pp. 6517–6525 (2017). https://doi.org/10.1109/CVPR.2017.690. ISBN: 9781538604571. eprint: 1612.08242
11. Redmon, J., Farhadi, A.: YOLOv3: An Incremental Improvement (2018). https://arxiv.org/pdf/1804.02767.pdf. http://arxiv.org/abs/1804.02767. eprint: 1804.02767
12. Bochkovskiy, A., Wang, C.Y., Liao, H.Y.M.: YOLOv4: optimal speed and accuracy of object detection (2020). https://arxiv.org/pdf/2004.10934.pdf. eprint: 2004.10934
13. Wang, C.-Y., Yeh, I.-H., Liao, H.-Y.M.: You Only Learn One Representation: Unified Network for Multiple Tasks pp. 1–11 (2021). https://arxiv.org/pdf/2105.04206.pdf. eprint: 2105.04206
14. Ge, Z., Liu, S., Wang, F., Li, Z., Sun, J.: YOLOX: Exceeding YOLO Series in 2021, pp. 1–7 (2021). https://arxiv.org/pdf/2107.08430.pdf. eprint: 2107.08430
15. Zhu, Y., Li, S., Du, W., Du, Y.: Identification of table grapes in the natural environment based on an improved yolov5 and localization of picking points. Precision Agric. **24**(4), 1–22 (2023)
16. Li, C., Li, L., Jiang, H., et al.: "YOLOv6: A Single-Stage Object Detection Framework for Industrial Applications. (2022). https://arxiv.org/pdf/2209.02976.pdf. eprint: 2209.02976
17. Wu, M., Cui, G., Lv, S., et al.: Deep convolutional neural networks for multiple histologic types of ovarian tumors classification in ultrasound images. Front. Oncol. **13**, 1154200 (2023)
18. Reis, D., Kupec, J., Hong, J., Daoudi, A.: Real-Time Flying Object Detection with YOLOv8. (2023). https://arxiv.org/pdf/2305.09972.pdf. eprint: 2305.09972
19. YOLO-NAS. A Next-Generation, Object Detection Foundational Model generated by Deci's Neural Architecture Search Technology (2023). https://github.com/Deci-AI/supe-gradients/blob/master/YOLONAS.md. Accessed 19 Sept. 2023
20. Lin, T.Y., Maire, M., Belongie, S., et al.: Microsoft COCO: common objects in context. In: Lecture Notes in Computer Science (including subseries Lecture Notes in Artificial Intelligence and Lecture Notes in Bioinformatics). LNCS, vol. 8693, pp. 740–755 (2014). https://doi.org/10.1007/978-3-319-10602-148. eprint: 1405.0312
21. Liu, Y., Liu, Y., Jiang, C., et al.: HOI4D: a 4D egocentric dataset for category-level human-object interaction. In: Proceedings of the IEEE Computer Society Conference on Computer Vision and Pattern Recognition, vol. 2022, pp. 20 981–20 990 (2022). https://doi.org/10.1109/CVPR52688.2022.02034. ISBN: 9781665469463. eprint: 2203.01577
22. Fan, G., Carlson, K.D., Thomas, R.D.: Individual differences in cognitive constructs: a comparison between American and Chinese culture groups. Front. Psychol. **12**, 1–12 (2021). https://doi.org/10.3389/fpsyg.2021.614280. ISSN:16641078
23. Goli, A., Ala, A., Mirjalili, S.: A robust possibilistic programming framework for designing an organ transplant supply chain under uncertainty. Ann. Oper. Res. **328**, 493–530 (2023)
24. Goli, A., Ala, A., Hajiaghaei-Keshteli, M.: Efficient multi-objective meta-heuristic algorithms for energy-aware non-permutation flow-shop scheduling problem. Expert Syst. Appl. **213**(Part B), 119077 (2023)
25. Goli, A., Tirkolaee, E.B.: Designing a portfolio-based closed-loop supply chain network for dairy products with a financial approach: accelerated benders decomposition algorithm. Comput. Oper. Res. **155**, 106244 (2023)
26. Mohammed, A.A.Q., Lv, J., Islam, M.D.: A deep learning-based end-to-end composite system for hand detection and gesture recognition. Sensors (Switzerland) **19**(23), 5282 (2019). https://doi.org/10.3390/s19235282. ISSN: 14248220
27. Haji Mohd, M.N., Mohd Asaari, M.S., Lay Ping, O., Rosdi, B.A.: Vision-based hand detection and tracking using fusion of kernelized correlation filter and single-shot detection. Appl. Sci. (Switzerland) **13**(13), 7433 (2023). https://doi.org/10.3390/app13137433. ISSN: 20763417

28. Terven, J., Cordova-Esparza, D.: A Comprehensive Review of YOLO: From YOLOv1 and Beyond, pp. 1–34 (2023). eprint: 2304.00501
29. Jin, P., Rathod, V., Zhu, X.: Pooling Pyramid Network for Object Detection, pp. 2–4 (2018). https://arxiv.org/pdf/1807.03284.pdf. eprint: 1807.03284
30. Girshick, R., Donahue, J., Darrell, T., Malik, J.: Rich feature hierarchies for accurate object detection and semantic segmentation. In: Proceedings of the IEEE Conference on Computer Vision and Pattern Recognition, pp. 580–587 (2014)
31. Girshick, R.: Fast r-cnn. In: Proceedings of the IEEE International Conference on Computer Vision, pp. 1440–1448 (2015)
32. Ren, S., He, K., Girshick, R., Sun, J.: Faster r-cnn: towards real-time object detection with region proposal networks. Adv. Neural Inf. Process. Syst. **28** (2015)
33. Dai, J., Li, Y., He, K., Sun, J.: R-fcn: object detection via region-based fully convolutional networks. Adv. Neural Inf. Process. Syst. **29** (2016)
34. He, K., Gkioxari, G., Dollar, P., Girshick, R.: Mask r-cnn. In: Proceedings of the IEEE International Conference on Computer Vision, pp. 2961–2969 (2017)
35. Zeren, M.T., Aytulun, S.K., Kirelli, Y.: Comparison of SSD and faster R-CNN algorithms to detect the airports with data set which obtained from unmanned aerial vehicles and satellite images. Eur. J. Sci. Technol. **19**, 643–658 (2020). https://doi.org/10.31590/ejosat.742789
36. Chen, J., Ni, Z., Sang, N.: Multi-scale YOLOv2 for hand detection in complex scenes. In: 2018 15th International Conference on Control, Automation, Robotics and Vision, ICARCV 2018, pp. 1525–1530 (2018). https://doi.org/10.1109/ICARCV.2018.8581090
37. Van Staden, J., Brown, D.: An evaluation of YOLO-based algorithms for hand detection in the kitchen. In: Proceedings of the 4th International Conference on Artificial Intelligence, Big Data, Computing and Data Communication Systems (icABCD 2021) (2021). https://doi.org/10.1109/icABCD51485.2021.9519307
38. Simonyan, K., Zisserman, A.: Very deep convolutional networks for large-scale image recognition. In: Proceedings of the 3rd International Conference on Learning Representations, ICLR 2015 - Conference Track, pp. 1–14 (2015). eprint: 1409.1556
39. Gao, Q., Liu, J., Ju, Z., Zhang, L., Li, Y., Liu, Y.: Hand detection and location based on improved SSD for space human-robot interaction. In: Lecture Notes in Computer Science (including subseries Lecture Notes in Artificial Intelligence and Lecture Notes in Bioinformatics). LNAI, vol. 10984, pp. 164–175 (2018). https://doi.org/10.1007/978-3-319-97586-315. ISBN: 9783319975856
40. Bandini, A., Zariffa, J.: Analysis of the hands in egocentric vision: a survey. In: IEEE Transactions on Pattern Analysis and Machine Intelligence, p. 1 (2020). https://doi.org/10.1109/tpami.2020.2986648. ISSN: 0162-8828. eprint: 1912.10867
41. Roy, K., Mohanty, A., Sahay, R.R.: Deep learning based hand detection in cluttered environment using skin segmentation. In: Proceedings of the 2017 IEEE International Conference on Computer Vision Workshops, ICCVW 2017, pp. 640–649 (2017). https://doi.org/10.1109/ICCVW.2017.81. ISBN: 9781538610343
42. Adiguna, R., Soelistio, Y.E.: CNN based posture-free hand detection. In: Proceedings of 2018 10th International Conference on Information Technology and Electrical Engineering: Smart Technology for Better Society, ICITEE 2018, vol. 2018, pp. 276–279 (2018). https://doi.org/10.1109/ICITEED.2018.8534743
43. Yang, L., Qi, Z., Liu, Z., et al.: A light CNN-based method for hand detection and orientation estimation. In: Proceedings of the International Conference on Pattern Recognition, vol. 2018, pp. 2050–2055 (2018). https://doi.org/10.1109/ICPR.2018.8545493, ISSN: 10514651
44. Yang, L., Qi, Z., Liu, Z., et al.: An embedded implementation of CNN-based hand detection and orientation estimation algorithm. Mach. Vision Appl. **30**(6), 1071–1082 (2019). https://doi.org/10.1007/s00138-019-01038-4. ISSN: 14321769

45. Zhang, M., Cheng, X., Copeland, D., et al.: Using computer vision to automate hand detection and tracking of surgeon movements in videos of open surgery. In: AMIA ... Annual Symposium Proceedings. AMIA Symposium, vol. 2020, pp. 1373–1382 (2020). ISSN: 1942597X. eprint: 2012.06948
46. Xu, C., Cai, W., Li, Y., Zhou, J., Wei, L.: Accurate hand detection from single-color images by reconstructing hand appearances. Sensors (Switzerland) **20**(1), 1–21 (2020). ISSN: 14248220. https://doi.org/10.3390/s20010192
47. Long, X., Deng, K., Wang, G., et al.: PP-YOLO: an effective and efficient implementation of object detector (2020). https://arxiv.org/pdf/2007.12099.pdf. eprint: 2007.12099
48. Couturier, R., Noura, H.N., Salman, O., Sider, A.: A deep learning object detection method for an efficient clusters initialization. arXiv preprint arXiv:2104.13634 (2021). eprint: 2104.13634
49. Fang, Y., Liao, B., Wang, X., et al.: You only look at one sequence: rethinking transformer in vision through object detection. Adv. Neural Inf. Process. Syst. **31**, 26183–26197 (2021). ISSN: 10495258. eprint: 2106.00666
50. Huang, X., Wang, X., Lv, W., et al.: PP-YOLOv2: A Practical Object Detector, pp. 1–7 (2021). https://arxiv.org/pdf/2104.10419.pdf. eprint: 2104.10419
51. Wang, C.-Y., Bochkovskiy, A., Liao, H.-Y.M.: YOLOv7: trainable bag-of-freebies sets new state-of-the-art for real-time object detectors, pp. 7464–7475 (2022). https://arxiv.org/pdf/2207.02696.pdf. eprint: 2207.02696
52. Rath, S.: YOLOv8 Ultralytics: State-of-the-Art YOLO Models (2023). https://learnopencv.com/ultralytics-yolov8/. Accessed 19 July 2023
53. Liu, W., Anguelov, D., Erhan, D., et al.: SSD: single shot multi-box detector. In: Proceedings of the Computer Vision–ECCV 2016: 14th European Conference, Amsterdam, 11–14 October 2016, Part I 14, pp. 21–37. Springer (2016)
54. Zhang, F., Bazarevsky, V., Vakunov, A., et al.: MediaPipe hands: on-device real-time hand tracking. In: CVPR Workshop on Computer Vision for Augmented and Virtual Reality (2020). http://arxiv.org/abs/2006.10214. eprint: 2006.10214
55. Chu, G., Chen, L.-C., Chen, B., et al.: Searching for MobileNetV3. In: International Conference on Computer Vision, pp. 1314–1324 (2019)

Fine-Tuning of 3D Hand Pose Estimation on HOI4D Dataset by Convolutional Neural Networks

Van-Dinh Do[1] and Van-Hung Le[2(✉)]

[1] Electrical Department, Sao Do University, Chi Linh, Vietnam
[2] Tan Trao University, Tuyen Quang, Vietnam
van-hung.le@mica.edu.vn

Abstract. To build a support system and guide blind people to grasp objects intuitively, the hand pose, hand opening size, and grasping direction need to be determined accurately. To do this, the 3D hand pose needs to be estimated accurately and quickly to suit the requirements of blind people in reality, especially when the camera is mounted on the person's chest/EVC and data is obscured by other objects and gaze directions. In the study, we fine-tune some 3D right-hand skeleton estimation (3D-R-HSE) models based on high-performance CNNs such as P2PR PointNet (PPN), Hand PointNet (HPN), V2V-PoseNet (VPN), and Hand-FoldingNet (HFN). The model fine-tuning is performed on the HOI4D dataset. CNNs to regress 3D hand pose/3D hand keypoints/3D hand skeleton from hand region data obtained from EVC. HOI4D is a benchmark dataset published at CVPR2022. 3D-HSE estimation results of PPN, HPN, VPN, and HFN with Err_a are *32.71 mm, 35.12 mm, 26.32 mm, and 20.49 mm*, respectively. HandFoldingNet has the best results and can be applied to further studies such as determining hand shape/type and size for object grasping or hand activity recognition. 3D HSE best result when using HFN on cam^{4th} data, distance error is *12.41 mm*. This helps us choose a good method for 3D-HSE. The time to estimate 3D HS is 5.4 fps.

Keywords: Fine-tune model · 3D Hand Skeleton Estimation · HOI4D dataset · Egocentric Vision · CNNs

1 Introduction

Nowadays there are many applications developed from 3D-HSE to design a robotic hand that works like a human hand [1], HMI [2], VR [3], etc. To estimate the 3D skeleton of the hand one can, one of the input data types: RGB, depth map or point cloud data (PCD).

Previously, Research on 3D hand pose estimation often used data obtained from second or third-person cameras or surveillance cameras. However, to research and develop systems to support visually impaired people or people practicing hand grasping, the camera needs to be mounted on the person at the head or chest, this camera setup is called egocentric vision. (EVC). Our previous studies on 3D HSE were performed on the FPHAB dataset [4]. Although deep learning (DL) has achieved many convincing results in computer vision, DL requires a large amount of training data and the FPHAB dataset does not fully represent the types of real-world hand grasps. Therefore, enriching the data for learning the HSE model obtained from EVC is a necessary issue that needs to be researched and tested. In the study [5], the input data to train the 3D HSE model are RGB, depth map, and PCD. CNNs with PCD input data will estimate the hand skeleton more intuitively. In particular, Bandini et al. [6] have systematized the studies on 3D HSE according to the applications. Among them, some CNNs have outstanding results in estimating the hand skeleton [7–9]. While some CNNs have outstanding HSE results, studies often use predefined hand models, depth images, or RGB images of multiple views. Recently, the HOI4D dataset [10] was published at CVPR 2022 for the research community to train and evaluate models on 3D HSE and hand activity recognition, HOI4D is collected using four EVCs in an indoor environment. Currently, there are not many experimental studies on HOI4D.

In this study, we perform fine-tuning the 3D HSE models based on several CNNs models: PPN [11], HPN [11], VPN [12], and HFN [13] on the image data of the right hand of the HOI4D dataset. To train and evaluate the estimated model, we perform ground truth normalization of the 3D R-HSE and construct the model training and testing datasets.

Our paper includes the following contributions:

- Standardizing the ground truth data of 3D-R-HS and building the training data set, test data of 3D R-HSE model on HOI4D database.
- Normalizing and constructing the 3D PCD data of the hand data region based on the ground truth of the 2D hand skeleton.
- Fine-tuning 3D hand skeleton estimation model on HOI4D database with several CNNs: PPN, HPN, VPN, and HFN.

Our articles are written in a structured manner. Related studies on 3D HSE are presented in Sect. 2. The architecture of the CNNs for 3D HSE is presented in Sect. 3. The experiments, evaluation metrics, and results of 3D HSE are presented in Sect. 4. Finally, some conclusions and future research are presented in Sect. 5.

2 Related Works

The problem of estimating 3D HS is not a new problem, but estimating 3D HS from EVC data is a problem that has not received much research attention. In this study, we present two aspects: 3D HSE methods and datasets collected from EVC.

Over the past decade, there have been many studies performed on 3D HSE, including some studies examining the value [14–17]. Li et al. [14] presented a comprehensive survey study on data collection methods, 3D HSE estimation methods, assessment metrics, results, and challenges of 3D HSE.

Huang et al. [15] also presented a survey on 3D HSE methods and challenges from two types of input data, RGB image and depth map. At the same with methods using depth maps as input data, it is possible to follow some methods: generative approaches, regression-based approaches, detection-based approaches, structural constraint approaches, multi-stage prediction approaches, ensemble prediction approaches, synthetic data approaches, and other CNN approaches. Based on the generative approaches, the geometry models and constraints on the hand used to represent the 3D model: skeleton model [18, 19], sphere model [20, 21], triangulated mesh model [22], cylindrical model [23], etc. Based on the regression-based methods, DL models are used to train the 3D HSE model according to two directions: using 3D CNNs [12, 24, 25] and point-set networks [26, 27]. Based on the detection-based approaches, the depth map is used to build 2D heatmaps and 3D heatmaps by using the predicted dense, point sets, or voxel sets. The dense probability map for each joint [11, 12, 28] is built. Other CNNs are presented in [15]. Based on RGB image as input data, 3D-HSE be done according to some methods: generative approaches [29], 2D-to-3D lifting approaches [30], cross-modality approaches [31], disentanglement approaches [32], model-based 3D hand reconstruction approaches [33], model-free 3D hand reconstruction approaches [34], weakly-supervised and semi-supervised learning CNNs approaches [35], sequential modeling and tracking approaches [36].

Ohkawa et al. conducted a survey study on the ground truth construction methods of 3D HS for evaluating 3D HSE models, The methods are classified into four types as follows: manual, synthetic-model-based, hand-sensor-based, and computational.

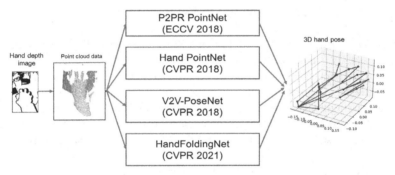

Fig. 1. Illustration of some fine-tuned models for 3D HSE.

The benchmark datasets for 3D HSE assessment published in the last three years are also presented as follows: GRAB dataset [37], YouTube3Dhands dataset [38], some datasets collected from EVC as HO-3D dataset [39], DexYCB dataset [40], H2O dataset [41], InterHand2.6M dataset [34], AssemblyHands dataset, Assembly101 dataset [16] also presented.

About 3D HSE research on EVC datasets, Le et al. [42] proposed a deep framework for 3D HS estimation and hand activity recognition based on the estimated 3D HS. The 3D HSE step is performed based on the combination of HopeNet [43] and GraphCNN,

experiments and model evaluation were performed on the FPHAB dataset with a minimum error of 36.6 mm. Le [44] also conducted a study on automatic 3D HS estimation from hand region data detected from YOLOv7. The author used the HFN to make the estimation, the best estimation result with a distance error of 19.98 mm was obtained when performing the estimation on the hand data region of the ground truth of the hand-bounding box. Prakash et al. [45] published the WildHands database collected from EVC in a kitchen environment to evaluate 3D HSE models, the collected data includes RGB images and the ground truth of hand-annotated using the FrankMocap system. Plizzari et al. [46] presented the applications that can be developed from the data collected from EVC. In the study, Pramanick et al. [47] published the EgoVLPv2 dataset, which was collected from EVC for sign language assessment. The study by Zhang et al. [48] published the EgoHOS database for evaluating object segmentation and hand interaction models, which contains 11,243 RGB images. The ground truth data for object segmentation evaluation is annotated at the pixel-level. In Gong et al. [49] study, the MMG-Ego4D dataset containing four types of data was published: PCD, video, audio, and inertial motion sensor modalities, and a generalizable database using supervised machine learning models. In the study, Khaleghi et al. [50] published the MuViHand database which was constructed and synthesized from 12 EVCs to evaluate the 3D HSE model. It contains 402,000 synthetic hand RGB images synthesized from 4,560 videos. The MIXAMO was used to annotate the ground truth of the 3D HS, including 21 joints. Plizzari et al. [51] have published the N-EPIC-Kitchens database collected from EVC for evaluating hand gesture recognition models. The collected data includes RGB images and optical flow.

3 Fine-Tuning 3D HSE Using CNNs

As surveyed in [17], The CNNs model can use input data such as RGB, depth map, and PCD. In this study, we propose to fine-tune some CNN models for 3D HSE with PCD input data. The framework for fine-tuning is presented in Fig. 1. We rely on the research of Le [44], The hand data area is obtained from the hand detection step on the RGB image and correspondingly taken on the depth map, the PCD data is then constructed based on the combination of the hand region on the depth map and the internal parameters of the camera. The result of the 3D HSE model is a 3D skeleton consisting of 21 joints. In this study, we fine-tuned the 3D HSE model using the HOI4D database, with 16 hand-grasping activities and 2.4 million RGB-D frames. From there, the learning model can learn many hand poses based on the learning data in 3D space. Next, we present CNNs to estimate 3D HS.

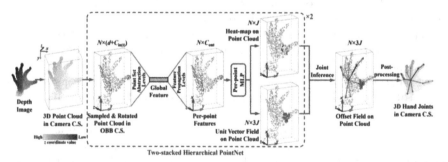

Fig. 2. Illustration of P2PR PointNet architechture for 3D-HSE [11]. From the depth map data is converted to PCD and normalized in the base coordinate system and the estimation is performed.

3.1 P2PR PointNet (PPN)

P2PR PointNet [11] based on the input data is the normalized PCD and performs regression of the position of the keypoints of the 3D skeleton, the P2PR PointNet architecture is presented in Fig. 2. PCD input data is normalized to the base coordinate system (CS) and the data is reduced in three levels: 1024 points, 512 points, and 128 points.

The normalized data of the hand lies in a 3D bounding box (OBB) with the origin at the wrist and the orientation of the hand in the direction of the x-axis as shown in Fig. 2, this data will be used as input for the regression network. The value of the normalized data in the x, y, and z directions is from 0.5 and 0.5. P2PR PointNet uses then two-stacked hierarchical PointNet [27] to regress the 3D heatmap on the global feature set extracted from PCD from abstraction levels. A unit vector is generated which is the 3D regression heatmaps, and the output of the regression model is 3D HSE based on 3D heatmaps. The MLP is used to regress 3D heatmaps on the per-point feature as shown in Fig. 3. The 3D heatmaps are used to infer 3D HS on the unit vector fields using the last hierarchical PointNet module. The final step is to post-process the estimated 3D-HS from the normalized coordinate system to the real coordinate system with two problems to solve as follows: (1) unreliable estimation in large space, and (2) no constraints between hand joints.

The first PPN model uses a threshold to replace the 3D heatmap candidate estimation with the direct regression result of the 3D HS. Second, PPN used the PCA space to construct constraints and estimate space.

3.2 Hand PointNet (HPN)

HPN [11] uses a regression method to estimate the 3D joints of the 3D HS based on the PointNet [27] backbone. The structure of the HPN model is shown in Fig. 4.

The input data is the depth map of the detected hand region converted to PCD data when combined with the camera's internal parameter set. The data is then normalized to the base coordinate system in a 3D BB or 3D OBB using the PCA space. The normalized data is then extracted into three subsets (abstraction set - AS) with sizes 512 points, 128 points, and 64 points, respectively. 3D joints are regressed on three AS using the hierarchical PointNet, and three FC layers are used to combine the regression results at three AS for 3D HSE. The final step, fingertip refinement is based on the structure of fingers. The final step is that the fingertips are refined using the hand structure, this study added a 3D random offset with a spherical radius of $r = 15\ mm$ to find k points neighboring to the estimated fingertip point location on the ground truth of the hand with PCD data and calculate the angle of the joints. This step is applied to the model training process to increase the accuracy of the 3D HSE process. To find the neighboring points of the fingertips, HPN used the processing algorithms on PCD and calculated the distance and angle between the joints.

3.3 V2V-PoseNet (VPN)

VPN [12] is a CNN that estimates 3D HS with depth image input and converts to 3D volumetric (Vol.) data. Each candidate 3D is estimated per-voxel likelihood using the

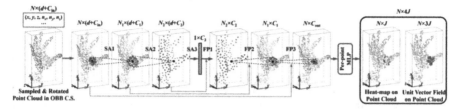

Fig. 3. The process of creating the 3D heatmaps and unit vector fields [11] of PPN.

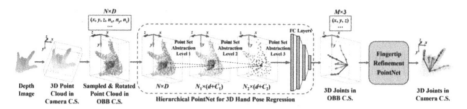

Fig. 4. Illustration of HPN architecture.

encoding and decoding of features on the 3D voxelized. The architecture of VPN is illustrated in Fig. 5.

The VPN model consists of four building blocks: the first is the Vol. basic block including a Vol. Conv., ReLU, and Vol. batch normalization, this block is placed at the first and last of the VPN. The second block is the Vol. residual block, it was developed from the 2D residual block. The third block is the Vol. down-sampling block, including a Vol. max pooling layer. The fourth block is the volumetric up-sampling block is final and includes a volumetric de-convolution layer, a Vol. batch normalization layer, and the activation function. To reduce complexity in training, VPN can add the ReLU block.

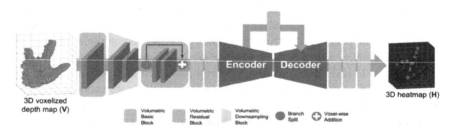

Fig. 5. Illustration of VPN architechture [12].

3.4 HandFoldingNet (HFN)

HFN [13] is a CNN used to train a 3D HSE model based on input data which is a depth map of the hand region taken from hand detection on RGB images converted to PCD based on intrinsic camera parameters. The model of HFN is presented in Fig. 6 and contains

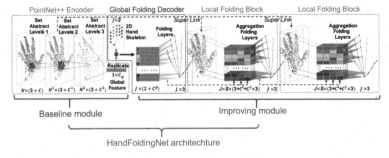

Fig. 6. HFN architecture [13].

four blocks: (1) the PointNet++ encoder, (2) the global folding (GF) decoder, and (3), (4) two local folding blocks (FBs). HFN uses the underlying network as a PointNet++ encoder. To improve accuracy, HandFoldingNet added additional information such as the 2D hand pose, 3D surfaces normal vector (f_i^{nor}) is the corresponding normalized PCD (p_i^{nor}). To extract LFs with different levels (the size of abstract 1st level, 2nd level, and 3rd level are $N * (3 + C)$, $N_1 * (3 + C_1)$, $N_2 * (3 + C_2)$, respectively), HandFoldingNet used the hierarchical PointNet++ decoder on a single GF.

GF is used in the GF decoder, the initial keypoint coordinates allow for redefining 2D HS. Each LFB uses an MLP (1×1) and a max pooling to regress to a matrix with the size ($J \times 3$). The calculation process of each joint-wise local feature-based FB is presented in Fig. 7, including three inputs: previous estimated joint coordinates, the previous FB containing m intermediate layers, and the previous AS extracted from the local feature map. Two LFBs estimate 3D joint coordinates by grouping the local features close to the initialized 3D coordinates of the 3D keypoint.

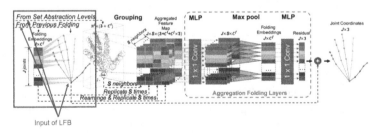

Fig. 7. Illustration of joint-wise local feature-based FB [13].

4 Experimental Results

4.1 Data Collection

The HOI4D dataset [10] is a large-scale 4D radial benchmark dataset. The HOI4D dataset contains 2.4 million RGB-D frames. This dataset uses four cameras to collect data from indoor environments. In this study, we divide the HOI4D database by each data collection camera. The hand actions in the HOI4D dataset consist of 16 hand activities (grasping ToyCar, grasping Mug, grasping Laptop, grasping StorageFurniture, grasping Bottle, grasping Safe, grasping Bowl, grasping Bucket, grasping Scissors, grasping Pliers, grasping Kettle, grasping Knife, grasping TrashCan, grasping Lamp, grasping Stapler, grasping Chair) and are shown in Fig. 8. The GT hand data of the HOI4D dataset includes a 2D hand BB, 3D HS annotation, and hand action label. In the study, we use RGB images, depth map and 2D BB annotation of hand, 3D HS annotation to prepare for training the 3D-HSE model. 2D hand BB annotation standardized according to the BB format of the COCO 2017 dataset. The data is collected from 4 ECVs labeled cam^{1st}, cam^{2nd}, cam^{3rd}, cam^{4th}.

We only use right-hand data for the hand detection model and evaluation model from 603,332 frames. We divide the data of the HOI4D database by left-hand and right-hand activities, by camera label of data collection, and by 16-hand object grasping activities. We use a ratio (7:3), with 70% of the number of frames for the training model, and 30% number of frames for the testing model, in which the data is divided randomly. After that, the 2D HS annotation is also obtained corresponding to the RGB image that has been divided into the training and testing data. The ground truth of 3D HS is also extracted from 2D HS annotation combined with camera intrinsic parameters to obtain 3D HS annotation for evaluating the estimated model.

In this study, to prepare the PCD data as input for fine-tuning the 3D HSE model, we use the data area of the 2D BB annotation on the RGB images to get the corresponding data area on the depth map (because the two types of data, RGB image, and deth map, are a pair of data collected simultaneously from the camera and have been calibrated to the same center). The PCD data area of he hand includes the points $P_i(x,y,z)$. The coordinates of P_i in 3D space are calculated as in Eq. 1.

$$x_{P_i} = \frac{(x_d - c_x) * D_{va}}{f_x}$$
$$y_{P_i} = \frac{(y_d - c_y) * D_{va}}{f_y} \qquad (1)$$
$$z_{P_i} = Dva$$

where (fx, fy) are the focal length along the x and y axes of the camera, and (cx, cy) are the coordinates of the center of the image along the x and y axes. These are called the camera's internal parameters. Dva is the depth map value at the pixel (x_d, y_d).

The process of constructing the PCD data of the hand as input for training the 3D HSE model is also calculated as in Eq. 1 and presented in Fig. 9. The generated PCD data includes the PCD data of the hand and additional data of the object, because the 2D hand BB is a rectangle, so the data of the hand and the object in the image cannot be segmented. To segment the PCD data of hands and objects, we use the Euclidean distance [52] and keep the largest data area (with the largest number of points) as the PCD data of the hand.

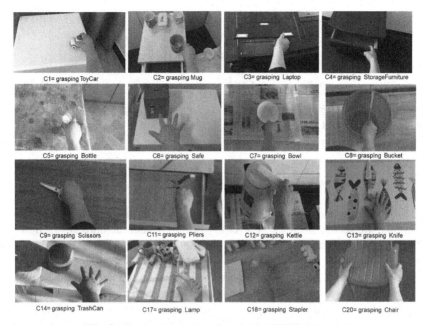

Fig. 8. Illustrating hand actions in the HOI4D dataset.

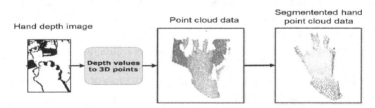

Fig. 9. PCD data construction process of hand for training and testing of 3D HSE model.

4.2 Metrics

Similar to the 3D HSE studies, we use the average error distance (Err_a) between the joints of the 3DHS ground truth and the estimated 3D HS to evaluate the performance of the 3D HSE model (as calculated in Eq. 2) fine-tuned based on PPN, HPN, VPN, and HFN on the HOI4D database.

$$Err_a = \frac{1}{Num_s} \sum_{n=1}^{Num_s} \frac{1}{k} \sum_{k=1}^{K} DIS(p_g, p_e) \qquad (2)$$

where $DIS(pg, pe)$ is the Euclidean distance of a 3D joint annotation p_g and an estimated 3D joint p_e (mm); Num_s is the number of hands tested; $K = 21$ is the number of joints on the predefined 3D hand.

To ensure peer comparison, the fine-tuning of the 3D HSE model was performed with an epoch number is 50 epochs, a batch size is 32, and a value of learning rate is

Table 1. Average distance error of 3D hand joints on 3D hand skeleton (Err_a) on the HOI4D database.

Methods	Average error distance (Err_a) (mm)			
	Cam^{1st}	Cam^{2nd}	Cam^{3rd}	Cam^{4th}
PPN [11]	18.2	35.9	50.44	26.33
HPN [11]	**13.1**	61.07	49.68	16.65
VPN [12]	15.2	28.9	45.7	15.5
HFN [13]	14	**22.72**	**32.85**	**12.41**

10E-04. We use the number of PCD data point clouds at the first abstraction level is 1024 points, the second abstraction level is 512 points, and the third abstraction level is 128 points. The number of components of the PCA method is 42, The number of neighbor points for finding the nearest neighbor using the KNN algorithm is 64, The radius of the nearest neighbor search sphere is 0.015 at level 1 and level 2, and 0.04 at level 3, respectively. Adam's [53] method is used to optimize the training process of the 3D HSE model. In addition, other parameters are tested and selected by us as the best. Matlab programming language is used to develop data preprocessing source codes such as building PCD data, using the PCA method for data reduction, and data normalization. The source code for developing the HSE 3D model to train and test the application is written in the Python programming language and is also supported by several libraries such as Tqdm, PyTorch, etc.

In the study, to perform model training and model evaluation, we experiment on a server with the following GPU configuration: RTX 2080 Ti, 12 GB, installed the CUDA 11.2/cuDNN 8.1.0 libraries and CPU: 12th Gen Intel i9-12900K (24) @ 6.500 GHz. The experimental source code is built in Python language ($\geq$3.9 version) with the support of some libraries of AI programming and computer vision. The source code for training and testing the 3D HSE model is inherited and reused by us as follows: PPN[1], HPN[2], VPN[3], and HFN[4].

4.3 Results and Discussions

Table 1 shows the 3D HSE results of the right hand on the HOI4D database when fine-tuning on CNN models such as PPN, HPN, VPN, and HFN. The error distance when fine-tuning with the model of PPN is 18.2 mm on the cam^{1st} dataset, same as above HPN is 13.1 mm, VPN is 15.2 mm), and HFN is 14 mm. The error distance when fine-tuning with the model on the cam^{2nd} dataset of PPN, HPN, VPN, and HFN is 35.9 mm, 61.07 mm, 28.9 mm, and 22.72 mm, respectively. The error distance when fine-tuning with the model on the cam^{3rd} dataset of PPN, HPN, VPN, and HFN is 50.44 mm,

---
[1] https://github.com/ataata107/Point-to-Point-egressionand-Pose [Accessed October 10, 2023].
[2] https://github.com/3huo/Hand-Pointnet [Accessed October 10, 2023].
[3] https://github.com/mks0601/V2V-PoseNet_RELEASE [Accessed October 10, 2023].
[4] https://github.com/cwc1260/HandFold [Accessed October 10, 2023].

49.68 mm, 45.7 mm, and 32.85 mm, respectively. The error distance when fine-tuning with the model on the cam^{4th} dataset of PPN, HPN, VPN, and HFN is 26.33 mm, 16.65 mm, 15.5 mm, and 12.41 mm, respectively.

The 3D HSE results are shown in Table 1, with the best estimation result of HandFoldingNet, with a distance error (Err_a) of 12.41 mm. This is a small distance error compared to the size of the hand moving and grasping the object on the HOI4D database. At the same time, this is an acceptable error distance for the 3D HS estimation system with data obtained from ECV containing many challenges due to being obscured by the camera's view direction, the object, and the ambiguity of the data. Especially the speed of movement of the hand in grasping objects is fast.

This distance error is acceptable when building robotic arms with fingers that can perform complex operations like human hands, or building applications that guide visually impaired people to find and grasp objects in the environment, especially unfamiliar environments. The distance error distribution of 3D HSE when performing model fine-tuning with HFN on test data at cameras is shown in Fig. 10. The results show that the distance error of model HandFoldingNet when tested on cam^{4th} data is the largest, which also shows that the 3d HSE results in Table 1 correctly represent the nature of the problem.

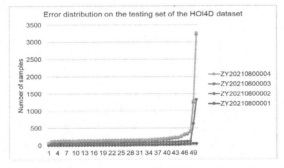

Fig. 10. Distance error distribution when using the HFN model to train and test the estimated model on the test data of the HOI4D database.

The 3D HSE test results of 16 right-hand object grasping actions using HFN to train the 3D HSE model are shown in Fig. 11. The red 3D HSE is the estimated result of the HandFoldingNet model, and the blue 3D HS is the 3D HS annotation. The results are visually displayed according to each hand grip action.

Table 2. Processing speed for 3D HSE with each frame having a hand performing an object grasping action

Methods	PPN	HPN	VPN	HFN
Processing time (fps)	3.5	4.2	3.8	5

Table 2 shows the computation time/processing speed for 3D HSE on the HOI4D database when using CNNs to train the 3D HS estimation model: PPN, HPN, VPN, and HFN.

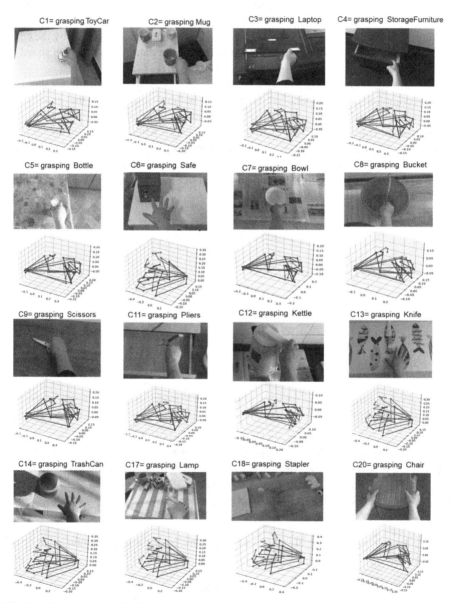

Fig. 11. The estimated 3D HS and ground truth of 3D HS when using HFN to fine-tune the estimated model on the HOI4D database. The red 3D HS is the estimated 3D HS, the blue 3D HS is the ground truth of 3D HS, and 16 actions are shown. (Color figure online)

5 Conclusions and Future Works

The problem of estimating hand pose from image data is not new in computer vision and robotics, but the problem of estimating 3D HS on data obtained from EVC is currently facing many challenges such as the hand being obscured by the viewing direction, by the object being grasped, and the speed of the hand moving, making the data blurry and containing a lot of noise. The results of 3D HSE can be applied to build robotic arms with fingers that can perform complex operations like human hands using simulation technology and learned posture control. Or it is possible to build systems to help the visually impaired find and grasp objects in new environments, especially to grasp them safely to avoid burns. In this study, we conduct fine-tuning of CNNs models (PPN, HPN, VPN, and HFN) for 3D HSE from PCD data of the hand combined from the hand region on the depth map and the internal parameter set of the camera, we only experiment on the right hand of the HOI4D database. To perform fine-tuning of the 3D HS model, we normalize the data of the HOI4D database to obtain the ground truth of the 3D HD for model evaluation.

The estimated distance error of 3D hand joints is the Euclidean distance of PPN, HPN, VPN, and HFN is *32.71 mm, 35.12 mm, 26.32 mm,* and *20.49 mm*, respectively. The results of HFN's 3D HSE are the best, with a distance error of *12.4 mm* when performed on the cam[4th] dataset. These results show that this method can be applied to build robot hands that can perform complex activities like human hands or to build applications that support and guide visually impaired people to find and grasp objects in new environments. In the short term, will test on newer 3D HSE models and make

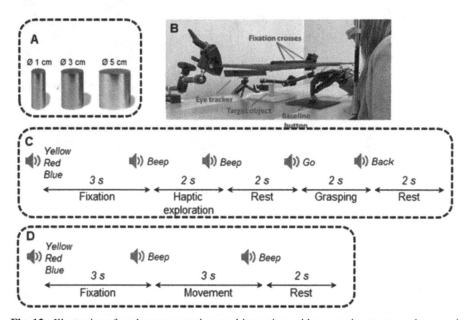

Fig. 12. Illustration of a robot arm grasping an object using guidance on the structure, shape, and orientation of the hand [54].

improvements to the estimation model to improve the estimation results. Apply the 3D HSE module to the model to build a system that helps and guides blind people or a robot arm to grasp objects, as illustrated in Fig. 12 [54]. In further plans, we will conduct research on integrating the 3D-HPE module into the support system, evaluating whether people practice hand rehabilitation as shown in Fig. 13 [55], or apply it to the control system using table gestures as shown in Fig. 14 [56].

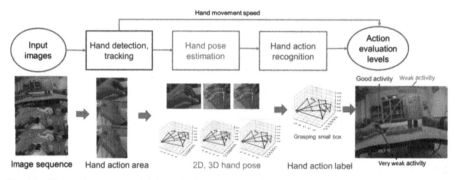

Fig. 13. Illustration of a model for building a system to support practitioners in recovering hand function [55].

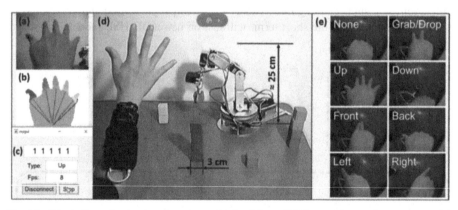

Fig. 14. Illustration of using hand gestures obtained from ECV to perform robot arm grasping instruction [56].

Acknowledgment. This research is supported by Tan Trao University, Vietnam.

References

1. Gomez-Donoso, F., Orts-Escolano, S., Cazorla, M.: Accurate and efficient 3D hand pose regression for robot hand teleoperation using a monocular RGB camera. Expert Syst. Appl. **136**, 327–337 (2019). https://doi.org/10.1016/j.eswa.2019.06.055. ISSN: 09574174

2. Bandi, C., Thomas, U.: Regression-based 3D Hand Pose Estimation for Human-Robot Interaction. Communications in Computer and Information Science, CCIS, vol. 1474, no. Visigrapp, pp. 507–529 (2022). https://doi.org/10.1007/978-3-030-94893-124. ISSN: 18650937
3. Kaeser-Chen, C., Guleryuz, O.G.: Fast lifting for 3D hand pose estimation in AR/VR applications. In: 2018 25th IEEE International Conference on Image Processing (ICIP), 2018, pp. 106–110. ISBN: 9781479970612
4. Garcia-Hernando, G., Yuan, S., Baek, S., Kim, T.-K.: First-person hand action benchmark with rgb-d videos and 3d hand pose annotations. In: Proceedings of Computer Vision and Pattern Recognition (CVPR) (2018)
5. Le, V.H., Nguyen, H.C.: A survey on 3D hand skeleton and pose estimation by the convolutional neural network. Adv. Sci. Technol. Eng. Syst. **5**(4), 144–159 (2020). https://doi.org/10.25046/aj050418. ISSN: 24156698
6. Bandini, A., Zariffa, J.: Analysis of the hands in egocentric vision: a survey. IEEE Trans. Pattern Anal. Mach. Intell. **45**(6), 6846–6866 (2023). https://doi.org/10.1109/TPAMI.2020.2986648. ISSN: 19393539. eprint: 1912.10867
7. Fan, L., Rao, H., Yang, W.: 3D hand pose estimation based on five layer ensemble CNN. Sensors (Switzerland) **21**(2), 1–16 (2021). https://doi.org/10.3390/s21020649. ISSN: 14248220
8. Isaac, J.H., Manivannan, M., Ravindran, B.: Single shot corrective CNN for anatomically correct 3D hand pose estimation. Front. Artif. Intell. **5**, 1–11 (2022). https://doi.org/10.3389/frai.2022.759255. ISSN:26248212
9. Cheng, J., Wan, Y., Zuo, D., et al.: Efficient virtual view selection for 3D hand pose estimation. In: Proceedings of the 36th AAAI Conference on Artificial Intelligence, AAAI 2022, vol. 36, pp. 419–426 (2022). https://doi.org/10.1609/aaai.v36i1.19919. ISSN: 2159-5399. eprint: 2203.15458
10. Liu, Y., Liu, Y., Jiang, C., et al.: HOI4D: a 4D egocentric dataset for category-level human-object interaction. In: Proceedings of the IEEE Computer Society Conference on Computer Vision and Pattern Recognition, vol. 2022, pp. 20981–20990. https://doi.org/10.1109/CVPR52688.2022.02034, ISBN: 9781665469463. eprint: 2203.01577
11. Ge, L., Ren, Z., Yuan, J.: Point-to-point regression point net for 3D hand pose estimation. In: European Conference on Computer Vision, LNCS, vol. 11217, pp. 489–505 (2018). https://doi.org/10.1007/978-3-030-01261-829. ISBN: 9783030012601
12. Moon, G., Chang, J.Y., Lee, K.M.: V2V-PoseNet: voxel-to-voxel prediction network for accurate 3D hand and human pose estimation from a single depth map. In: IEEE/CVF Conference on Computer Vision and Pattern Recognition (CVPR, 2018), pp. 5079–5088 (2018)
13. Cheng, W., Park, J.H., Ko, J.H.: Handfoldingnet: a 3d hand pose estimation network using multiscale-feature guided folding of a 2d hand skeleton. In: Proceedings of the IEEE/CVF International Conference on Computer Vision, pp. 11260–11269 (2021)
14. Li, R., Liu, Z., Tan, J.: A survey on 3D hand pose estimation: cameras, methods, and datasets. Pattern Recogn. **93**, 251–272 (2019). https://doi.org/10.1016/j.patcog.2019.04.026. ISSN: 00313203
15. Huang, L., Zhang, B., Guo, Z., Xiao, Y., Cao, Z., Yuan, J.: Survey on depth and RGB image-based 3D hand shape and pose estimation. Virt. Reality Intell. Hardw. **3**(3), 207–234 (2021). https://doi.org/10.1016/j.vrih.2021.05.002. ISSN: 26661209
16. Ohkawa, T., Furuta, R., Sato, Y.: Efficient annotation and learning for 3D hand pose estimation: a survey. Int. J. Comput. Vis. 1–18 (2023). https://doi.org/10.1007/s11263-023-01856-0. ISSN: 15731405. print: 2206.02257
17. Le, V.H., Nguyen, H.C.: A survey on 3D hand skeleton and pose estimation by convolutional neural network. Adv. Sci. Technol. Eng. Syst. **5**(4), 144–159 (2020). https://doi.org/10.25046/aj050418. ISSN: 24156698

18. Tompson, J., Stein, M., Lecun, Y., Perlin, K.: Real-time continuous pose recovery of human hands using convolutional networks. ACM Trans. Graph. **33**(5) (2014). https://doi.org/10.1145/2629500. ISSN: 15577368
19. Zhou, X., Wan, Q., Wei, Z., Xue, X., Wei, Y.: Model-based deep hand pose estimation. In: IJCAI International Joint Conference on Artificial Intelligence, vol. 2016, pp. 2421–2427 (2016). ISSN: 10450823. eprint: 1606.06854
20. Oikonomidis, I., Kyriazis, N., Argyros, A.A.: Full DOF tracking of a hand interacting with an object by modeling occlusions and physical constraints. In: Proceedings of the IEEE International Conference on Computer Vision, pp. 2088–2095 (2011). https://doi.org/10.1109/ICCV.2011.6126483. ISBN: 9781457711015
21. Qian, C., Sun, X., Wei, Y., Tang, X., Sun, J.: Realtime and robust hand tracking from depth. In: Proceedings of the IEEE Computer Society Conference on Computer Vision and Pattern Recognition, pp. 1106–1113 (2014). https://doi.org/10.1109/CVPR.2014.145. ISBN: 9781479951178
22. Gorce, M.D.L., Fleet, D.J., Paragios, N.: Model-based 3D hand pose estimation from monocular video. In: IEEE Transactions on Pattern Analysis Machine Intelligence (PAMI), pp. 1–15 (2009)
23. Lamberti, L., Camastra, F.: Real-time hand gesture recognition using a color glove. Lecture Notes in Computer Science (including subseries Lecture Notes in Artificial Intelligence and Lecture Notes in Bioinformatics). LNCS, vol. 6978, no. PART 1, pp. 365–373 (2011). https://doi.org/10.1007/978-3-642-24085-038. issn: 03029743
24. Ge, L., Liang, H., Yuan, J., Thalmann, D.: 3D convolutional neural networks for efficient and robust hand pose estimation from single depth images. In: Proceedings of the 30th IEEE Conference on Computer Vision and Pattern Recognition, CVPR 2017, vol. 2017, pp. 5679–5688 (2017). https://doi.org/10.1109/CVPR.2017.602. ISBN: 9781538604571
25. Deng, X., Yang, S., Zhang, Y., Tan, P., Chang, L., Wang, H.: Hand3D: hand pose estimation using 3D neural network. arXiv preprint https://arxiv.org/abs/1502.06807 (2017). eprint: 1704.02224. [Online]. Available: http://arxiv.org/abs/1704.02224
26. Qi, C.R., Su, H., Mo, K., Guibas, L.J.: PointNet: deep learning on point sets for 3D classification and segmentation. In: Proceedings of the 30th IEEE Conference on Computer Vision and Pattern Recognition, CVPR 2017, vol. 2017, pp. 77–85 (2017). https://doi.org/10.1109/CVPR.2017.16. ISBN: 9781538604571. eprint: 1612.00593
27. Qi, C.R., Yi, L., Su, H., Guibas, L.J.: PointNet++: deep hierarchical feature learning on point sets in a metric space. Adv. Neural Inf. Process. Syst., **2017**, 5100–5109 (2017)
28. Pavlakos, G., Zhou, X., Derpanis, K.G., Daniilidis, K.: Coarse-to-fine volumetric prediction for single-image 3D human pose. In: Proceedings of the 30th IEEE Conference on Computer Vision and Pattern Recognition, CVPR 2017, vol. 2017, pp. 1263–1272 (2017). https://doi.org/10.1109/CVPR.2017.139. eprint: 1611.07828
29. Mueller, F., Bernard, F., Sotnychenko, O., et al.: GANerated hands for real-time 3D hand tracking from monocular RGB. In: CVPR, pp. 49–59 (2017)
30. Zimmermann, C., Brox, T.: Learning to estimate 3D hand pose from single RGB images. In: CVPR, pp. 4903–4911 (2017)
31. Theodoridis, T., Chatzis, T., Solachidis, V., Dimitropoulos, K., Daras, P.: Cross-modal variational alignment of latent spaces. In: IEEE Computer Society Conference on Computer Vision and Pattern Recognition Workshops, vol. 2020, pp. 4127–4136 (2020). https://doi.org/10.1109/CVPRW50498.2020.00488. ISBN: 9781728193601
32. Yang, L., Yao, A.: Disentangling latent hands for image synthesis and pose estimation. In: Proceedings of the IEEE Computer Society Conference on Computer Vision and Pattern Recognition, vol. 2019, pp. 9869–9878 (2019). https://doi.org/10.1109/CVPR.2019.01011. ISBN: 9781728132938. eprint: 1812.01002

33. Zhou, Y., Habermann, M., Xu, W., Habibie, I., Theobalt, C., Xu, F.: Monocular real-time hand shape and motion capture using multimodal data. In: Proceedings of the IEEE Computer Society Conference on Computer Vision and Pattern Recognition, pp. 5345–5354 (2020). https://doi.org/10.1109/CVPR42600.2020.00539. eprint: 2003.09572
34. Moon, G., Lee, K.M.: I2L-MeshNet: Image-to-Lixel Prediction Network for Accurate 3D Human Pose and Mesh Estimation from a Single RGB Image. Lecture Notes in Computer Science (including subseries Lecture Notes in Artificial Intelligence and Lecture Notes in Bioinformatics). LNCS, vol. 12352, pp. 752–768 (2020). https://doi.org/10.1007/978-3-030-58571-644. ISSN: 16113349. eprint: 2008.03713
35. Spurr, A., Iqbal, U., Molchanov, P., Hilliges, O., Kautz, J.: Weakly Supervised 3D Hand Pose Estimation via Biomechanical Constraints," Lecture Notes in Computer Science (including subseries Lecture Notes in Artificial Intelligence and Lecture Notes in Bioinformatics). LNCS, vol. 12362, pp. 211–228 (2020). https://doi.org/10.1007/978-3-030-58520-413. ISSN: 16113349. eprint: 2003.09282
36. Fan, Z., Liu, J., Wang, Y.: Adaptive Computationally Efficient Network for Monocular 3D Hand Pose Estimation. Lecture Notes in Computer Science (including subseries Lecture Notes in Artificial Intelligence and Lecture Notes in Bioinformatics). LNCS, vol. 12349, pp. 127–144 (2020). https://doi.org/10.1007/978-3-030-58548-88. ISSN: 16113349
37. Taheri, O., Ghorbani, N., Black, M.J., Tzionas, D.: GRAB: A Dataset of Whole-Body Human Grasping of Objects. Lecture Notes in Computer Science (including subseries Lecture Notes in Artificial Intelligence and Lecture Notes in Bioinformatics). LNCS, vol. 12349, pp. 581–600 (2020). https://doi.org/10.1007/978-3-030-58548-834. ISSN: 16113349. eprint: 2008.11200
38. Kulon, D., Guler, R.A., Kokkinos, I., Bronstein, M., Zafeiriou, S.: Weakly supervised mesh-convolutional hand reconstruction in the wild. In: Proceedings of the IEEE Computer Society Conference on Computer Vision and Pattern Recognition, pp. 4989–4999 (2020). https://doi.org/10.1109/CVPR42600.2020.00504. ISSN: 10636919. eprint: 2004.01946
39. Hampali, S., Rad, M., Oberweger, M., Lepetit, V.: Honnotate: a method for 3D annotation of hand and object poses. In: Proceedings of the IEEE Computer Society Conference on Computer Vision and Pattern Recognition, pp. 3193–3203 (2020). https://doi.org/10.1109/CVPR42600.2020.00326. ISSN: 10636919. eprint: 1907.01481
40. Chao, Y.W., Yang, W., Xiang, Y., et al.: DexYCB: a benchmark for capturing hand grasping of objects. In: Proceedings of the IEEE Computer Society Conference on Computer Vision and Pattern Recognition, pp. 9040–9049 (2021). https://doi.org/10.1109/CVPR46437.2021.00893. ISSN: 10636919. eprint: 2104.04631
41. Kwon, T., Tekin, B., Stuhmer, J., Bogo, F., Pollefeys, M.:H2O: two hands manipulating objects for first person interaction recognition. In: Proceedings of the IEEE International Conference on Computer Vision, pp. 10118–10128 (2021). https://doi.org/10.1109/ICCV48922.2021.00998, isbn: 9781665428125. eprint: 2104.11181
42. Le, V.D., Hoang, V.N., Nguyen, T.T., et al.: Hand activity recognition from automatic estimated egocentric skeletons combining slow fast and graphical neural networks. Vietnam J. Comput. Sci. **10**(1), 75–100 (2023). https://doi.org/10.1142/S219688882250035X. ISSN: 21968896
43. Doosti, B., Naha, S., Mirbagheri, M., Crandall, D.J.: Hope-net: a graph-based model for hand-object pose estimation. In: Proceedings of the IEEE Computer Society Conference on Computer Vision and Pattern Recognition, pp. 6607–6616 (2020). https://doi.org/10.1109/CVPR42600.2020.00664. eprint: 2004.00060
44. Le, V.H.: Automatic 3D hand pose estimation based on YOLOv7 and HandFoldingNet from egocentric videos. In: Proceedings of the 2022 RIVF International Conference on Computing and Communication Technologies, RIVF 2022, pp. 161–166 (2022). https://doi.org/10.1109/RIVF55975.2022.10013903. ISBN: 9781665461665

45. Prakash, A., Tu, R., Chang, M., Gupta, S.: 3D hand pose estimation in egocentric images in the wild. arXiv preprint https://arxiv.org/abs/2312.06583 (2023). eprint: 2312.06583
46. Plizzari, C., Goletto, G., Furnari, A., et al.: An outlook into the future of egocentric vision. arXiv preprint https://arxiv.org/abs/2308.07123 (2023). eprint: 2308.07123
47. Pramanick, S., Song, Y., Nag, S., et al.: EgoVLPv2: egocentric videolanguage pre-training with fusion in the backbone, pp. 5262–5274. arXiv preprint https://arxiv.org/abs/2307.05463 (2024). https://doi.org/10.1109/iccv51070.2023.00487. eprint: 2307.05463
48. Zhang, L., Zhou, S., Stent, S., Shi, J.: Fine-Grained Egocentric HandObject Segmentation: Dataset, Model, and Applications. Lecture Notes in Computer Science (including subseries Lecture Notes in Artifcial Intelligence and Lecture Notes in Bioinformatics). LNCS, vol. 13689, pp. 127–145 (2022). https://doi.org/10.1007/978-3-031-19818-28. issn: 16113349. eprint: 2208.03826
49. Gong, X., Mohan, S., Dhingra, N., et al.: MMG-Ego4D: multi-modal generalization in egocentric action recognition, pp. 6481–6491. arXiv preprint https://arxiv.org/abs/2305.07214 (2023)
50. Khaleghi, L., Sepas-Moghaddam, A., Marshall, J., Etemad, A.: MultiView video-based 3D hand pose estimation. In: IEEE Transactions on Artificial Intelligence, pp. 1–14 (2022). https://doi.org/10.1109/TAI.2022.3195968. issn: 26914581. eprint: 2109.11747
51. Plizzari, C., Planamente, M., Goletto, G., et al.: E2(GO) MOTION: motion augmented event stream for egocentric action recognition, vol. 2, pp. 19935–19947. arXiv preprint https://arxiv.org/abs/2112.03596 (2021)
52. Liu, H., Song, R., Zhang, X., Liu, H.: Point cloud segmentation based on Euclidean clustering and multi-plane extraction in rugged field. Measur. Sci. Technol. **32**(9), 095106 (2021). https://doi.org/10.1088/1361-6501/abead3
53. Diederik, P.K., Jimmy, B.: Adam: a method for stochastic optimization. In: ICLR (2015)
54. Pirruccio, M., Monaco, S., Della Libera, C., Cattaneo, L.: Gaze direction influences grasping actions towards unseen, haptically explored, objects. Sci. Rep. **10**(1), 1–10 (2020). https://doi.org/10.1038/s41598-020-72554-x. issn: 20452322
55. Nguyen, H.C., Nguyen, T.H., Scherer, R., Le, V.H.: YOLO series for human hand action detection and classification from egocentric videos. Sensors (Basel, Switzerland) **23**(6), 1–24 (2023). https://doi.org/10.3390/s23063255. issn: 14248220
56. Chen, F., Deng, J., Pang, Z., Nejad, M.B., Yang, H., Yang, G.: Finger angle-based hand gesture recognition for smart infrastructure using wearable wrist-worn camera. Appl. Sci. (Switzerland) **8**(3) (2018). https://doi.org/10.3390/app8030369. issn: 20763417

Prediction of Ethereum Prices Based on Blockchain Information in an Industrial Finance System Using Machine Learning Techniques

Syrine Ben Romdhane[1](✉), Fahmi Ben Rejab[1], and Khadija Mnasri[2]

[1] University of Tunis, Higher Institute of Management of Tunis, BESTMOD Lab., Tunis, Tunisia
`srnbenromdhane@gmail.com`
[2] University of Lorraine, CEREFIGE Lab., Nancy, France
`Khadija.mnasri-mazek@univ-lorraine.fr`

Abstract. Ethereum is one of the most established Blockchains. It has managed to hold on to second place in terms of market capitalization, after Bitcoin. Like all other cryptocurrencies, Ethereum is subject to normal price fluctuations and is equally affected by bear and bull markets. However, predicting the price of Ethereum remains a challenging task. This paper discusses the main drivers of Ethereum prices, including its attractiveness, macroeconomic and financial factors, with a particular focus on the use of Blockchain information in prediction models. We apply time series to daily data for the period from 30/07/2015 to 30/09/2023. We used Python and TensorFlow library version 2.11.0. Price prediction is performed with three machine learning techniques: support vector machines (SVMs), decision trees (DTs) and multilayer perceptron technique (MLP), for time-series analysis. The proposed approaches are used for price prediction in an industrial finance system and exhibit suitable accuracy scores. Two results are worth noting: When using the proposed model, the SVM algorithm provides better results than the MLP and the decision tree algorithms. The accuracy of the proposed model can be increased by adding features to the SVM algorithm. Also, Ethereum-specific Blockchain information is the most important variable in predicting Ethereum prices. This study highlights the importance of integrating Blockchain factors into cryptocurrency price prediction models. This conclusion improves investor decision-making and provides a reference for governments to design the best regulatory policies.

Keywords: Ethereum Price Prediction · Blockchain Information · Machine Learning Techniques

1 Introduction

Cryptocurrency markets embody an intricate financial system. Conventional asset pricing models and standard risk factors fall short in elucidating cryptocurrency prices and returns. Additionally, the absence of essential data such as earnings, dividends, and cash

© The Author(s), under exclusive license to Springer Nature Switzerland AG 2025
A. Mirzazadeh et al. (Eds.): ODSIE 2023, CCIS 2204, pp. 189–211, 2025.
https://doi.org/10.1007/978-3-031-81455-6_12

flows adds to the unpredictability of cryptocurrencies' dynamics (Ben-Ahmed et al., 2023).

Transactions via Ethereum, as the second most renowned cryptocurrency after Bitcoin, have become even more important in the wake of the DeFi explosion. By 2023, Ethereum, or its token ETH, is one of the major digital decentralized currencies today, with a market capitalization of over $200 billion, making it one of the most valuable cryptocurrencies (Coinmarketcap, 2022). Ethereum price forecasting enables investors to make informed decisions when buying and selling Ethereum. After a significant price increase in 2021, Ethereum consolidated its place among other cryptocurrencies, and for the first time, new investors began buying Ethereum instead of Bitcoin.

Cryptocurrencies are based on Blockchain technology, which assures users of transaction trust and transparency. The technology, exemplified by cryptocurrencies, is widely recognized as a major innovation with profound implications for the future of finance (Liu et al., 2022). The rapid advancement of Blockchain technology has brought about remarkable transparency for its users. Furthermore, this technology can guarantee security and transparency even in the event of IoT objects or devices being compromised by intruders. Capable of tracking, organizing, and supporting communications, Blockchain technology stores data from numerous devices and facilitates peer interactions without reliance on a centralized cloud (Poongodi et al., 2020).

Ethereum stands as a prominent public Blockchain, not only securing the position of its native cryptocurrency Ether as the second-largest by market capitalization, but also serving as the underlying infrastructure for Web3 and decentralized applications (DApps), powered by Smart Contracts (Ferenczi & Bădică, 2023). Beyond facilitating transactions, Ethereum serves as a ledger for processing information, laying the groundwork for the creation of decentralized applications by others. Data is communicated through transactions, and decentralized applications utilize the Blockchain as a back-end, executing their operations through smart contracts (Peter & Styppa, 2023). Also, Ethereum serves as a cryptocurrency payment platform where clients carry out transactions, with machines executing the requested operations. More broadly, ether plays a vital role as an incentive, ensuring developers maintain and uphold quality throughout the transaction process (Poongodi et al., 2020).

The value of Ethereum fluctuates and is influenced by various factors, including the cryptocurrency market dynamics, supply and demand, technological advances, internal competition, market dynamics, economic conditions, security concerns, political factors, and the adoption of innovative strategies (Monish et al., 2022). Ethereum holds significant value due to its versatility as a form of currency; it can be used for transactions, and partial payments can be made using Ethereum. Predicting the price of Ethereum accurately can offer significant opportunities for traders and investors to make informed decisions and maximize their profits. Compared to Bitcoin, Ethereum is characterized by distinct information on the Ethereum Blockchain (Kim et al., 2021). In fact, compared with Bitcoin, Ethereum has a unique Blockchain system (Antonopoulos & Wood, 2018). Although Ethereum is the second generation Blockchain after Bitcoin, very few studies have examined the relationships between Ethereum Blockchain information and Ethereum prices. Therefore, through this study, we will attempt to fill this first gap by emphasizing the need for using information from the Ethereum Blockchain to predict the value of Ethereum (Kim et al., 2021). Also, several traditional statistical time series models exist for price prediction and have proven to be capable and effective. However, with the advancements in Machine Learning (ML) and Deep Learning (DL), the ability to capture long-term information for improved prediction has significantly enhanced, yielding superior results. While ML has shown success in predicting stock market prices using various time models, its effectiveness in predicting cryptocurrency prices has been less pronounced. This is primarily due to the inherent complexity of cryptocurrency price determinants. The increased use of ML techniques to predict cryptocurrency prices motivated this study to seek more effective methods for better accurate Ethereum price prediction. Therefore, to fill this second gap, we used time-series analyses and advanced ML techniques to forecast daily movements of the price of Ethereum.

In total, there is a lack of conclusive findings regarding the predictors of cryptocurrency prices. The results are further complicated by contradictions, attributed to variations in samples and diverse methodologies across studies. Consequently, there is a compelling need to gain deeper insights into the behavior of cryptocurrency prices. The current study aims to address these gaps by examining the key factors that determine the price of Ethereum, including macroeconomic factors, technical indicators, and Blockchain-specific factors and developing a ML models that can accurately predict the future price of Ethereum. We use the Support Vector Machines (SVMs), which is well-suited for handling high-dimensional and nonlinear data. We also apply the Multilayer Perceptrons (MLP) that excels in learning complex patterns from large datasets. Finally, we use Decision Tree (DTs) that offers interpretability and simplicity in decision-making.

This study seeks to assess the dual research questions: 1) Which variable the most affects Ethereum prices? 2) What is the most effective ML method for predicting their future movements? The contributions of this research are three-fold. First, we demonstrate that ML techniques have a great capability for forecasting cryptocurrency prices. Otherwise, the aim of this study is to compare the predictive capabilities of the SVMs, DTs, and MLP algorithms in the case of the cryptocurrency markets, and to investigate whether the ML method is more advanced in financial time series forecasting. Second, we check the most factors that determine the price of Ethereum. Third, we provide insights into the complex nature of the cryptocurrency market that can be used by investors, traders, and regulators to make informed decisions about the future of the market. For this, we have retained the following forecast statistics: Mean Absolute Error (MAE), Mean Square Error (MSE), and Root Mean Square Error (RMSE).

The remainder of the paper is structured as follows: Sect. 1 delves into a review of recent literature, while Sect. 2 outlines the methodology, and Sect. 3 presents the comparative analysis, experimental results and discussion, followed by the concluding remarks.

2 A Literature Review

Traders and investors seek precise predictions of cryptocurrency prices to enhance returns and mitigate risks. The dynamic nature of cryptocurrency prices poses a challenge in accurately predicting their future values. Additionally, cryptocurrencies exhibit high volatility, experiencing sudden rises and dips over time influenced by various factors (Hansun et al., 2022). Consequently, the trading community requires an accurate prediction method to aid strategic decision-making and capitalize on their investments. While researchers have suggested predictors using statistical, ML, and DL approaches, the available literature remains limited. This section investigates how Ethereum prices are predicted, and what are the main determinants of Ethereum prices. As already stated, Ethereum holds the second position in cryptocurrency market capitalization. However, there is a scarcity of studies in this area.

Murray et al. (2023) conduct a comprehensive analysis comparing the performance of commonly used statistical, ML and DL methods. Additionally, they explore hybrid models and ensembles to evaluate the potential enhancement in prediction accuracy through model combination. Their findings indicate that DL approaches, particularly LSTM, consistently emerge as the most effective predictors across all examined cryptocurrencies. Kolokotronis et al. (2021) explore the influence of technological factors and sentiment analysis on Ethereum's closing price predictions. The study assesses whether DL models, such as LSTM, outperform ML techniques, like XGBoost. It also investigates whether temporal stability of a model contributes to more accurate results and examines the applicability of DL without assuming stationarity. The key findings highlight the greater significance of technological variables over financial ones, and LSTM demonstrates overall superior performance compared to XGBoost. The study of Angela and Sun (2020) aims to study the determinants of Ethereum prices. Based on the ARDL (Autoregressive Distributed Lag) test model, and considering weekly data over the period 2016–2018, the results show that EUR/USD only affects Ethereum prices in the short term, while the price of gold shows no effect on Ethereum prices. Another study by Hansson (2022) focuses on the volatility of Ethereum and the factors that influence it. Over the period 2017–2022, and by applying a GARCH $(1,1)$ model, it turned out that the variables Google trends, hash rate, S&P 500, number of addresses and volume of transactions affect the volatility of Ethereum. In the same context, and using an extension of the autoregressive conditional heteroskedasticity model over the period 2015–2018, Zhamharyan (2018) looked at Bitcoin and Ethereum. The results of his study show that the price returns of S&P, SSE, Nikkei, Cyber 15, gold and oil are significant for the models and that, unlike Ethereum, Bitcoin is also used for diversification and hedging purposes.

However, some studies have pointed out the importance of integrating special variables into the Blockchain information of other coins (Kim et al., 2021). So, using time series analysis and advanced ML techniques, Kim et al. (2021) examine the relationship between information inherent in the Ethereum Blockchain and Ethereum prices. They collected data concerning Ethereum from various sources between August 11, 2015 and November 28, 2018. Two main results are worth highlighting: Bitcoin Blockchain information is significantly related to Ethereum prices, and Ethereum-specific Blockchain information factors are significantly associated with Ethereum prices. Thus, according to these authors, practitioners should actively use macroeconomic factors and Ethereum and Bitcoin Blockchain information to accurately forecast Ethereum prices.

It should be noted that several studies have focused on specifying the benefit of using ML to predict asset prices (Chevallier et al., 2021; Atsalakis et al., 2019; Jang & Lee,

2017; Mallqui & Fernandes, 2019). Atsalakis et al. (2019), Jang and Lee (2017) use artificial neural network (ANN) and Bayesian neural network (BNN)-based methodology for Bitcoin prices, respectively. These authors show that ML techniques outperform other analytical methods in predicting Blockchain coin prices. This result was confirmed, on the one hand, by Mallqui and Fernandes (2019) who show the effectiveness of the ANN, SVM and k-Means clustering method compared to other studies, and on the other hand, by the study of Chevallier et al. (2021) which led to the result that the six ML algorithms (ANN, SVM, RF, k-Nearest Neighbors, AdaBoost, Ridge regression) perform better and that this performance largely depends on several design choices. Recently, wishing to explore the exponential evolution of Ethereum, Likhitha et al. (2023) use a long-term and short-term memory (LSTM) ML model, and show its reliability in predicting Ethereum prices. Also, the contribution of Bouteska et al. (2024) is significant by focusing on various cryptocurrencies and providing a comprehensive comparative analysis of DL forecasting models. The obtained results show the effectiveness of complex ML methods.

Table 1 summarizes the various studies that have been conducted, using different methods, to identify potential factors affecting Ethereum prices, as documented in the literature.

Table 1. Recent studies using various approaches and main determinants for Ethereum Prices

References	Methodology	Study Period	Key Factors	Results
Murray et al. (2023)	Machine learning (ML) and deep learning (DL) approaches	1 June 2017 to 31 May 2022	Opening price, highest price, lowest price, and trading volume of five popular cryptocurrencies, i.e., XRP, Bitcoin (BTC), Litecoin (LTC), Ethereum (ETH), and Monero (XMR)	DL approaches are the best predictors, particularly the LSTM
Hansun et al. (2022)	Recurrent neural networks (RNNs), Long Short-Term Memory (LSTM), Bidirectional LSTM (Bi-LSTM), and Gated Recurrent Unit (GRU)	The start date of available data for each cryptocurrency to October 2021	Daily recorded data of Bitcoin (BTC), Ethereum (ETH), Cardano (ADA), Tether (USDT), and Binance Coin (BNB)	Bi-LSTM and GRU exhibit comparable accuracy performance. When considering execution time, both LSTM and GRU yield similar results, with GRU showing a slight advantage and lower variation on average

(*continued*)

Table 1. (*continued*)

References	Methodology	Study Period	Key Factors	Results
Monish et al. (2022)	Recurrent Neural Networks (RNNs), Long Short-Term Memory (LSTM) and Bi-directional Long Short-Term Memory (Bi-LSTM)	Not mentioned	Closing price for the last 2000 days that is used to predict both short-term (30 days) and long-term (90 days) Ethereum prices	Bidirectional LSTM is the best model among RNN, LSTM and Bi-LSTM to forecast the price of Ethereum
Nimbark (2022)	Deep Coin Cap Algorithm	April 2020 to April 2021	BTC dataset, Coinmarketcap dataset	The proposed methodology can provide the optimum precision for Ethereum price prediction using incidence logging
Gurkan and Palandoken (2021)	Recurrent Neural Network (RNN), Long Short-Term Memory (LSTM), Gated Recurrent Unit (GRU), Bi-directional Long Short Term Memory (Bi-LSTM), Bidirectional Gated Recurrent Unit (Bi-GRU)	August 07, 2015 to April 15, 2021	Opening price, highest price, lowest price, and trading volume of Ethereum	The best forecasting performance among these models has been achieved by the BiGRU model
Kim et al. (2021)	Time-series analyses and advanced machine-learning techniques	August 11, 2015 to November 28, 2018	Macro-economic development indices, global currency ratios, generic Blockchain information (on Ethereum, Bitcoin, Litecoin, and Dashcoin), and Ethereum-specific Blockchain information	Ethereum prices are strongly determined by macro-economy factors, Ethereum-specific Blockchain information, and the Blockchain information of other cryptocurrency

(*continued*)

Table 1. (*continued*)

References	Methodology	Study Period	Key Factors	Results
Kolokotronis et al. (2021)	LSTM and XGBoost models	July 2015 to 31 of December 2020	Financial, technological and sentiment features	Technological variables are more important than financial, LSTM has overall a higher performance than XGBoost
Angela and Sun (2020)	Autoregressive Distributed Lag (ARDL) test model	Weekly data during the 2016–2018 period	Macroeconomic aspects, EUR/USD exchange rate and the price of gold, as well as Bitcoin and other altcoins prices	EUR/USD affects the Ethereum prices in the short term. While Bitcoin, Litecoin, and Monero significantly affect the Ethereum prices, the price of gold, Ripple and Stellar have no effect on Ethereum prices
Zoumpekas et al. (2020)	CNN-2L, CNN-3L, LSTM, SLSTM, BiLSTM and GRU	August 8, 2015 to May 28, 2018	Endogenous price data	LSTM achieved the best values in the performance metrics, while the GRU is less computational expensive than the previous
Poongodi et al. (2020)	Linear Regression (LR) and Support Vector Machine (SVM)	July 2015 to April 2019	Opening price, highest price, lowest price, and trading volume of Ethereum	SVM method has a higher accuracy than the LR method
Chen et al. (2017)	Logistic Regression, Support Vector Machine, Random Forest, Naive Bayes, ARIMA models	August 30, 2015 to December 2, 2017	Ether price	All methods demonstrated accuracy surpassing 50%, with the ARIMA model outperforming others

Source: Authors' Contribution

3 Methodology

3.1 Dataset and Experiments

Table 2 shows and describes the full set of variables used in the empirical analysis. The different sources of data are reported in the last column. The predictive models are trained and tested over a time window of 9 years. For our methodology, we collected daily Ethereum data from Factset Database, a professional financial data platform. The period covers the days from July 30, 2015 (birth date of Ethereum) to September 30, 2023. In fact, the first version of Ethereum, dubbed Frontier, was released on July 30, 2015, marking the cryptocurrency's official launch. The dependant variable is the Ethereum price. Ethereum-specific Blockchain information, Ethereum Blockchain information, global currency ratio, Macro-economic development indices, attractiveness and generic Blockchain information on Bitcoin, Doge Coin and XRP, are used as the predictors. We chose to integrate these three others cryptocurrencies with Blockchain information. Our choice is explained by their high market transaction volumes and the temporal overlap of their Blockchain information with that of Ethereum. We used Kaggle for the data pre-pocessing and for the experiments (60% training set and 40% test set). We applied three algorithms of ML. These latter consist of the Support Vector Machines (SVMs) algorithm, the Decision Trees (DTs) algorithm, and the Multilayer Perceptron (MLP) algorithm. These ML algorithms are capable of producing the most accurate and fastest predictions (Siva et al., 2024). In order to determine the SVM optimal parameters (i.e. the weight W, the bias b, the cost C and the KernelType), we used the Grid Search technique proposed by Chen and Lin in (Chen & Lin, 2005). For MLP and DTs algorithms, the cross validation has been applied to determine the appropriate model.

According to the literature review, these methods worked better for several types of time series data (Chen et al., 2017; Poongodi et al., 2020; Kim et al., 2021; Murray et al., 2023). ML algorithms have the potential to find natural patterns in data to improve decision making (Chevallier et al., 2021). In recent years, the use of ML techniques has attracted increasing interest in different scientific fields, particularly in cryptocurrency price prediction (Ashayer, 2019). The methods underlying these models and their assumptions are briefly summarized below, with references to Chen et al. (2017), Friedman et al. (2001), James et al. (2013), and Chevallier et al. (2021).

Support Vector Machines. SVMs are a set of supervised learning techniques designed to solve discrimination and regressive problems. SVMs algorithm is a non-parametric methodology introduced by Vapnik in 1995. In a regression approach, SVMs perform a linear regression in a high-dimensional feature using an e-insensitive loss (Chen et al., 2017). The basic idea is to divide a p-dimensional space into two halves (for more details, see Friedman et al., 2001; James et al., 2013).

Considering a dataset (x_i, y_i), $i = 1,...,n$, the linear support vector classifier can be represented as:

$$f(x) = \beta_0 + \sum_{i=1}^{n} \beta_i \sum_{j=1}^{n} x_{ij} y_{ij} \tag{1}$$

Decision Trees. DTs learning are a non-parametric supervised learning algorithm used as a predictive model to draw conclusions about a set of observations. In decision analysis, DTs can be used to visually and explicitly represent decisions and decision making. First proposed by Ho (1988) and further developed by Breiman (2001), the main objective of this technique is to create a model that predicts the value of a target variable by considering simple decision rules determined from data characteristics. While several authors are keen to use the DTs due to the fact that they work well with noisy or missing data, can easily be regrouped to form more robust predictors, and give more accurate results, others authors prefer to use the Random Forests (RFs) algorithm proposed by Breiman (2001) (example Chevallier et al., 2021; Chen et al., 2017). RFs algorithm approximate the following expectation (for more details, see Geurts et al., 2006; Biau, 2012; Genuer et al., 2010).

$$\hat{f}_{rf} = E_\theta T(x; \theta) = \lim_{\beta \to \infty} \widehat{f(x)}_{rf}^\beta \qquad (2)$$

Multilayer Perceptron. MLP, or Multilayer Neural Network, is the most popular classic neural network architectures. This technique can evaluate any continuous function mapping from one-finite dimensional discrete space to another (Guresen et al., 2011). The classical neural network layer performs a convolution on a given sequence X, outputting another sequence Y whose value at time t is:

$$\hat{y} = \sum_{j=1}^{p} f(\beta_j, x_j(t)) + \varepsilon(t) \qquad (3)$$

where β_j are the parameters of the layer trained by back-propagation (for more details on neural networks, see Maclin & Shavlik, 1995; Vapnik, 1999; Scholkopf, 1998).

For a more complete study of the three ML algorithms used in our research, Table 3 presents the advantages and disadvantages, extracted from the literature, of each method (Ciaburro & Venkateswaran, 2017; Pineda-Jaramillo 2019; Quinlan, 1992; Lantz, 2015).

Table 4 summarizes the various studies that have been conducted, using SVMs, DTs and MLP algorithms to predict cryptocurrency prices, as documented in the literature.

3.2 Experimental Protocol and Evaluation Criteria

In this study, a detailed evaluation is conducted using mean square error (MSE), root mean square error (RMSE), and mean absolute error (MAE) to evaluate the used algorithms with cryptocurrency datasets. These measurements offer vital information about how the model predicts the future, as it relates to Ethereum price forecasting (Likhitha et al., 2023). We apply 50 epochs that correspond to the total number of iterations used with our datasets in order to follow the evolution of the evaluation criteria all over the time. These evaluation criteria are presented by the following formulas:

$$MSE = \frac{1}{N} \sum_{i=1}^{n} (y_i^{real} - y_i^{pred})^2$$

$$RMSE = \sqrt{MSE} = \sqrt{\frac{1}{N}\sum_{i=1}^{n}(y_i^{real} - y_i^{pred})^2} \qquad (4)$$

Table 2. Data for empirical study.

Data category	Research Variables	Definition	Data Source
Dependent variable	Ethereum price	The daily US dollar price of one unit of Ethereum on the Bitstamp exchange	https://coinmarketcap.com/
Independent variables			
Macro-economic Development Index	Standard & Poor's 500 index (S&P 500) Stock Index of Eurozone (Euro Stoxx 50) National Association of Securities Dealers Automated Quotations (NASDAQ) Crude Oil Gold Volatility index of S&P 500 Dow Jones Industrial Average Financial Times Stock Exchange 100 Index (FTSE100)		Factset database
Global Currency Ratio	British Currency Sterling (GBP)/US Dollar (USD) Japanese Yen (JPY)/US Dollar (USD) Swiss Franc (CHF)/US Dollar (USD) Euro (EUR)/US Dollar (USD)		Factset database
Ethereum-Specific Blockchain information	Number of transactions (Demand for Ethereum)	The daily total volume of Ethereum transactions validated and recorded by a Blockchain ledger. This variable is used as a proxy for the demand side of the Ethereum market	https://www.blockchain.com/

(*continued*)

Table 2. (*continued*)

Data category	Research Variables	Definition	Data Source
	Number of Ethereum mined (Supply for Ethereum)	The daily total amount of Ethereum units currently in circulation. This variable is introduced as a proxy for the supply side of the Ethereum market	https://www.blockchain.com/
	Ethereum Average Gas Price	The daily value of units traded on the Bitstamp platform, expressed in US dollars. This variable is employed as a measure of the Ethereum market activity	https://www.blockchain.com/
	Ethereum Daily Gas Used	The Ethereum Daily Gas Used Chart shows the historical total daily gas used of the Ethereum network	https://www.blockchain.com/
	Ethereum Average Gas Limit	The Ethereum Average Gas Limit is the maximum number of units of gas you are willing to pay for in order to carry out a transaction or Ethereum Virtual Machine (EVM) operation	https://www.blockchain.com/
	Uncle Block	Uncle Block refers to a block that was not registered as a formal block in the Blockchain network	https://www.blockchain.com/
Ethereum Blockchain information	Average Txn Fees (USD)	Transaction fees are the difference between the amount of Ethereum sent and the amount received. Fees are employed as an incentive for miners to add transactions to blocks	https://www.blockchain.com/
	Ethereum Block Count	Ethereum Block Count shows the historical number of blocks produced daily on the Ethereum network	https://www.blockchain.com/

(*continued*)

Table 2. (*continued*)

Data category	Research Variables	Definition	Data Source
	Ethereum Network Hash Rate (TH/s)	The daily average exa-hashes per second (1 EH/s = 1018 hashes) is an indicator of the processing capability of high-powered mining hardware that individual miners use to unlock new Ethereum units. The higher the hash rate is, the more resilient the network is to malicious cyber-attacks. We utilize this variable to represent the security aspect of the Ethereum network	https://www.blockchain.com/
	Address Count	The number of addresses which fulfills the defined activity parameter on a given Blockchain. This variable is employed to measure how active a given Blockchain is, and can be more representative compared to tracking number of transactions	https://www.blockchain.com/
	Block Reward	When miners successfully mine a Block into existence on the Ethereum Blockchain, they receive a reward in ETH	https://www.blockchain.com/
	Network Utilization	Ethereum Network Utilization shows the average gas used over the gas limit in percentage	https://www.blockchain.com/
	Block Size Average Block Size	The size of a block equals the amount of data it stores. And just like any other container, a block can only hold so much information	https://www.blockchain.com/

(*continued*)

Table 2. (*continued*)

Data category	Research Variables	Definition	Data Source
	Bloc Difficulty (Network Difficulty)	The difficulty is a measure of how difficult it is to mine an Ethereum block	https://www.blockchain.com/
Generic Blockchain Information (Bitcoin, Doge Coin, XRP)	Price	The daily US dollar price of one unit of cryptocurrency on the Bitstamp exchange	https://coinmarketcap.com/
	Number of transactions	The daily total volume of cryptocurrency transactions validated and recorded by a Blockchain ledger	https://coinmarketcap.com/
	Number of cryptocurrency mined	The daily total amount of cryptocurrency units currently in circulation	https://coinmarketcap.com/

Source: Authors' Contribution

Table 3. The advantages and disadvantages of SVMs, DTs and MLP algorithms.

	Advantages	Disadvantages
Support Vector Machines (SVMs)	Interpretability, power, flexibility, high quality results	Not suitable for nonlinear problems, not the best choice for a large number of more complex features
Multilayer Perceptron (MLP)	MLP is relatively robust to noisy or missing input data, and can still produce accurate output even when some input data is missing or corrupted MLP with a single hidden layer is able to approximate any continuous function to any desired level of accuracy. This property is known as the universal approximation theorem	MLP is susceptible to overfitting, especially when the number of parameters in the model is high. This means that the network may learn to fit the training data too closely, resulting in poor generalization performance on new data
Decision Trees (DTs)	Powerful and precise, suitable for linear and nonlinear complex problems	No interpretability, the number of trees must be chosen manually

Source: Authors' Contribution

Table 4. Studies conducted for cryptocurrency price prediction using SVMs, DTs and MLP algorithms.

ML Technique	References	Dataset	Main Results
SVMs	Chen et al. (2017)	Ethereum	The ARIMA algorithm had the best performance. The price of Ether is not stationary, it is volatile and has an upward or downward trend
	Sebastiao and Godinho (2021)	Ethereum Litecoin	The best results were determined by the linear models, RFs, and SVMs on Ethereum prices
	Kang et al. (2022)	Bitcoin Ethereum Ripple	The LSTM achieved the optimal RMSE for all three cryptocurrencies, with RMSE of 928.62 on Bitcoin, 11.69 on Ethereum, and 0.16 on Ripple
DTs or RFs	Chevallier et al. (2021)	Bitcoin	RFs algorithms work as the best ML models and could be implemented in the internal IT system of a banking institution for Bitcoin predictability
	Derbentsev et al. (2020)	Bitcoin Ethereum Ripple	Gradient boosting machine is able to better forecast the price compared to RFs
	Chen et al. (2017)	Ethereum	The underperformance of the random forest-based classifier is explained by the continuous nature of the dataset. Also, it is impossible for the algorithm to explore the entire feature space, since this is time series data, with price features that do not necessarily repeat

(*continued*)

Table 4. (*continued*)

ML Technique	References	Dataset	Main Results
MLP	Jay et al. (2020)	Bitcoin Ethereum Litecoin	The stochastic neural networks showed an average improvement of 4.84% for Bitcoin, 4.15% for Ethereum, and 4.74% for Litecoin, compared to deterministic MLP and LSTM
	Jiang (2020)	Bitcoin	The MLP model achieved the best result, which had the minimal RMSE of 19.020
	Sin and Wang (2017)	Bitcoin	The incorporation of MLP into ANN increased the Bitcoin price prediction accuracy
	Atlan and Pence (2021)	Bitcoin Ethereum Ripple	As a result of the cross-validation test, estimation results of the Multilayer Perceptron technique are found to be successful and important. With the successfully developed model, it is possible to estimate price for BTC, Ethereum and Ripple with instant data in the future
	Albariqi and Winarko (2020)	Bitcoin	Long-term prediction has a better result than short-term prediction, with the best accuracy in MLP and RNNs. Also, MLP outperforms RNNs with accuracy of 81.3%.
	Alonso-Monsalve et al. (2020)	Bitcoin Dash Ether Litecoin, Monero Ripple	LSTM significantly outperformed all other models. CNNs are effective, especially for Bitcoin, Ether and Litecoin

(*continued*)

Table 4. (*continued*)

ML Technique	References	Dataset	Main Results
	Mallqui and Fernandes (2019)	Bitcoin	RNNs and MLPs achieve the best results in price trend and closing price prediction, respectively
	Adcock and Gradojevic (2019)	Bitcoin	MLP is suitable for predicting Bitcoin returns

Source: Authors' Contribution

$$MAE = \frac{1}{N}\sum_{i=1}^{n}\left|y_i^{real} - y_i^{pred}\right|$$

where N is the number of samples, y_i^{real} is the actual value, and y_i^{pred} is the estimated value.

4 Comparative Analysis, Results and Discussion

The descriptive statistics of Ethereum Price and Ethereum-Specific Blockchain information are shown in Table 5. We report the descriptive statistics of Global Currency Ratio and others cryptocurrencies Blockchain information variables in Table 6.

It is obvious from Table 5, that the values of the mean and Standard Deviation are very close confirming that we do not have noisy data.

For this research, we refer to the study of Kim et al. (2021) and we conduct a stepwise analysis from Models 1 to 8 (Table 7). Our analyses show that SVM method works better than the other methods across all models. In fact, considering all inputs in our model, the findings show that SVM performs very well and it is better than DTs and MLP methods. This method achieves the lowest RMSE (0.1353765) and MAE (0.5702575). This means that it performs the best in forecasting the price of Ethereum. MLP performs the second best and the DTs perform the worst. This result proves that SVM is a relevant Ethereum price forecasting tool, compared to others ML methods. Concerning the different models considered, Model 1 includes all variables (MSE = 0.0194054, RMSE = 0.1353765, MAE = 0.5702575). Model 2 includes variables with iteration of only macro-economic factors and we found that the error doubled; MSE, RMSE and MAE did not significantly improve the results of the analysis (MSE = 0.0349733, RMSE = 0.2558478 =, MAE = 1.0333336). These findings are in line with previous studies that have shown a significant positive link between macroeconomic factors and Ethereum price (Kim et al., 2021; Chaigusin, 2014; Derbentsev et al., 2020). Therefore, we recommend considering the macro-economic factors for the prediction of Ethereum prices. Model 3 with iteration of Ethereum-Blockchain information presented the worst performance, as shown in Table 7 (MSE = 0.3005836, RMSE = 1.3021723, MAE = 1.1319348). Then, we find that Ethereum-Blockchain information is the most important

Table 5. Descriptive Statistics of Ethereum Price and Ethereum-Specific Blockchain Information.

	Ethereum Price	Number of Ethereum mined (Supply)	Number of transactions (Demand)	Ethereum Average Gas Price	Ethereum Daily Gas Used	Uncle Block	Ethereum Average Gas Limit
Count	2384.000000	2384.000000	2.384000e+03	2384.000000	2.384000e+03	2.384000e+03	2384.000000
Mean	0.368410	666.747391	1.002292e+08	75497.956916	4.737260e+10	4.085093e+10	515.478607
Standard Deviation	0.393488	1076.679462	1.349590e+07	127187.703053	5.971831e+10	3.281063e+10	323.133796
Min	0.004090	0.440000	7.227992e+07	32.644000	7.320701e+09	3.886300e+07	126.000000
25%	0.009537	43.735000	9.009561e+07	3961.938740	1.575086e+10	3.058952e+09	331.000000
50%	0.269000	220.650000	1.031340e+08	23169.375686	2.354762e+10	4.057771e+10	399.000000
75%	0.486450	550.407500	1.115459e+08	56215.822896	5.505893e+10	7.228083e+10	494.000000
Max	3.380000	4810.970000	1.195831e+08	568966.160087	9.395883e+11	1.015230e+11	2096.000000

Source: Authors' estimations

Table 6. Descriptive Statistics of Global Currency Ratio and others cryptocurrencies Blockchain Information.

	GPB-USD	JPY-USD	EURO-USD	CHF-USD	Bitcoin Price	Doge Coin Price	XRP Price
Count	2384.000000	2384.000000	2384.000000	2384.000000	2384.000000	2384.000000	2384.000000
Mean	4691.866661	0.919365	0.786440	40.017932	0.716279	12759.847282	0.036019
Standard Deviation	3272.140191	0.620882	0.529591	29.263458	0.482151	16576.764693	0.089607
Min	0.000000	0.000000	0.000000	0.000000	0.000000	210.490000	0.000114
25%	0.000000	0.000000	0.000000	0.000000	0.000000	1175.920000	0.000324
50%	6531.695000	1.288350	1.112925	48.710000	1.011800	7120.940000	0.002521
75%	7253.067500	1.344450	1.158950	63.877500	1.041075	11323.417500	0.004156
Max	7877.450000	1.577400	1.250300	96.480000	1.141400	67566.830000	0.684800

Source: Authors' estimations

variable in predicting Ethereum prices. Our research confirms the findings of several studies which examined the use of Blockchain information in predicting cryptocurrency prices and which found a positive long-term correlation between cryptocurrency

Blockchain information and cryptocurrency prices (Kristoufek, 2015; Kubal & Kristoufek, 2022; Jang & Lee, 2017; Saad & Mohaisen, 2018). Also, Model 4 with iteration of Blockchain information of Bitcoin does not significantly improve the results of the analysis (MSE = 0.0319806 =, RMSE = 0.2978149, MAE = 0.9417674). This result illustrates convincing evidence of the benefits of integrating Bitcoin-Blockchain data into Ethereum price forecasts, since we note that Bitcoin-Blockchain information contributes significantly in predicting Ethereum prices. The same result is demonstrated for the two others cryptocurrencies: Doge Coin and XRP. In fact, we find that MSE, RMSE and MAE for both Models 5 and 6 do not improve the analysis result, respectively with iteration of Doge Coin-Blockchain information and XRP-Blockchain information. These results show that there is a relationship between the Blockchain information of Doge Coin and XRP and the Ethereum price, thus playing an important and determining role in predicting Ethereum prices. In this study, it appears that Bitcoin, XRP and Doge Coin prices contribute significantly toward predicting Ethereum prices. This result is in line with some authors' works confirming the existence of direct relationships between Blockchain information and cryptocurrency prices (Mallqui & Fernandes, 2019; Guo et al., 2021). Model 7, without Attractiveness variable, improves slightly the analysis result (MSE = 0.0224258, RMSE = 0.1601084, MAE = 0.6574294). Based on these values, our findings show that our sentiment indicator (Attractiveness) improves the efficiency of our forecasting model. So, Ethereum price may be affected by its attractiveness as an investment opportunity. This result is in line with studies conducted by Kristoufek (2015), Abraham et al. (2018), Hansson (2022), and Ciaian et al. (2016), showing that attractiveness variable is closely related to Ethereum prices. Thus, we can conclude that Ethereum price fluctuations are influenced by the publication of positive or negative news. This result implies that supervisors, analysts and researchers should consider the attractiveness variable and integrate it into their forecasting models, since this variable constitutes an important determinant of Ethereum's price. Finally, we find that MSE, RMSE and MAE are slightly improved in Model 8, with iteration of global currency ratio (MSE = 0.0104607, RMSE = 0.0721287, MAE = 0.3067682), suggesting that global currency ratios determine the price of the Ethereum in the short term as well as in the long term. The results found in our study could provide good references for researchers and practitioners wishing to predict Ethereum prices. The aim of this research is to discover new variables that could influence Ethereum prices and to broaden theoretical perspectives in this field. Finally, it should be emphasized that our study incorporates the majority of variables likely to influence Ethereum prices, taking into account macroeconomic factors, Ethereum-specific Blockchain information, global currency ratio, Blockchain information of other cryptocurrencies and social variables to predict Ethereum prices, which has not been considered in previous studies.

It's essential to note that the superiority of SVMs over MLP and DTs depends on the nature of the data, the choice of hyperparameters, and the complexity of the problem. When predicting cryptocurrency prices using datasets with a high number of attributes, SVMs may outperform DTs and MLP due to their capability to effectively handle high-dimensional spaces. This advantage stems from SVMs' focus on support vectors, which are data points closest to the decision boundary, thus avoiding the curse of dimensionality. In contrast, MLP might struggle with high dimensionality due to increased complexity and potential overfitting, while DTs could become excessively large and prone to overfitting.

Table 7. The Results of Data Analysis.

	Model	Train			Validation		
		MSE	RMSE	MAE	MSE	RMSE	MAE
Model 1: All Inputs	MLP	0,26398046	4,3403796	1,4942290	0,0095944	2,6326893	1,5653828
	DTs	0,36546002	6,5164863	10,9096074	0,3098404	4,3426833	9,7306258
	SVMs	0,02227104	0,2580683	0,6351600	0,0194054	0,1353765	0,5702575
Model 2: Iteration of Macro-economic Development Index	MLP	0,03439821	5,6744298	1,9534922	0,0141982	3,6279142	2,0465157
	DTs	0,46687194	7,8282783	13,8210737	0,3965685	5,0622271	12,2990865
	SVMs	0,04067678	0,4850824	1,1571704	0,0349733	0,2558478	1,0333336
Model 3: Iteration of Blockchain information of Ethereum	MLP	0,30100479	1,2387913	1,0445742	0,3028745	1,8959452	1,3326705
	DTs	0,38705765	1,1489422	5,0403741	0,3824187	1,1028213	4,9199101
	SVMs	**0,30061688**	**0,3019526**	**0,3326926**	**0,3005836**	**1,3021723**	**1,1319348**
Model 4: Iteration of Blockchain information of Bitcoin	MLP	0,04127754	6,1265364	2,7028837	0,0259639	4,8651907	2,8830759
	DTs	0,04871092	0,8940676	1,4638042	0,0414505	0,6085071	1,3068897
	SVMs	0,03718821	0,4622519	1,0522531	0,0319806	0,2978149	0,9417674
Model 5: Iteration of Blockchain information of Doge Coin	MLP	0,00157394	0,2580515	0,0860172	0,0005675	0,1548309	0,0903180
	DTs	0,18367132	3,5613481	5,7241881	0,1568042	2,5911550	5,1945979
	SVMs	0,00095083	0,0105940	0,0269405	0,0008271	0,0059893	0,0241720
Model 6: Iteration of Blockchain information of XRP	MLP	0,03689408	6,0761328	2,1913922	0,0183848	4,2831756	2,4902184
	DTs	0,35480062	6,3278865	10,5658367	0,3034950	4,2469015	9,4591166
	SVMs	0,03441316	0,3815059	0,9600628	0,0302852	0,2211986	0,8610967
Model 7: Iteration of Attractiveness	MLP	0,03913351	6,2578649	2,1852862	0,0163469	4,0725788	2,3839485
	DTs	0,25944233	4,6327131	7,8104133	0,2209785	3,1209065	6,9815861
	SVMs	0,02593632	0,2957779	0,7347960	0,0224258	0,1601084	0,6574294
Model 8: Iteration of Global Currency Ratio	MLP	0,02734571	4,3817590	1,5548177	0,0119785	2,8976148	1,6961648
	DTs	0,17445857	3,0748708	5,2216320	0,1484787	2,0675306	4,6637808
	SVMs	0,01201402	0,1365652	0,3418526	0,0104607	0,0721287	0,3067682

Source: Authors' estimations

5 Conclusion

Predicting the price of Ethereum using ML techniques is a fascinating endeavor. By training a model with historical data and employing various algorithms and techniques, we can gain insights into the future price movements of this popular cryptocurrency. This study aims to address the lack of conclusive findings in cryptocurrency price forecasting, by examining the key factors driving Ethereum's price, including macroeconomic factors, technical indicators and Blockchain-specific factors, and developing a ML model capable of accurately predicting Ethereum's future price. The goal of this research is to explore which variables affect Ethereum price level. In order to achieve this objective, we applied time series to daily data for the period from 30/07/2015 to 30/09/2023. We used Python and TensorFlow library version 2.11.0. Price prediction was performed with three ML techniques: support vector machines (SVMs), decision trees (DTs) and multilayer perceptron technique (MLP), for time-series analysis. The proposed approaches are used for price prediction in an industrial finance system and exhibit suitable accuracy scores. We followed Kim et al. (2021) method conducting a stepwise analysis. Two results are worth noting: when using the proposed model, the SVM algorithm provides better results than the MLP and the decision tree algorithms. Then, SVMs turns out to be the best method of predicting Ethereum prices. The performance of the two other models was quite limited for Ethereum. Also, Ethereum-specific Blockchain information is the most important variable in predicting Ethereum prices. This study highlights the importance of integrating Blockchain factors into cryptocurrency price prediction models.

However, in practice, investors are more interested in predicting the volatility of future Ethereum prices, something which is not considered in our study. Therefore, the main limitation of this research is that it does not aim to predict the volatility of future Ethereum prices. The second limitation is that ML techniques have some drawbacks since they depend on feature engineering and are hampered by model complexity when lots of training data is available. Thus, we believe that it is necessary to consider other more effective methods for predicting cryptocurrency prices.

Therefore, as an avenue of research, we recommend introducing other revealing variables to predict deeply and correctly cryptocurrency prices, examining more frequent data, and proposing model which can be increased by adding features to the SVM algorithm. Further research into predicting Ethereum price volatility is needed. Finally, we hope that our results will have served to provide a theoretical basis for investors, regulators, and researchers to discover additional variables, and expand knowledge in the field of cryptocurrency research.

References

Abraham, J., Higdon, D., Nelson, J., Ibarra, J.: Cryptocurrency price prediction using tweet volumes and sentiment analysis. SMU Data Sci. Rev. **1**(3) (2018)

Adcock, R., Gradojevic, N.: Non-fundamental, non-parametric Bitcoin forecasting. Physica A **531**, 121727 (2019)

Albariqi, R., Winarko, E.: Prediction of bitcoin price change using neural networks. In: 2020 International Conference on Smart Technology and Applications (ICoSTA) (2020)

Alonso-Monsalve, S., Suárez-Cetrulo, A.L., Cervantes, A., Quintana, D.: Convolution on neural networks for high-frequency trend prediction of cryptocurrency exchange rates using technical indicators. Expert Syst. Appl. **149**, 113250 (2020)

Angela, O., Sun, Y.: Factors affecting cryptocurrency prices: evidence from ethereum. In: 2020 International Conference on Information Management and Technology (ICIMTech), 13–14 August 2020. IEEE (2020)

Antonopoulos, A.M., Wood, G.: Mastering Ethereum. The Ethereum Book LLC, Gavin Wood, vol. 384. O'Reilly Media, Inc., Printed in the United States of America (2018)

Ashayer, A.: Modeling and Prediction of Cryptocurrency Prices Using Machine Learning Techniques. East Carolina University (2019)

Atlan, F., Pence, I.: Forecasting of bitcoin price using the multilayer perceptron technique. Techno-Science **4**(2), 68–74 (2021)

Atsalakis, G.S., Atsalaki, I.G., Pasiouras, F., Zopounidis, C.: Bitcoin price forecasting with neuro-fuzzy techniques. Eur. J. Oper. Res. **276**, 770–780 (2019)

Ben-Ahmed, K., Theiri, S., Kasraoui, N.: Short-term effect of COVID-19 pandemic on cryptocurrency markets: a DCC-GARCH model analysis. Heliyon **9**(8), e18847 (2023)

Biau, G.: Analysis of a random forests model. J. Mach. Learn. Res. **13**, 1063–1095 (2012)

Bouteska, A., Zoynul Abedin, M., Hajek, P., Yuan, K.: Cryptocurrency price forecasting – A comparative analysis of ensemble learning and deep learning methods. Int. Rev. Financ. Anal. **92**, 103055 (2024)

Breiman, L.: Random forests. J. Mach. Learn. **45**, 5–32 (2001)

Chaigusin, S.: An application of decision tree for stock trading rules: a case of the stock exchange of Thailand. In: Proceedings of Eurasia Business Research Conference (2014)

Chen, Y.W., Lin, C.J.: Combining SVMs with various feature selection strategies. In: Studies in Fuzziness and Soft Computing, Taiwan University, vol. 207, pp. 315–324. Springer (2005)

Chen, M., Narwal, N., Schultz, M.: Predicting Price Changes in Ethereum. Cs229. Stanford Education, pp. 1–6 (2017)

Chevallier, J., Guégan, D., Goutte, S.: Is it possible to forecast the price of bitcoin? Forecasting **3**, 377–420 (2021)

Ciaburro, G., Venkateswaran, B.: Neural Networks with R: Smart Models Using CNN, RNN, Deep Learning, and Artificial Intelligence Principles, Livre numérique, Edition. Packt Publishing, Birmingham (2017)

Ciaian, P., Miroslava, R., D'Artis, K.: The economics of BitCoin price formation. Appl. Econ. **48**(19), 1799–1815 (2016)

Derbentsev, V., Semerikov, S., Serdyuk, O., Solovieva, V., Soloviev, V.: Recurrence based entropies for sustainability indices. E3S Web Conf. **166**, 1–7 (2020)

Ferenczi, A., Bădică, C.: Prediction of ethereum gas prices using DeepAR and probabilistic forecasting. J. Inf. Telecommun. (2023). https://doi.org/10.1080/24751839.2023.2250113

Friedman, J., Hastie, T., Tibshirani, R.: The Elements of Statistical Learning, vol. 1, no. 10. Springer, New York (2001)

Genuer, R., Poggi, J.M., Tuleau-Malot, C.: Variable selection using random forests. Pattern Recognit. Lett. **31**, 2225–2236 (2010)

Geurts, P., Ernst, D., Wehenkel, L.: Extremely randomized trees. J. Mach. Learn. **63**, 3–42 (2006)

Guo, H., Zhang, D., Liu, S., Wang, L., Ding, Y.: Bitcoin price forecasting: a perspective of underlying blockchain transactions. Decis. Supp. Syst. **151**, 113650 (2021)

Guresen, E., Kayakutlu, G., Daim, T.U.: Using artificial neural network models in stock market index prediction. Expert Syst. Appl. **38**, 10389 (2011)

Gurkan, C., Palandoken, M.: Time series forecasting of ethereum prices using deep learning methods. In: Conference: The Fifth International Conference on Computational Mathematics and Engineering Sciences (2021)

Hansson, P.: The Underlying Factors of Ethereum Price Stability: An Investigation on What Underlying Factors Influence the Volatility of the Returns of Ethereum. Independent thesis Advanced level, degree of Master (2022)

Hansun, S., Wicaksana, A., Khaliq, A.Q.M.: Multivariate cryptocurrency prediction: comparative analysis of three recurrent neural networks approaches. J. Big Data **9**(1) (2022)

Ho, T.K.: The random subspace method for constructing decision forests. IEEE Trans. Pattern Anal. Mach. Intell. **20**, 832–844 (1988)

James, G., Witten, D., Hastie, T., Tibshirani, R.: An Introduction to Statistical Learning, vol. 112. Springer, Heidelberg (2013)

Jang, H., Lee, J.: An empirical study on modeling and prediction of bitcoin prices with Bayesian neural networks based on blockchain information. IEEE Access **6**, 5427–5437 (2017)

Jay, P., Kalariya, V., Parmar, P., Tanwar, S., Kumar, N., Alazab, M.: Stochastic neural networks for cryptocurrency price prediction. IEEE Access **8**, 82804–82818 (2020)

Jiang, X.: Bitcoin price prediction based on deep learning methods. J. Math. Financ. **10**, 132–139 (2020)

Kang, C.Y., Chin, P.L., Kiang, M.L.: Cryptocurrency price prediction with convolutional neural network and stacked gated recurrent unit. **7**(11), 149 (2022)

Kim, H.M., Bock, G.W., Lee, G.: Predicting ethereum prices with machine learning based on blockchain information. Exp. Syst. Appl. **184** (2021)

Kolokotronis, D., Van, D.E.J.M., Boriš Cule, W.D.: Ethereum forecasting by utilizing machine and deep learning. Thesis submitted in partial fulfillment of the requirements for the degree of Master of Science in Data Science & Society, Department of Cognitive Science & Artificial Intelligence, School of Humanities and Digital Sciences, Tilburg University (2021)

Kristoufek, L.: What are the main drivers of the bitcoin price? Evidence from wavelet coherence analysis. PLoS ONE **10**(4), e0123923 (2015)

Kubal, J., Kristoufek, L.: Exploring the relationship between Bitcoin price and network's Hashrate within endogenous system. Int. Rev. Financ. Anal. **84**, 102375 (2022)

Lantz, B.: Machine Learning with R. Packt Publishing, Birmingham (2015)

Likhitha, B.B., Raj, A., Salim Ul Islam, M.: Unveiling ethereum's future: LSTM-based price prediction and a systematic blockchain analysis. E3S Web Conf. **453**, 01043, ICSDG (2023)

Liu, Y., Tsyvinski, A., Wu, X., Liu, Y., Tsyvinski, A., Wu, X.: Common risk factors in cryptocurrency. J. Financ. **77**(2), 1133–1177 (2022)

Maclin, R., Shavlik, J.W.: Combining the predictions of multiple classifiers: using competitive learning to initialize neural networks. IJCAI **95**, 524–531 (1995)

Mallqui, D.C.A., Fernandes, R.A.S.: Predicting the direction, maximum, minimum and closing prices of daily bitcoin exchange rate using machine learning techniques. Appl. Soft Comput. **75**, 596–606 (2019)

Monish, S., Mohta, M., Rangaswamy, S.: Ethereum price prediction using machine learning techniques – a comparative study (2022). https://doi.org/10.33564/IJEAST.2022.v07i02.018

Murray, K., Rossi, A., Carraro, D., Visentin, A.: On forecasting cryptocurrency prices: a comparison of machine learning, deep learning, and ensembles. Forecasting **5**(1), 196–209 (2023)

Nimbark, H.: Development of strategy for ethereum price analysis using deep learning based on time series data. Bus. Manag. Econ. Eng. **20**(2) (2022)

Peter, F.J., Styppa, K.: Predicting millionaires from Ethereum transaction histories using node embeddings and artificial neural nets. Expert Syst. Appl. **223**, 119834 (2023)

Pineda-Jaramillo, J.D.: A review of Machine Learning (ML) algorithms used for modeling travel mode choice. DYNA **86**(211), 32–41 (2019)

Poongodi, M., et al.: Prediction of the price of Ethereum Blockchain cryptocurrency in an industrial finance system. Comput. Electr. Eng. **81**, 106527 (2020)

Quinlan, J.C.: Programs for Machine Learning. Morgan Kaufmann Publishers Inc., San Francisco (1992)

Saad, M., Mohaisen, A.: Towards characterizing blockchain-based cryptocurrencies for highly-accurate predictions. In: IEEE Infocom 2018 - IEEE Conference on Computer Communications Workshops (Infocom Wkshps) (2018)

Scholkopf, B.: Support vector machines: a practical consequence of learning theory. IEEE Intell. Syst. **13**, 4 (1998)

Sebastião, H., Godinho, P.: Forecasting and trading cryptocurrencies with machine learning under changing market conditions. Financ. Innov. **7**(1), 1–30 (2021)

Sin, E., Wang, L.: Bitcoin price prediction using ensembles of neural networks. In: 13th International Conference on Natural Computation, Fuzzy Systems and Knowledge Discovery (ICNC-FSKD), pp. 666–671 (2017)

Siva, R., Subrahmanian, B., Chaturya, K.: Analyzing the machine learning methods to predict Bitcoin pricing. World J. Adv. Res. Rev. **21**(01), 1288–1294 (2024)

Vapnik, V.N.: The Nature of Statistical Learning Theory. Data Mining and Knowledge Discovery, Editor. Springer (1995)

Vapnik, V.N.: An overview of statistical learning theory. IEEE Trans. Neural Netw. **10**, 988–999 (1999)

Zhamharyan, A.: Price Determinants and Prediction of Bitcoin and Ethereum. Study Submitted to American University of Armenia Manoogian Simone College of Business and Economics in Partial Fulfillment of the Requirements for the Degree of BA in Business, Yerevan (2018)

Zoumpekas, T., Houstis, E., Vavalis, M.: ETH analysis and predictions utilizing deep learning. Expert Syst. Appl. **162**, 113866 (2020)

Email Classification of Text Data Using Machine Learning and Natural Language Processing Technique

Oluwaseyi Ijogun[1], Hayden Wimmer[1], and Carl Rebman Jr.[2(✉)]

[1] Georgia Southern University, Statesboro, GA, USA
{oi00326,hwimmer}@georgiasouthern.edu
[2] University of San Diego, San Diego, CA, USA
carlr@sandiego.edu

Abstract. Spam and Phishing emails are the most crucial in social networks, many issues arise through emails such as cost of dealing with spam and phishing emails due to their large quantities, privacy resulting in loss of sensitive information, time taken to identify spam and phishing emails, and cyber security threat due to malicious content. Using a spam and phishing detection approach, a model can quickly recognize spam and phishing emails and classify them before they become a threat to the organization. In this study, a machine learning and Natural Language processing-based supervised learning approach was used and plays an effective role in improving email classification. The dataset was prepared and dynamically classified into 3 categories namely spam-ham, spam-phishing, and ham-phishing. Different methods for effective classification were performed such as data preprocessing, feature selection, model training, model testing, and classification result and performance evaluation. There were 5 machine learning algorithms used, and the result was evaluated using 8 performance indexes. The result shows that the XGBoost classifier out-performed other machine learning algorithms used, Results show that XGBoost machine learning algorithms outperformed other algorithms using the datasets. This research would help to improve categorizing emails into different folders based on their content, intent, or relevance, improve user experience, and better manage email inboxes by automatically filtering, sorting, and prioritizing messages.

Keywords: Phishing · Spam · NLP · Classification

1 Introduction

Increase in the number of internet users has significantly made email communication the most extensive use for individuals and businesses, however this has led to the emergence of unsolicited emails and information leakage caused by spam and phishing emails [1]. Phishing emails are considered more dangerous because they target sensitive information from their users such as usernames, passwords, card numbers or pins, from unsuspecting employees or individuals. Meanwhile spam messages could lead to burden on email users due to their volume and frequency, high network bandwidth, large memory space and sometimes malware product attachments. The global average total cost of a data breach was $4.35 million in 2022, were stolen and compromised information are responsible for 19% of breaches and phishing was responsible for 16% [2]. Different research has been developed for the classification of spam emails systems using supervised machine learning techniques, the objective of this research is to solve class imbalance problems, develop a novel spam-ham-phishing dataset and use the principles of machine learning and natural language processing to improve the performance of the three class of dataset developed namely (spam-ham, ham-phishing, spam-phishing). To avoid bias due to the imbalanced dataset SMOTE (Synthetic over-sampling technique) was used to over-sample the minority, various data pre-processing method was performed on the dataset such as Tokenization using the word and sentence tokenizer for word count analysis, removal of stop words, stemming using the porter stemmer and lemmatization as the data cleaning process. Then different features extraction and selection was performed before training of the dataset using 5 different machine learning algorithms. The dataset used to train the model consists of 70% of the entire data, while 30% was used to test the model accuracy and performance. The result of the 5-machine learning (SVM, XGBoost, Random Forest, Multi-nominal and Gaussian Naïve bayes) shows how SMOTE improved the performance index of the classification techniques used. Consequently, the impact of solving the imbalance dataset using the SMOTE technique greatly improved the performance of the result, it shows that XGBoost performs better using the three-dataset developed. The spam-ham dataset had an accuracy of 0.99, precision of 1.00, recall of 0.98, and f1-score of 0.99, the spam-phishing likewise has an accuracy of 0.95, precision of 0.94, recall of 0.95, f1-score of 0.95 and ham-phishing dataset having an accuracy of 0.99, precision of 1.00 recall of 0.98 and f1-score of 0.94. Hence the result shows that XGBoost machine learning algorithms outperformed other algorithms using the datasets.

The remainder of this paper is organized as follows: next we conduct a review of relevant literature, followed by the methodology, results, and finally a conclusion.

2 Literature Review

This section presents a review of relevant research studies that helped form the foundation of this research study. These areas include machine learning, email spam detection, phishing impacts on networks, natural language processing and algorithms for filtering. We identify a gap which exists on a general comparison of NLP methods applied to both spam, ham, and phishing filtering.

Nandhini and Marseline KS [3] analyzed the effectiveness of various machine learning algorithms for email spam detection. During the investigation, it was observed that a significant volume of unsolicited emails posed potential threats to users and their organizations' security. The primary focus of their study was to leverage established algorithms to build a machine learning model capable of classifying emails as either spam or ham. The dataset utilized for this research was sourced from the Spam Base dataset available in the UCI Machine Learning repository. Several machine learning methods for data classification were employed, including Logistic Regression, Decision Trees, Naive Bayes,

KNN, and SVM. To assess the research findings comprehensively, performance metrics such as Classification Accuracy, Confusion Matrix, Precision, Recall, and F1 Score were utilized. Among the five machine learning models investigated, the Random Tree model exhibited the highest performance index when applied to the UCI Machine Learning dataset. The results underscored the Random Tree model's exceptional performance, achieving accuracy, precision, recall, and F1 Score scores of 99.9%, 99.9%, and 99%, respectively [3].

Kumar and Sonowal [4] studied machine learning methods for email spam detection. They clarified that emails are deemed spam when they are delivered to a large number of recipients in an unsolicited manner or with an advertising message attached, wasting space, time, and transmission speed. The study used machine learning to detect spam as an efficient method. For example, the study first preprocessed the data by eliminating any missing values from the data set before utilizing text analysis to evaluate email messages. Using several machine learning traditional classifiers, ensemble learning techniques, boosting and adaboost classifiers, data transformation, data reduction by deleting stop words, tokenizing each word, and producing a bag of words for feature selection are all steps in the process. Naive Bayes, Support Vector Machine, Decision Tree, K-Nearest Neighbor, Random Forest, Adaboost Classifier, and Bagging Classifier are some of the machine learning methods employed by the researcher. The algorithm with the highest accuracy, 98%, according to the results, was the Multinomial Naive Bayes classifier Kumar and Sonowal [4].

Junnarkar, et al. [5] employed machine learning and natural language processing to classify spam emails. They noted that the amount of spam communications has rapidly expanded with the rate of information transmission via email. In order to lower the rate of unsolicited bulk emails or spam, the research underlined the need to develop a thorough system for spam classification based on semantic text classification utilizing NLP. As part of the innovation of the work, the researcher also categorize the URL present utilizing a three-step filtering and analysis procedure. The dataset was first preprocessed using various text classification approaches, and once it had been trained using machine learning algorithms, URL filtering was carried out to safeguard users from emails containing harmful URLs. The method was applied to two distinct datasets, namely the Enron spam dataset and spam.csv from Kaggle. Utilizing feature engineering, text was preprocessed using naive bayes, KNN, decision trees, random forests, and support vector machines, among five other machine learning algorithms. The most effective machine learning algorithm, according to the results, was support vector machines, which had performance indices of 91% accuracy, 94% precision, 93% recall, and 91% F1 score [5].

Olatunji [6] examined the significance of spam detection as a first stage in the email filtering process in the wake of an increased influx of unsolicited messages, which now make up the bulk of inbox messages. The researched focused on the application of support vector machines and extreme learning machine models for email spam identification. Olatunji [6] found that there has been a lot of studies comparing ELM with SVM for classification and regression difficulties. Hence, this investigation focused on the usage of SVM and ELM for email spam detection and the requirement for an efficient detection method to identify and separate unsolicited mail. This study's findings demonstrated that the SVM provided a better performance index than the ELM, with an accuracy score of

94% compared to 93% for the latter. The ELM did well in terms of speed, with a training time of 0.94 s, while the SVM fared better in terms of accuracy, despite the fact that both models produced good results. In addition, the author also contrasted the speed of the operation of the two models [6].

Feng, et al. [7] explored support vector machine-based naive bayes algorithm for spam filtering. In their paper, they highlighted the use of naive bayes classifiers for spam emails and also pointed out shortcomings, such as having strong independence assumptions across features. The study used a naive bayes-SVM-NB filtering system as a spam detection method. A hyperplane was used to divide the training sample into two sections after the classification and training were first carried out using naive bayes classifiers. The training set was reduced and the independence of the samples in each category was improved by using the trimming approach offered by SVM to remove poor examples from the training space. When compared to a pure SVM-based solution, the results indicated that SVM-NB delivers improved precision and recall rates for detecting spam [7].

Verma, et al. [8] characterized phishing as a form of network theft in which attackers produce phony web pages or websites that seem quite legitimate with the intention of tricking unsuspecting users. Erma, et al. [8] continued by stating that emails are becoming a very popular medium where attackers transmit nefarious links or pop-up ads that recipients unintentionally open, leaving them entirely exposed. The goal of the study was to compare the various accuracy rates and provide a full description of how phishing is classified using machine learning and natural language processing techniques. For better results, various data preprocessing techniques were used, including the removal of stop words, punctuation, tokenization, and stemming. They used seven different machine learning algorithms for classification which produced the following results; K Nearest Neighbors with accuracy of 94.75, Decision Tree with accuracy of 97.55, Random Forest with accuracy of 98.42, Logistic Regression Classifier with accuracy of 98.56, SGD with accuracy of 98.34, Nave Bayes classifier with accuracy of 98.70, and SVM Linear classifier with accuracy of 98.77. The study concluded that the SVM linear was recommended as the best ML algorithm for the used dataset because it had the best accuracy [8].

Fette, et al. [9] used machine learning to study phishing email detection. They stressed that emails with links to websites that gather information are the most prevalent way for phishing attacks to start. The report noted that because hackers are becoming more skilled every day, it is crucial to recognize phishing assaults. The study's goal was to use machine learning to identify email phishing assaults. 860 phishing emails and 6950 non-phishing emails made up the dataset. Over 96% of phishing emails were successfully recognized by the model, with only 0.1% misclassified. The technique employed was a machine learning strategy known as PILFER, including decision trees, random forests, SVM rule-based strategies, and Bayesian as the classifier. The difference was not statistically significant, according to the researcher's analysis of total accuracy. The result shows that the PILFER approach gave an overall accuracy of 99.5% [9].

Bahgat, et al. [10] noted that monitoring and categorizing a large number of emails is a significant difficulty since there has been a constant increase in email users, which has led to unsolicited emails. They devised an effective email classification strategy based on semantic methodologies. They established a productive email filtering system based on semantic techniques. The approach taken in his research uses the wordnet ontology together with several semantic-based approaches and similarity measurements to reduce the complexity of textual properties over time and space. Various feature selection methods, including PCA and correlation feature selection, were employed. To test the accuracy, the researchers employed a variety of machine learning techniques, including logistic regression, naive bayes, support vector machines, random forests, and radial basis function networks. They also demonstrated that, when compared to other machine learning techniques, logistic regression provided the highest accuracy. The study showed that, with the addition of feature selection techniques, the outcome provided an average of 90% accuracy with shorter execution times [10].

Toolan and Carthy [11] described methods for phishing and spam feature selection. Unsolicited Bulk Email (UBE), as the researcher called spam emails, continuously evolves and gets past some junk mail filters, which is why his study is important. The researcher investigated the usefulness of 40 traits by calculating information gained over spam, phishing, and ham corpora. The results with the highest information gain created the best classifier after each dataset's evaluation was tested using its information gain. With the help of the C5.0 decision tree learning method, each dataset attribute was assessed. The findings demonstrated that the classifier developed using the best features outperformed those created using the best IG values, median IG values, and lastly the worst IG values [11].

Magdy, et al. [12] described a deep learning-based spam and phishing email filtering method. Magdy, et al. [8] stated that spam emails pose a serious threat to the internet, consume a lot of server storage space and cause delays in surfing unwanted bulk emails (UBE), which ultimately costs people and businesses money. A deep learning model that performed better than comparable experiments was presented in the research report. The research's classifier supported three different classes: phishing, spam, and ham. A predictive ANN model with two hidden layers and a tolerable training time was introduced. The outcome compared the accuracy of deep learning ANN to that of machine learning random forest and SVM. When compared to conventional machine learning models, the deep learning model's accuracy was 99.9% [12].

Zhang and Yuan [13] employed a neural network to identify phishing in emails. In light of the increasingly complex attacks, they emphasized the significance of efficient email filters for phishing emails. The goal of the research was to identify phishing emails using multilayer feedforward neural networks. Large real-world samples of 4202 spam emails and 4560 phishing emails make up the dataset used. The technique adopted included feature selection, phishing dataset processing, neural network system implementation, and performance comparisons between NN and machine learning techniques. The outcome compares the performance of the neural network with five machine learning models for the dataset in use. The researcher claimed that while the decision tree algorithm had the highest accuracy (96%), neural networks had the highest recall (95% + precision). NNs are therefore more effective at spotting phishing emails, as stated in the study [13].

Mujtaba, et al. [1] utilized machine learning to study email classification for forensic analysis, focused on the impact of an enormous increase in email data and the difficulties in email management. They emphasized the need for email detection and classification based on content and other data elements. The strategy suggested using multiple labels to classify emails in order to arrange them and aid in the forensic analysis of large amounts of email data. With the use of a unique blend of TF-IDF properties, the emails were divided into four classes. Five distinct machine learning algorithms—logistic regression, naive Bayes, stochastic gradient descent, random forest, and support vector machine—were used to assess the effectiveness of the technique. Comparing several machine learning methods used for classification, the logistic regression method produced the highest accuracy. Their outcome results demonstrated that logistic regression outperformed alternative techniques with an accuracy of 91.9% [1].

Bagui, et al. [14] examined the effectiveness of deep learning and machine learning for classifying phishing emails using one-hot encoding. They stressed the significance of creating anti-phishing technology to aid people and companies in avoiding losing enormous sums of money to criminals. To determine the characteristics of the text body, the researchers used a semantic analysis. Phishing emails made up 3,416 of the 18,366 tagged records in the sample, while regular emails made up 14,950. Data cleansing, lemmatization, and vectorization utilizing one-hot encoding were some data pretreatment techniques employed to get the dataset ready for machine learning and deep learning model. The Convolutional neural networks (CNN) and long memory DL techniques were applied with 30% of the dataset utilized to validate and test the model's performance, while 70% was used for training. (LSTM) [14]. The study results demonstrated the power of semantic analysis in classifying phishing emails as they determined that the CNN model with Word Embedding performed with 96.34% accuracy which was better than CNN with one-hot encoding (95.97%).

Chakravarty and Manikandan [15] proposed clever techniques for identifying email spam using machine learning techniques. They underlined the importance of an efficient anti-spam filtering system and described spam as one of the greatest threats to contemporary internet usage. The authors stated that one of the largest issues in a supervised learning system is anticipating the class labels in a personalized mailbox; as a result, a reliable detection system is required. The research methodology included an NLP component that can separate spam from non-spam emails and further categorize them. On the SpamAssassin spam corpus dataset, various well-known techniques, including Bayesian classification, KNN, ANN, SVM, counterfeit safe framework, and unpleasant sets, were applied. The naive bayes gave and the unpleasant sets performed adequately well according to the results of the six machine learning algorithms utilized, but the NB gave the best performance index when compared to the other methods. Naive Bayes provided accuracy, precision, and recall values of 99.48%, 99.68%, and 99.48% respectively. The algorithm's output was then put into practice, and the potency of the categorization method was established. An overall dataset of 6000 spam and non-spam emails is used by the author [15].

Given the most successful techniques for filtering spam and phishing are still based in heuristics, at least as far as commercial products, expanding to understanding the ability of NLP methods need to be investigated. While NLP has been applied, it has been heavily applied to SPAM and not phishing and few studies attempt to classify spam, ham, and phishing and no studies reviewed examined the efficacy of the machine learning techniques employed in this work.

3 Methodology

The research methodology comprises five core stages. Firstly, data acquisition involved obtaining phishing data from an educational institute and ham-spam data from Kaggle. In the second stage, dataset pre-processing encompassed tasks like handling empty rows and columns, converting email text to lowercase, removing punctuation to eliminate special characters, excluding non-meaningful or non-English words through WordNet, lemmatization using WordNet lemmatizer, and eliminating stop words to emphasize crucial terms. In the third stage, the pre-processed data underwent the feature selection and extraction stage. The fourth step involved employing five machine learning classifiers, namely Support Vector Machine (SVM), XGBoost, Naïve Bayes (Multinomial and Gaussian), and Random Forest, on the chosen attributes. The entire experiment was conducted using Google Colab through the Jupyter notebook IDE. The outcomes and evaluation of the machine learning algorithm are detailed in the subsequent sections.

3.1 Dataset

Getting the right dataset for effective spam-ham-phishing classification is of paramount importance. The dataset employed in this study originates from two distinct sources. Firstly, data was sourced from Tanusree Sharma's GitHub repository, encompassing educational institute phishing data (https://github.com/TanusreeSharma/phishingdataAnalysis/blob/master/1st%20data/PhishingEmailData.csv). Additionally, spam-ham email data was obtained from Kaggle (https://www.kaggle.com/datasets/shantanudhakadd/email-spam-detection-dataset-classification). To create the dataset, these sources were amalgamated, followed by cleaning and segmentation into three distinct categories: spam-ham, spam-phishing, and ham-phishing. Consequently, the dataset employed for this research comprises three primary classes: ham, spam, and phishing data. The dataset features two columns: "V1" denoting the email class (spam, ham, or phishing) and "V2" representing the email body in supervised learning. Further examination of the dataset reveals the following class combinations:

- Spam-ham: There were 5,572 instances, comprising 4,825 instances of ham and 747 instances of spam. This accounts for 13.4% of spam and 86.6% of ham.
- Spam-phishing: There were 936 instances, with 747 instances classified as spam and 189 instances classified as phishing. This represents 79.8% of spam instances and 20.2% of phishing instances.
- Ham-phishing: There were 5,014 instances, with 4,825 instances classified as ham and 189 instances classified as phishing. This accounts for 3.77% of the phishing dataset and 96.23% of the ham dataset.

3.2 Exploratory Data Analysis

Exploratory data analysis (EDA), which involves examining and understanding the dataset before modeling, is an essential part of machine learning and natural language processing (NLP). Two columns and 5,762 rows make up the dataset, where each row stands for a single data instance. EDA uses a variety of statistical and graphical approaches to extract insightful information from the dataset. Word frequency analysis, sentence length analysis, average word length analysis, and word cloud production are a few of these methods. These techniques give a deeper comprehension of the traits and patterns in the dataset. Another useful tool in EDA is the word cloud, a graphic representation of text data. With word size and color indicating their prominence, it visualizes the frequency and significance of terms in the dataset. EDA, in general, acts as a foundational step to get deeper understanding of the properties of the dataset, permitting deeper conclusions during later machine learning and NLP activities Fig. 1.

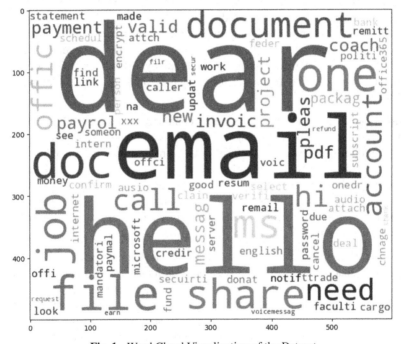

Fig. 1. Word Cloud Visualization of the Dataset

3.3 Data Preprocessing

Data extracted from the real world are usually inadequate, consisting of noise and missing values. Hence, the need for data pre-processing is imperative to transform them from a dirty or incomplete stage into a clean, usable, and organized form. Inconsistencies in a dataset can include typos, missing data, and data with different scales. The dataset was first prepared for use in the model. Examples of data pre-processing done on the email dataset include Text cleaning, Conversation to lowercase, Tokenization, Removal of stop words, stemming/ lemmatization, and feature extraction. Skipping this important stage in a machine learning model would affect the result because most models can't handle missing values, while some are affected by outliers, high dimensionality, and noisy data preprocessing makes the dataset completer and more accurate.

Tokenization. Tokenization plays a pivotal role in Natural Language Processing (NLP), being indispensable for effectively handling text data. It involves breaking down lengthy text into smaller units referred to as tokens, as explained by Pai [16]. These tokens can encompass words, characters, or sub-words (n-gram characters), depending on the specific tokenization method employed. Among these methods, Word Tokenization stands out as a widely utilized algorithm, where text is segmented into individual words. This approach is exemplified by pre-trained word embeddings such as Word2Vec and GloVe. Conversely, Character Tokenization dissects text into sets of individual characters, adept at handling Out of Vocabulary (OOV) terms while preserving word information. Sub-word Tokenization, on the other hand, disassembles text into sub-word components or N-gram characters, separated by spaces or punctuation characters (referred to as delimiters). Traditional NLP techniques like Count Vectorizer and TF-IDF leverage vocabulary as features, enhancing model performance. In this study, word tokenization was adopted to segment the text data into tokens, followed by the calculation of the total word count for subsequent analysis.

Stemming. Stemming is a natural language technique, that reduces and modifies words into their root forms, hence improving text preprocessing according to Nebojsa Bacanin [17] To build a robust model it is necessary to normalize texts by removing repetitive words and transform words into their base form through stemming. It can be applied to different forms of information retrieval, text mining, as well as email classification. There are different types of stemming such as Porter stemmer, Snowball stemmer, Lancaster stemmer, and Regex stemmer. Porter stemmer gives a resultant stem in a shorter word with the same root meaning, the porter stemmer was subsequently used in this research.

Stop Words. Stop words, in the context of Natural Language Processing (NLP), refer to English words that contribute minimally to the overall meaning of a sentence. Typically, these words are excluded from Natural Language data processing, as they often constitute the most common and less meaningful terms in a language from the perspective of machine learning models. Common examples of stop words include "the," "is," "at," "which," and "on". In contrast to tasks like language translation, stop words offer limited utility for our model, thus warranting their exclusion from the corpus. One notable advantage of removing stop words is the reduction in training time, with little impact on model accuracy. Moreover, this process enhances performance by retaining a more focused set of significant tokens, ultimately leading to improved classification accuracy. However, it's essential to exercise caution when removing stop words, as improper selection and removal can alter the text's intended meaning. Python offers several libraries, such as NLTK, SpaCy, and Gensim packages, which facilitate the removal of stop words while maintaining the text's integrity and improving the efficiency of NLP tasks.

Label encoding. Label encoding is a technique used to convert categorical variables into numerical variables suitable for the machine learning model. It converts all the columns in each table from a categorical column into a numerical column which can be fitted into the model. It is a vital preprocessing stage in a machine learning project. In this research, the email class was converted into a numerical column where each email category is denoted by 0 and 1. In this research, the categorical value was replaced with a numerical value between 0 and 1, where 0 stands for ham, and 1 stands for spam for the spam/ham dataset. 0 stands for ham, and 1 for phishing for the ham/phishing dataset. 0 for phishing and 1 stands for spam in the spam/phishing dataset.

Text vectorization TF-IDF. Highly frequent words tend to dominate the document hence leading to larger score, but they may not contain the needed information to the model, Term Frequency-Inverse Document Frequency is an approach used to rescale the frequency of words based on how often they appear in all documents. While Term Frequency refers to scoring of words in the current document, the weight of a term that occurs in a document is proportional to the term frequency and is illustrated below.

$$\text{Tf}(t,d) = \text{count of t in d /number of words in d.} \tag{1}$$

Document frequency tests the meaning of the text and it's very similar to TF, in the corpus. While term frequency is the frequency counter for a term t, df is the number of occurrences in the document set N. Df(t) - occurrence of t in documents. Inverse Frequency scores how rare and relevant the word is across the documents. The aim is to locate the appropriate records that fit the demand. It is computed using the TfidfVectorizer() method in Sklearn. The figure below shows an array of the frequency of the cleaned email dataset. Frequently occurring words often dominate a document, potentially inflating their importance, although they might not necessarily convey essential information to the model. To address this issue, Term Frequency-Inverse Document Frequency (TF-IDF) is a valuable approach for rescaling word frequencies based on their prevalence across all documents. Term Frequency (TF) quantifies word occurrences within a specific document, calculated as follows:

$$TF(t,d) = \frac{count\ of\ t\ in d}{number\ of\ words\ in\ d} \tag{2}$$

Document Frequency (DF) assesses the significance of a term in the corpus by counting its occurrences in the document set DF(t) = occurrence of t in documents. Inverse Frequency (IDF) gauges a term's rarity and relevance across documents, facilitating the identification of pertinent records. This computation is typically achieved using the TfidfVectorizer() method available in libraries like Scikit-Learn. The relationship between TF and IDF enables the generation of TF-IDF scores, rebalancing word importance in text data. These concepts are illustrated in Eq. 2 above, showcasing the frequency distribution within the cleaned email dataset.

Feature selection. In machine learning, feature selection is the act of selecting a subset of useful characteristics from a dataset while removing unimportant or redundant ones. This improvement increases computational effectiveness, lowers overfitting, and improves model performance. Filter methods (such as correlation analysis), wrapper methods (such as forward selection), and embedding methods (such as the feature importance of Random Forest) are the three basic techniques for feature selection. While wrapper techniques and embedded methods incorporate feature selection into the model training process, filter approaches assess feature relevance separately [18].

Feature Importance. Feature importance measures how much each input feature affects model predictions. Each aspect is given a score, with higher values indicating more effect. By exposing feature-target linkages and enhancing models through dimensionality reduction, this method helps users better understand their data and make model conclusions more understandable. By keeping the most informative properties and enhancing their functionality, it aids in model optimization [19].

Imbalanced data (SMOTE). Machine learning models frequently struggle with datasets exhibiting notable class imbalances. The Synthetic Minority Oversampling Technique, or SMOTE, is one efficient method for resolving this problem. By creating synthetic samples for the minority class inside the training dataset, SMOTE tackles the issue of class imbalance. SMOTE achieves this by duplicating existing examples from the minority class, allowing the creation of as many synthetic instances as needed [17]. These synthetic instances are strategically crafted to closely resemble the characteristics of the original minority class examples. The implementation of SMOTE was facilitated through the Python library "imbalanced-learn" integrated into the development environment. This technique is instrumental in rebalancing class distributions, enhancing model performance, and ensuring that the classification model effectively learns from both majority and minority class instances.

3.4 Machine Learning Algorithms

This research aims to create a robust email classifier capable of accurately categorizing incoming emails as either spam, ham (legitimate), or phishing. To assess the effectiveness of various machine learning algorithms for this email classification task, computational experiments were conducted. Three distinct algorithms were employed namely Multinomial Naive Bayes, Support Vector Machine, XGBoost. Performance evaluation was carried out using a confusion matrix, enabling the calculation of essential metrics such as accuracy, precision, recall, and F-measure. These metrics collectively gauge the efficiency and reliability of the developed email classification model and are shown in Table 1.

Table 1. Confusion Matrix showing the Positives and Negatives

Predicted class	Actual Class	
	YES	NO
YES	TP	FN
NO	FP	TN

3.5 Evaluation

In machine learning, model evaluation is a crucial step to determine how well a trained model performs in real-world scenarios. To gauge the effectiveness of a proposed method, a comprehensive set of eight metrics is employed. The F1 score, a balanced measure of precision and recall, provides insights into overall model performance. Accuracy quantifies the model's correctness, while precision assesses its ability to avoid false positives. Recall measures the model's sensitivity to detecting positive cases. Users can evaluate the model's capacity to differentiate between spam, ham, and phishing emails using this comprehensive evaluation approach. They are better able to understand the model's strengths and potential improvement areas by taking into account a variety of measures, which enables analysts to make adept choices about how to improve its performance.

True Positive rate (TP). It is a performance metric used to denote the percentage of spam messages that were classified by the machine learning model. it is illustrated as the total number of spam messages divided by the number of spam messages accurately classified.

$$TP = \frac{P}{S} \quad (3)$$

where S is the total number of spam messages, and P is the predicted spam messages.

True Negative Rate (TN). It is defined as the total number of non-spams divided by the number of non-spams predicted by the model. True Negatives denote the percentage of non-spam messages accurately predicted as non-spam by the machine learning model.

$$TN = \frac{Q}{N} \quad (4)$$

False Positive Rate (FP). False Positives occur in a model during classification where the machine learning algorithm misclassifies or wrongly categorizes non-spam messages as spam messages. E.g if non-spam messages are denoted N, and the misclassified non-spam messages as M. The illustration is highlighted below.

$$FN = \frac{M}{N} \quad (5)$$

False Negatives rate (FN). It indicates the proportion of spam messages that the machine learning algorithm misclassified as non-spam messages. False negatives misclassify the spam messages thereby wrongly classifying them as non-spam messages. The formula used to indicate the percentage of FN is illustrated in Eq. 6 below.

$$FN = \frac{T}{S} \quad (6)$$

Precision. It indicates the proportion of messages that the machine learning algorithm categorized as spam. It demonstrates the absolute correctness of the model. It is denoted by the formula below.

$$precision = \frac{TP}{TP + FP} \quad (7)$$

Recall. This shows the measure of completeness of the model. It denotes the percentage of messages that were spam and classified as spam.

$$Recall = \frac{TP}{TP + FN} \quad (8)$$

F1-score. It measures the harmonic means of precision and recall. The formula is illustrated in the equation below.

$$F1\ Score = 2 * \frac{Precision * Recall}{Precision + Recall} \quad (9)$$

Accuracy. Accuracy is a performance index used to measure classification algorithms. It is calculated as the ratio of correctly predicted samples to the total number of test samples. Accuracy is a good measure to determine how well the model classified the data. For an imbalanced dataset, they tend to Favor the dataset with the highest number of non-spam emails. An example is predicting a classification model where non-spam is 98 and spam is 2, the accuracy of the prediction would be 98% by predicting all samples as non-spam, while failing to effectively recognize spam.

4 Results

In the result below, text classification was initially performed using the Term Frequency Inverse Document Frequency approach. Predicting the outcome of the email classification technique, 5 different machine learning algorithms were implemented to perform a comparative analysis of its performance using different key performance metrics such as accuracy, precision, recall, and F1 score. A confusion matrix was also developed to highlight regions in which the model failed to correctly classify data during testing. The following is a summary of the machine learning algorithm utilized in this study: Support Vector Machine.

Gaussian Naive Bayes, Multinomial Naive Bayes, Random Forest, and XGBoost. The algorithm was implemented on the 3 different datasets namely spam-ham.csv, Spam-phishing.csv, and ham-phishing.csv. The result is divided into three sections. The first section shows the performance index for the non-smote for each of the machine learning algorithms used, the second section shows the performance of the model using the Smote technique and lastly, the third section shows the comparison between the results of spam-ham, spam-phishing, and ham-phishing using smote and non-smote technique for imbalanced dataset.

4.1 Section 1

Spam-Ham (Non-SMOTE) For SVC. In the result below, the performance index for the non-smote technique using a support vector machine shows the classification of spam to be 1 and ham denoted by 0. In the imbalance dataset, the model prediction tends to favor the spam class with precision, recall, F1 score showing value of 91%, 99%, 95% respectively, and an accuracy of 91%. For ham prediction, the model generally had a

Table 2. Classification Report for Spam-Ham (Non-Smote) SVC Algorithm

	Precision	Recall	F1-Score	Support
0	0.91	0.99	0.95	1345
1	0.85	0.35	0.50	206
Macro avg	0.88	0.67	0.72	1551

poor performance due to the majority class prediction of spam. However, the macro avg gives the average prediction of the model considering both factors as illustrated in Table 2 below.

Spam-Ham (Non-SMOTE) For XGBoost. The XGBoost classifier for spam-ham dataset using a non-smote technique, shows the macro avg performance index having precision of 87%, recall 61%, F1-score of 64% and accuracy of 89%, however, its prediction performed also favored the majority class consisting of spam set. Its prediction for ham gave a lesser value as compared to SVC and is shown in Table 3.

Table 3. Classification Report for Spam-Ham (Non-Smote) XGBoost

	Precision	Recall	F1-Score	Support
0	0.89	0.99	0.94	1345
1	0.85	0.22	0.35	206
Macro avg	0.87	0.61	0.64	1551

Spam-Ham (Non-SMOTE) For Multinomial NB. Using the Multinomial Naive Bayes classification algorithm, the model performed better in the spam class and under performed in the ham class category. In Table 4 below, the defaulter's class has a precision of 97%, recall of 19%, F1-score of 32%, and an overall accuracy of 89%. The macro avg shows the overall model performance to be 93% precision, 59% recall and 63% F1-score. Table 4 shows the classification report of Spam-Ham (Non-SMOTE) For Multinomial NB.

Table 4. Classification Report for Spam-Ham (Non-Smote) MNB.

	Precision	Recall	F1-Score	Support
0	0.89	1.00	0.94	1345
1	0.97	0.19	0.32	206
Macro avg	0.93	0.59	0.63	1551

Spam-Ham (Non-SMOTE) For Random Forest Classifier. The classification report of the random forest classifier is illustrated in Table 5. The table below shows the result of the classification for spam and ham class using non-smote. The model accuracy was 91% and the macro avg scores for precision, recall, F1-score was 88%, 67% and 72% respectively Table 5.

Table 5. Classification Report for Spam-Ham (Non-Smote) Random Forest Classifier

	Precision	Recall	F1-Score	Support
0	0.91	0.99	0.95	1345
1	0.85	0.35	0.50	206
Macro avg	0.88	0.67	0.72	1551

Spam-Ham (Non-SMOTE) For Gaussian NB. Table 6 shows the result for the Gaussian NB classification shown below denoted the macro avg for precision, recall and F1-score to be 57%, 63% 40% respectively.

Table 6. Classification Report for Spam-Ham (Non-SMOTE) GNB.

	Precision	Recall	F1-Score	Support
0	0.96	0.34	0.50	1345
1	0.17	0.91	0.29	206
Macro avg	0.57	0.63	0.40	1551

Spam-Phishing (Non-SMOTE) For SVC. In spam phishing classification report below, the macro avg has a precision value of 79%, recall 53%, F1-score 51% for the spam-phishing (non-smote) technique as shown in Table 7.

Table 7. Classification Report for Spam-Phishing (Non-SMOTE) SVC

	Precision	Recall	F1-Score	Support
0	0.75	0.06	0.12	47
1	0.82	1.00	0.90	202
Macro avg	0.79	0.53	0.51	249

Spam-Phishing (Non-SMOTE) For XGBoost. In the classification report below, the macro avg shows a lower performance index compared to the SVC for Non-smote technique. From the Table 8 below, the macro avg for precision, recall, and F1-score gave 66%, 51% and 47% respectively with an accuracy of 81%. The classification report further explains the model result of the spam and phishing class below.

Table 8. Classification Report for Spam-Phishing (Non-SMOTE) XGBoost

	Precision	Recall	F1-Score	Support
0	0.50	0.02	0.04	47
1	0.81	1.00	0.90	202
Macro avg	0.66	0.51	0.47	249

Spam-Phishing (Non-SMOTE) For Multinomial NB. The classification report for Spam-phishing (non-smote) for Multinomial NB is shown below. Table 9 compares the individual class and the macro avg.

Table 9. Classification Report for Spam-Phishing (Non-SMOTE) MNB

	Precision	Recall	F1-Score	Support
0	0.75	0.06	0.12	47
1	0.82	1.00	0.90	202
Macro avg	0.79	0.53	0.51	249

Spam-Phishing (Non-SMOTE) For Random Forest Classifier. The Random Forest classification for non-smote spam-phishing dataset shows the macro avg to have an improved classification result as shown in Table 10 below.

Table 10. Classification Report for Spam-Phishing (Non-SMOTE) Random Forest

	Precision	Recall	F1-Score	Support
0	0.91	0.21	0.34	47
1	0.84	1.00	0.91	202
Macro avg	0.88	0.60	0.63	249

Spam-Phishing (Non-SMOTE) For Gaussian NB. For Gaussian NB which is good at handling continuous values shows the performance index on a spam-phishing dataset. Table 11 shows the result of individual class performance during training as well as the macro avg with values of 63%, 70% and 53% for the precision, recall, and f1score respectively.

Ham-Phishing (Non-SMOTE) For SVC. The dataset used for ham-phishing classification has the majority class as ham, meanwhile the classification report shows how the accuracy of the prediction is using other performance indexes. The accuracy of model prediction using SVC denoted 96%, with a macro avg value for precision, recall, f1-score of 86%, 55% and 58% respectively (Table 12).

Table 11. Classification Report for Spam-Phishing (Non-SMOTE) GNB

	Precision	Recall	F1-Score	Support
0	0.29	0.96	0.44	47
1	0.98	0.45	0.62	202
Macro avg	0.63	0.70	0.53	249

Table 12. Classification Report for Ham-Phishing (Non-SMOTE) SVC

	Precision	Recall	F1-Score	Support
0	0.75	0.11	0.18	57
1	0.96	1.00	0.98	1351
Macro avg	0.86	0.55	0.58	1408

Ham-Phishing (Non-SMOTE) For Gaussian NB. Gaussian NB for Ham-phishing non-smote technique showed a reduced macro avg of 53% for precision, 68% for recall and 35% for recall. From Table 13 below the percentage of correctly predicting the phishing class is relatively low as compared to another ML model. The accuracy of prediction was also very low with 42% accuracy.

Table 13. Classification Report for Ham-Phishing (Non-SMOTE) GNB

	Precision	Recall	F1-Score	Support
0	0.06	0.96	0.12	57
1	1.00	0.40	0.57	1351
Macro avg	0.53	0.68	0.35	1408

Ham-Phishing (Non-SMOTE) For Random Forest Classifier. In the table below, the classification report shows the performance index using a random forest classifier. Table 14 shows the percentage of precision, recall f1-score using the ham-phishing data, the result shows the macro avg for each performance index to be 90%, 63% and 69% respectively with an accuracy of 97%.

Ham-Phishing (Non-SMOTE) For XGBoost. In the research experiment below, the XGBoost classifier shows a macro avg of 48% for precision, 50% recall and 49% f1score. The reason for this low performance shows the phishing classes was not giving a null value for the index highlighted below (Table 15).

Table 14. Classification Report for Ham-Phishing (Non-SMOTE) Random Forest

	Precision	Recall	F1-Score	Support
0	0.83	0.26	0.40	57
1	0.97	1.00	0.98	1351
Macro avg	0.90	0.63	0.69	1408

Table 15. Classification Report for Ham-Phishing (Non-SMOTE) XGBoost.

	Precision	Recall	F1-Score	Support
0	0.00	0.00	0.00	57
1	0.96	1.00	0.98	1351
Macro avg	0.48	0.50	0.49	1408

4.2 Section 2

Spam-Ham (SMOTE) For SVC.

Table 16 shows the results of the classification report for SVC for both the spam-ham emails. Using the macro average the precision, recall and F1_score all had a value of 96%.

Table 16. Classification Report for Spam-Ham (SMOTE) SVC

	Precision	Recall	F1-Score	Support
0	0.98	0.94	0.96	899
1	0.94	0.98	0.96	908
Macro avg	0.96	0.96	0.96	1807

Spam-Ham (SMOTE) For XGBoost. XGBoost was used for classifying spam and ham dataset using the smote technique, the result is illustrated in the Table 17.

Table 17. Classification Report for Spam-Ham (SMOTE) XGBoost.

	Precision	Recall	F1-Score	Support
0	0.98	1.00	0.99	899
1	1.00	0.98	0.99	908
Macro avg	0.99	0.99	0.99	1807

Spam-Ham (SMOTE) For Multinomial NB. Using the Multinomial Naïve Bayes approach, Table 18 illustrates the different classification report for both the Spam and ham class using the smote technique.

Table 18. Classification Report for Spam-Ham (SMOTE) MNB

	Precision	Recall	F1-Score	Support
0	0.82	0.98	0.89	899
1	0.98	0.78	0.87	908
Macro avg	0.90	0.88	0.88	1807

Spam-Ham (SMOTE) For Random Forest Classifier. Random Forest Classifier was also considered for this type of email class, and the result shown in Table 19 that the macro average has a precision, recall and f1-score of 96%.

Table 19. Classification Report for Spam-Ham (SMOTE) Random Forest.

	Precision	Recall	F1-Score	Support
0	0.97	0.95	0.96	899
1	0.95	0.97	0.96	908
Macro avg	0.96	0.96	0.96	1807

Spam-Ham (SMOTE) For Gaussian NB. GNB has a macro average classification of 95% from Table 20. The classification report shows the model has 823 True Positives and 894 True negatives classifying more customers as defaulters compared with other category.

Table 20. Classification Report for Spam-Ham (SMOTE) GNB

	Precision	Recall	F1-Score	Support
0	0.98	0.92	0.95	899
1	0.92	0.98	0.95	908
Macro avg	0.95	0.95	0.95	1807

Spam-phishing (SMOTE) for SVC. Considering the spam-phishing dataset using the smote technique, the SVC classifiers was used to determine the classification of spam-phishing dataset, in the Table 21 below, the classification report shows the precision is higher than recall and f1-score.

Table 21. Classification Report for Spam-Phishing (SMOTE) SVC

	Precision	Recall	F1-Score	Support
0	0.90	0.99	0.94	129
1	0.99	0.89	0.94	132
Macro avg	0.95	0.94	0.94	261

Spam-phishing (SMOTE) for XGBoost. In spam-phishing dataset, XGBoost shows a significantly lower classification metrics compared to the spam-ham dataset with a macro average of 95% (Table 22).

Table 22. Classification Report for Spam-Phishing (SMOTE) XGBoost.

	Precision	Recall	F1-Score	Support
0	0.94	0.95	0.95	129
1	0.95	0.94	0.95	132
Macro avg	0.95	0.95	0.95	261

Spam-phishing (SMOTE) for Multinomial NB. For spam-phishing dataset using MNB the model shows an improved classification report as compared against spam-ham data using the same classifier. Table 23 illustrates the classification report.

Table 23. Classification Report for Spam-Phishing (SMOTE) MNB

	Precision	Recall	F1-Score	Support
0	0.92	0.97	0.94	129
1	0.97	0.92	0.94	132
Macro avg	0.94	0.94	0.94	261

Spam-phishing (SMOTE) for Random Forest Classifier. Using the Random Forest Classifier for Spam-phishing dataset the macro average shows a precision of 95% and a recall and f1-score of 94% each (Table 24).

Table 24. Classification Report for Spam-Phishing (SMOTE) Random Forest.

	Precision	Recall	F1-Score	Support
0	0.91	0.98	0.94	129
1	0.98	0.90	0.94	132
Macro avg	0.95	0.94	0.94	261

Spam-phishing (SMOTE) for Gaussian NB. Gaussian was effective in classifying spam-phishing dataset using the smote technique, the result shows the macro average for precision, recall and f1-score to be 96% each (Table 25).

Table 25. Classification Report for Spam-Phishing (SMOTE) GNB

	Precision	Recall	F1-Score	Support
0	0.98	0.95	0.96	129
1	0.95	0.98	0.96	132
Macro avg	0.96	0.96	0.96	261

Ham-phishing (SMOTE) for SVC. Considering a Ham-phishing dataset using the smote technique, the SVC classifier shows a very effective model for classifying the dataset. With a precision, recall and f1-score value of 97%. The result shows that SVC is very effective for this dataset (Table 26).

Table 26. Classification Report for Ham-Phishing (SMOTE) for SVC

	Precision	Recall	F1-Score	Support
0	0.95	0.99	0.97	897
1	0.99	0.95	0.97	910
Macro avg	0.97	0.97	0.97	1807

Ham-phishing (SMOTE) for XGBoost. In this model category the result best algorithm for classifying ham-phishing dataset using the smote technique. Table 27 shows the precision, recall and f1-score all had a macro average of 99%.

Table 27. Classification Report for Ham-phishing (SMOTE) XGBoost

	Precision	Recall	F1-Score	Support
0	1.00	0.98	0.99	897
1	0.98	1.00	0.99	910
Macro avg	0.99	0.99	0.99	1807

Ham-phishing (SMOTE) for Multinomial NB. MNB algorithm for ham-phishing dataset in the Table 28 below shows the precision, recall and f1 score has a macro average of 98%. This also shows that the model was effective in classifying the result.

Ham-phishing (SMOTE) for Random Forest Classifier. With ham having a greater percentage of this dataset the Random Forest Classifier was effective at accurately classifying the dataset, the Table 29 shows the micro average of this classifier as 98%.

Table 28. Classification Report for Ham-Phishing (SMOTE) MNB

	Precision	Recall	F1-Score	Support
0	0.97	0.98	0.98	897
1	0.98	0.97	0.98	910
Macro avg	0.98	0.98	0.98	1807

Table 29. Classification Report for Ham-Phishing (SMOTE) Random Forest.

	Precision	Recall	F1-Score	Support
0	0.96	1.00	0.98	897
1	1.00	0.95	0.97	910
Macro avg	0.98	0.98	0.98	1807

Ham-phishing (SMOTE) for Gaussian NB. GNM for Ham-phishing dataset using SMOTE, shows the precision, recall and f1 value as 98% (Table 30).

Table 30. Classification Report for Ham-Phishing (SMOTE) GNB

	Precision	Recall	F1-Score	Support
0	0.98	0.99	0.98	897
1	0.99	0.98	0.98	910
Macro avg	0.98	0.98	0.98	1807

4.3 Section 3

Comparison of classification results across different dataset classes using machine learning algorithms with and without SMOTE (Synthetic Minority Over-sampling Technique) (Table 31).

The evaluation of machine learning algorithms using both SMOTE and non-SMOTE preprocessed data involved categorizing the data into three groups (Table 32) (spam-ham, spam-phishing, ham-phishing), and the results revealed that the SMOTE technique consistently demonstrated enhanced performance across all dataset categories, improving accuracy, precision, recall, and F1-score, making it a recommended approach for effectively addressing imbalanced datasets, particularly in the context of email classification (Table 33).

Spam-Ham

Table 31. Comparison between Smote and Non-Smote Performance Index for Spam-Ham.

Metric	SVC		Gaussian		XGBoost		Multinomial NB		Random Forest	
	S	NO/S	S	NO/S	S	NO/S	S	NO/S	S	NO/S
Accuracy	0.96	0.91	0.95	0.42	0.99	0.89	0.88	0.89	0.96	0.91
Precision	0.94	0.85	0.92	0.17	1.00	0.85	0.98	0.97	0.95	0.85
Recall	0.98	0.35	0.98	0.91	0.98	0.22	0.78	0.19	0.97	0.35
F1-Score	0.96	0.50	0.95	0.29	0.99	0.35	0.87	0.32	0.96	0.50

Spam-Phishing

Table 32. Comparison between Smote and Non-Smote Performance Index for Spam-Phishing.

Metric	SVC		Gaussian		XGBoost		Multinomial NB		Random Forest	
	S	NO/S	S	NO/S	S	NO/S	S	NO/S	S	NO/S
Accuracy	0.94	0.82	0.96	0.55	0.95	0.81	0.94	0.82	0.94	0.85
Precision	0.90	0.75	0.98	0.29	0.94	0.50	0.92	0.75	0.91	0.91
Recall	0.99	0.06	0.95	0.96	0.95	0.02	0.97	0.06	0.98	0.21
F1-Score	0.94	0.12	0.96	0.44	0.95	0.04	0.94	0.12	0.94	0.34

Ham-Phishing

Table 33. Comparison between Smote and Non-Smote Performance Index for Ham-phishing.

Metric	SVC		Gaussian		XGBoost		Multinomial NB		Random Forest	
	S	NO/S	S	NO/S	S	NO/S	S	NO/S	S	NO/S
Accuracy	0..97	0.96	0.98	0.42	0.99	0.96	0.98	0.96	0.98	0.97
Precision	0.95	0.75	0.98	0.06	1.00	0.00	0.97	0.00	0.96	0.83
Recall	0.99	0.11	0.99	0.96	0.98	0.00	0.98	0.00	1.00	0.26
F1-Score	0.97	0.18	0.98	0.12	0.99	0.00	0.98	0.00	0.98	0.40

5 Conclusion

Spam and phishing emails have become increasingly pervasive and problematic. Organizations and end-users are increasingly bombarded by spam. Phishing is even more problematic as a large percentage of network and cyber-security breaches come via phishing attacks as the point of entry. As it stands, heuristic methods are most commonly relying on blacklists, rules, and crowdsourcing to identify threats. As technology improves, we can apply more sophisticated techniques to the detection of these threats. With advances in technology and machine learning, specifically natural language processing, new methods can be explored and demonstrate effective amelioration of the impacts of spam and phishing. Using a NLP approach, this work seeks to compare the efficacy of various machine learning models. Specifically, NLP-based supervised learning approaches were explored, and their efficacy was reported. Results show all models were effective; however, we found XGBoost to be superior. This work can be employed by practitioners to develop more accurate filtering mechanisms. Future work will extend towards Large Language Models (LLMS) to compare LLM based approaches with traditional NLP approaches for email classification.

References

1. Mujtaba, G., Shuib, L., Raj, R.G., Majeed, N., Al-Garadi, M.A.: Email classification re-search trends: review and open issues. IEEE Access **5**, 9044–9064 (2017)
2. Bacanin, N., et al.: Application of natural language processing and machine learning boosted with swarm intelligence for spam email filtering. Mathematics **10**(22), 4173 (2022)
3. Nandhini, S., Marseline, KS.J.: Performance evaluation of machine learning algorithms for email spam detection. In: 2020 International Conference on Emerging Trends in Infor-mation Technology and Engineering (ic-ETITE), IEEE, pp. 1–4 (2020)
4. Kumar, N., Sonowal, S.: "Email spam detection using machine learning algorithms." In: 2020 Second International Conference on Inventive Research in Computing Applications (ICIRCA), IEEE, pp. 108–113 (2020)
5. Junnarkar, A., Adhikari, S., Fagania, J., Chimurkar, P., Karia, D.: "E-mail spam classifica-tion via machine learning and natural language processing." In: 2021 Third International Conference on Intelligent Communication Technologies and Virtual Mobile Networks (ICICV), IEEE, pp. 693–699 (2021)
6. S. O. Olatunji, "Extreme Learning machines and Support Vector Machines models for email spam detection," in 2017 IEEE 30th Canadian Conference on Electrical and Computer Engineering (CCECE), 2017: IEEE, pp. 1–6
7. Feng, W., Sun, J., Zhang, L., Cao, C., Yang, Q.: "A support vector machine based naive Bayes algorithm for spam filtering." In: 2016 IEEE 35th International Performance Compu-ting and Communications Conference (IPCCC), IEEE, pp. 1–8 (2016)
8. Verma, P., Goyal, A., Gigras, Y.: Email phishing: Text classification using natural lan-guage processing. Comput. Sci. Inf. Technol. **1**(1), 1–12 (2020)
9. Fette, I., Sadeh, N., Tomasic, A.: "Learning to detect phishing emails." In: Proceedings of the 16th international conference on World Wide Web, pp. 649–656 (2007)
10. Bahgat, E.M., Rady, S., Gad, W., Moawad, I.F.: Efficient email classification approach based on semantic methods. Ain Shams Eng. J. **9**(4), 3259–3269 (2018)
11. Toolan, F., Carthy, J.: "Feature selection for spam and phishing detection." In: 2010 eC-rime Researchers Summit, IEEE, pp. 1–12 (2010)

12. Magdy, S., Abouelseoud, Y., Mikhail, M.: Efficient spam and phishing emails filtering based on deep learning. Comput. Netw. **206**, 108826 (2022)
13. Zhang, N., Yuan, Y.: "Phishing detection using neural network." In: CS229 lecture notes, pp. 301 (2012)
14. Bagui, S., Nandi, D., Bagui, S., White, R.J.: Machine learning and deep learning for phishing email classification using one-hot encoding. J. Comput. Sci. **17**, 610–623 (2021)
15. Chakravarty, A., Manikandan, V.: "An Intelligent Model of Email Spam Classification." In: 2022 Fourth International Conference on Emerging Research in Electronics, Computer Science and Technology (ICERECT), IEEE, pp. 1–6 (2022)
16. Pai, A.: "What is tokenization in NLP? Here's All You Need To Know." In: Retrieved from Analytics Vidhya: https://www.analyticsvidhya.com/blog/2020~... (2020)
17. Bacanin, N., et al.: "Application of natural language processing and machine learning boosted with swarm intelligence for spam email filtering." In: MDPI vol. 2023 (2023)
18. Guyon, I., Elisseeff, A.: "An introduction to variable and feature selection." J. Mach. Learn. Res. **3**, pp. 1157–1182 (2003)
19. Hastie, T., Tibshirani, R., Friedman, J H.: The elements of statistical learning: data mining, inference, and prediction. Springer (2009)

The Fake News Detection Model Explanation and Infrastructure Aspects

Maksym Lupei[1](✉) and Myroslav Shliakhta[2]

[1] Pennsylvania State University, State College, University Park, PA 16801, USA
`maxim.lupey@gmail.com`
[2] Ivan Franko National University, Lviv 79000, Ukraine

Abstract. The spread of false information can significantly harm public opinion, underscoring the importance of accurately identifying untrustworthy news. This paper presents an innovative machine learning (ML) tool, TMining, designed to evaluate news credibility and facilitate various text-mining tasks. By examining a range of ML methodologies alongside preprocessing techniques, we aim to boost the system's effectiveness. Our research meticulously assesses different datasets, highlights the impact of applying stemming techniques, and employs Local Interpretable Model-Agnostic Explanations (LIMEs) to shed light on the rationale behind model predictions. The outcomes reveal a notable enhancement in both the precision and clarity of the news verification process. The ultimate version of the model has been made available as an Application Program Interface (API), and its source code has been shared openly, encouraging further exploration and collaboration within the scientific community. This initiative advances our ability to discern manipulative context from fictitious content and promotes transparency and understanding in the domain of ML applications.

Keywords: Text Mining · Fake News · Model Explanation · Machine Learning

1 Introduction

Today, traditional newspapers, television broadcasting, and even social media feeds on top of instant messaging platforms [2] have transformed how we learn about current affairs, giving people immediate access to world happenings and issues using gadgets. Worldwide, interconnectedness increases the risk of the fast dissemination of misinformation [1]. Fake news [3], a widely categorized phenomenon, has significant social, political, and economic impact, demanding that it be detected promptly [4].

Simply put, fake news can appear when news organizations do not adequately verify information or even artificially create it. In contrast to satire and parody news sites, fake news sites generate hoaxes, propaganda, and disinformation that are subsequently disseminated through social media to increase website views and impact. They can be motivated to influence voters' decisions to support a particular candidate, manipulate public perception to benefit them in one way or another or make money through online advertisement schemes. Such information often spreads very quickly, thus leading to incorrect conjectures among people, which also affects their actions and sometimes baffles state governments.

The conventional approach for tackling fake news is fact-checking [5]. When necessary, journalists and committed organizations check the accuracy of the information provided and make corrections or retractions. However, manual verification is not feasible due to the high volume of new online content added daily. Therefore, automatic mechanisms are needed to screen, assess, and filter out likely hoaxes.

Using supervised ML techniques, we cannot accurately ascertain the authenticity of the data. However, it is possible to detect manipulative contexts with these techniques. Verifying the authenticity of the data requires utilizing fact-checking tools and credible sources, as well as contextual analysis, which will be addressed in further research.

Constructing an ML model to identify news reliability is extremely difficult. This implies that the language can be pretty nuanced; false information may sometimes be not so explicit in its presentation, and fake news producers' techniques continue to evolve. There are instances where the black-box model, despite its accuracy, might be unable to instill the necessary trust for broad acceptance. This is where our research positions itself; we are building an accurate ML model and one that explains its decisions using visualizations.

Such a model and application, called TMining (text mining), was developed, as outlined in this paper, and it includes data retrieval, preprocessing, model training, testing, and final adjustments/optimizations. By combining the hands of different ML algorithms and local interpretable model-agnostic explainers (LIME) [6], the presented model is armed to fight against fraudulent news dissemination in social networks. The tool was developed and is publicly available for empirical testing and further scientific research.

This paper is a continuation of the author's previous research in text mining within the context of news media; it has predominantly utilized support vector machines (SVMs). For instance, "The identification of authorship of Ukrainian language texts of journalistic style using neural networks" [17]. This study, which specifically targeted the media outlet "Politeka", employed ML techniques for text mining. Another notable study, "Analyzing Ukrainian Media Texts by Means of SVM: Aspects of Language and Copyright" [18], expanded the scope of related research encompassing three distinct newspapers: "Ukrainska Pravda" and "Censor.NET", and "Zakarpattya Online".

2 Related Works

The growing threat of fake news has generated increased amounts of attention among researchers, resulting in a wealth of research focusing on the detection, origin, dissemination, and impact [7] of false data. This section surveys the modern literature, which outlines our understanding of the approaches to fake news detection through ML techniques [9].

Initially, efforts in fake news detection research gravitated toward understanding the production of fake news. Therefore, Castillo et al. (2011) provided insights into the reliability of tweets based on propaganda in important news [8], which shapes our understanding of social media platforms as information dissemination tools. Vlachos & Riedel (2014) provided a benchmarked fact-checking dataset, a tool for training patterns for early detection [5].

Early ML techniques for detecting fake news relied heavily on manual input. Zhou and Zafrani (2020) considered various factors that commonly reveal fake news [11] and identified linguists, images, and promoters' social network structures as contributing factors to fake news detection.

The advent of deep learning has completely changed how fake news is detected. Tang and other researchers (2015) [12], Zhang et al. (2019) [13], and Chen et al. (2019) [14] developed a model that used recurrent neural networks (RNNs), convolutional neural networks (CNNs) and transformer architectures to capture local and global context patterns in a text for pseudo information.

Acknowledging that false information is not created in a vacuum, Shu et al. (2019) emphasized the role of social context in its detection, a model that analyzes news content and user behavior on social media platforms [15].

The increasing complexity of the models has necessitated the investigation of model decision strategies. For example, Ribeiro et al. (2016) proposed the LIME method for a local surrogate model to predict any black-box classifier [6].

Monti and others admit the dynamic strategies employed by the media. (2019) investigated the effectiveness of adversarial training via graph neural networks (GNNs) in enhancing the detection performance for fraudulent cases [16].

As the above overview highlights, the literature on misinformation detection is rich and diverse.

3 Methodology

A total of three datasets were utilized: the LIAR dataset [10] (id 1), which is followed by a smaller set of parsed article titles from Politifact (id 2), and a bigger set of parsed article bodies from Politifact (id 3). Some Politifact subsets were sourced from social networks such as Twitter and Facebook, which presented more concise news. This section describes how these three datasets are employed in building a data reliability-checking ML model.

It is impossible to detect whether the information is true or false by supervised ML methods without a source of truth. For the first dataset (LIAR; id 1), the three classes "true", "half-true", and "mostly true" were rated among the "mostly reliable" sources, while the classes "false", "pants-fire", and "barely true" were among the "mostly unreliable". The second dataset (id 2) followed a similar pattern, simplifying the six classes into two categories: mostly reliable at one and mostly unreliable at zero. For the third dataset (id 3), we used a mostly reliable value of 1 and a mostly unreliable value of 0, simplifying the nine classes into two categories.

Every dataset includes all the normalized textual transformed data. This implies transforming all texts into small letters, excluding punctuation, unique signs, and digits, and reducing unnecessary gaps between words as much as possible Fig. 1.

The figure depicts the first step of the TMining application. It outlines four key steps: loading data from a CSV file, tokenization to break text into words or phrases, data cleaning to remove noise such as stopwords and unnecessary characters, and stemming and lemmatization to simplify words to their base or root forms. This process is essential for preparing raw text for ML models.

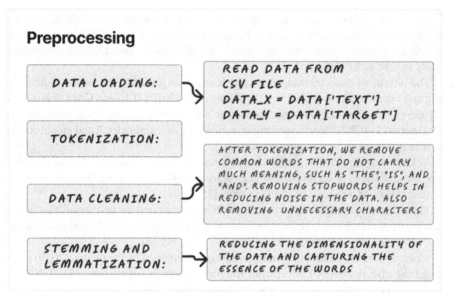

Fig. 1. The first step of the TMining application.

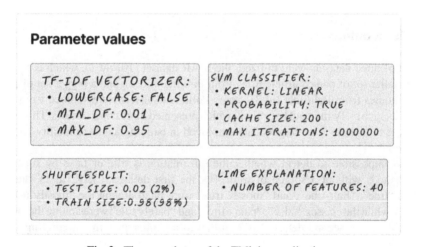

Fig. 2. The second step of the TMining application.

The figure depicts the second step of the TMining application. It shows a collection of parameter settings for various components of the ML pipeline. The TF-IDF Vectorizer is set to not convert to lowercase and to consider terms that appear in at least 0.1% but not more than 95% of the documents. For model evaluation, ShuffleSplit is configured with a test size of 2% and a training size of 98%. The SVM Classifier uses a linear kernel, and probabilities are enabled. It has a cache size of 200 MB and a maximum of 1,000,000 iterations. Lastly, for model interpretability, LIME is set to explain the results using 40 features.

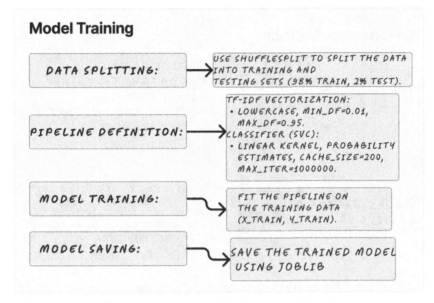

Fig. 3. The third step of the TMining application.

The figure depicts the third step of the TMining application. It starts with data splitting using ShuffleSplit for a 98% training set and a 2% testing set. The pipeline includes TF-IDF vectorization with specific parameters for document frequency and an SVM classifier with a linear kernel and probability estimates. After defining the pipeline, the model is trained with the specified data, and the final step is saving the trained model using the "joblib" library.

Figure 4 depicts the fourth step of the TMining application. It details the performance metrics for an ML model. It includes model evaluation, where predictions are made on the test data and metrics such as accuracy and F1 score are calculated. It also highlights the use of LIME for model interpretability, where explanations are generated for a subset of test instances and saved as HTML files for visualization purposes.

The NLTK was used to break down the articles into words (tokenization). To facilitate uniformity and reduce dimensionality, stemming was performed using LancasterStemmer, thereby changing words to their stems.

Further work was performed with the TfidfVectorizer modifier, a key component of the scikit-learn library. The idea behind the TfidfVectorizer is named TF-IDF. This TF controls the number of times a word appears in a document. When two statistical measures assess the importance of a word for a document in a corpus.

Documents differ in length, so the word frequency is naturally greater for longer documents. However, that does not mean that the position matters much. The IDF solves this problem by measuring the importance of the term. Common words in each document (such as 'is', 'an', and 'the') contain more specific information, so the IDF of these words is close to zero.

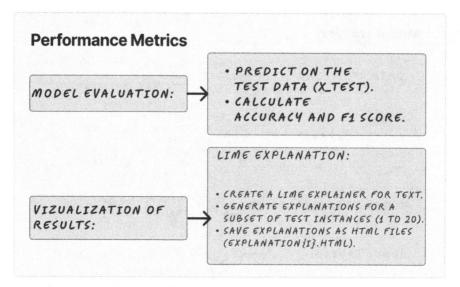

Fig. 4. The fourth step of the TMining application.

The TfidfVectorizer converts the text into feature vectors that can be used as input into the estimator. It converts a stack of documents into a TF-IDF object matrix. Thus, it provides a statistical measure of the importance of words, reducing the impact of tokens that appear frequently in a corpus and are not as informative as those that occur in a smaller proportion of the training corpus.

The models used were based on traditional ML algorithms:

- Support Vector Classifier (SVC): The SVC is a type of SVM used for classification problems. This method classifies data points in multidimensional space uniquely and divides the data into classes by finding the best hyperplane. The algorithm takes the input data and uses a line (2D space) or hyperplane (3D space and above) that separates the data into the classes with the greatest margin output.
- Support Vector Regression (SVR): Unlike SVC, SVR is used for regression problems. Instead of trying to fit the maximum street size possible between two classes while limiting margin violations, SVR tries to fit as many cases as possible in the street. The street width hyperparameter is called the epsilon and governs. SVR is a linear regression in a high (infinite) dimensional space.
- Logistic Regression: A logistic regression is a statistical model that uses a logistic function to model a binary dependent variable. In other words, the probability of an event is determined by fitting the data to the logistic curve. The results can be thought of as the probability of the given input point belonging to a class (mostly reliable or mostly unreliable).

K-fold cross-validation: K-fold cross-validation involves splitting a dataset into k subsets. The model is trained and tested k times through multiple iterations, using a different subset as the validation set each time. The performance metrics obtained from each iteration are averaged to gauge the model's overall predictive performance. We divided the datasets at a proportion of 80/20, accounting for 80% of the training data and 20% of the testing data. This methodology enhances the models' reliability.

We used the LIME approach to explain our model's predictions. This approach involves perturbing the input data it receives, noting its assumptions, and finally, training a simpler model to approximate what the complex model should do. This approach helped us understand which of the attributes was significant.

Model performance was measured using performance metrics such as accuracy, F1 score, and the area under the receiver operating characteristic curve (AUC-ROC).

We aim for a comprehensive methodology that covers many stages, starting with the collection of data and continuing until we explain how the model works. The basic dataset overview for the computational experiments is presented in Table 1.

Table 1. Dataset overview

ID	Source	Sample	Features	Target
1	LIAR dataset	10270	Article Body	Mostly Reliable - 1
2	Politifact parser	4472	Article Title	Mostly Unreliable - 0
3	Politifact parser	23640	Article Body	

The dataset was divided into two subsets: a training set and a test set. The main steps for preparation are as follows:

- reading and preparing the data;
- initialization and ML;
- data analysis.

4 Results of the Experiments

Following the completion of the empirical analysis, we employed three separate prototypes to evaluate each data cluster: the first dataset (id 1), the second dataset (id 2), and the third dataset (id 3). The quantified outputs of the TMining application are visualized in the subsequent tabular representation.

Table 2 shows the results for the three diverse datasets. Changes were made in the TMining application data processing applied for data processing, which significantly improved the outputs. The numerical data and accuracy of the results demonstrated a noticeable increase. In terms of confirming the effectiveness, the 5-fold cross-validation method is the most reliable. For each dataset, three different models: SVC, SVR, and Logistic Regression, were used, each yielding diverse performance metrics. In particular, the third dataset yielded satisfactory results with the SVC model after 5-fold cross-validation, with an average accuracy of 0.793 and an average error rate of $f1 = 0.883$.

The image appears to display a LIME (Local Interpretable Model-agnostic Explanations) visualization, which is a technique used to explain the predictions of a ML model. The visualization indicates that the model predicts the instance as "Mostly unreliable" with a high probability of 0.92. The words contributing to this classification, such as

Table 2. Results of the experiments.

ID	K-FOLD	Model	Values	Accuracy
1	5	**Logistic Regression**	**F1 = 0.67**	**0.593**
1	5	**SVC**	**F1 = 0.696**	**0.595**
1	5	**SVR***	**F1 = 0.691**	**0.595**
2	5	Logistic Regression	F1 = 0.845	0.158
2	5	SVC	F1 = 0.032	0.838
2	5	SVR*	F1 = 0.045	0.838
3	5	**Logistic Regression**	**F1 = 0.822**	**0.78**
3	5	**SVC**	**F1 = 0.883**	**0.793**
3	5	**SVR***	**F1 = 0.772**	**0.772**

* SVR prediction was rounded to an integer.

"false", "claim", and "Facebook", have weights shown, suggesting their influence on the prediction. "Mostly reliable" features like "factcheck" and "author" have lower weights and are less influential in this prediction.

Figure 5 presents varying coefficients assigned to a keyword bearing on the publication classification. The term "false" is allocated a coefficient of 0.14, whereas the keyword "evidence" is assigned a coefficient of 0.07, indicating a primary association with unreliable sources. Interestingly, for the term "Facebook", a coefficient of 0.06 is assigned to "mostly unreliable".

The image is a LIME (Local Interpretable Model-agnostic Explanations) visualization showing the prediction probabilities of a text being classified as "Mostly unreliable" or "Mostly reliable". The model predicts with a 70% probability that the text is reliable. The bar chart lists words from the text with their corresponding weights, indicating how each word influences the model's prediction of being reliable or unreliable. Words like "true", "vote", and "florida" are some of the features contributing to the text's reliability.

Figure 6 shows the explanation of another model decision. A "Mostly reliable" prediction with 0.7 probability is presented. Specifically, the term "mostly" has a coefficient of 0.06, paralleling the coefficient of the word "miss", which equals 0.06 as well.

As demonstrated in Fig. 2 and Fig. 3, we can classify the outcomes into two exhibited categories: 'positive', which corresponds to a condition of "mostly reliable", and 'negative', which identifies as "mostly unreliable". The weighting system significantly impacts the likelihood of predictions. Specifically, Fig. 2 suggests that, based on the calculations of word quantification within the given document, the system predicts that the given text is "mostly unreliable" with a probability equal to 0.92. In a similar context, Fig. 3 illustrates that the model predicts the given text as "mostly reliable" with a probability equal to 0.7.

This image is a bar chart from a LIME explanation showing the local feature importance for a text classified as "mostly reliable". The green bars indicate features that

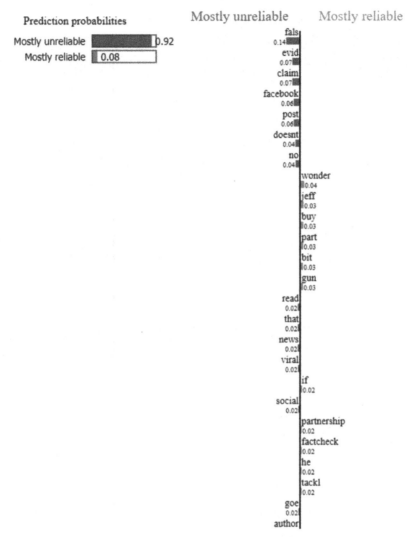

Fig. 5. Identification of the text fragment belonging to the "Mostly unreliable" group on the SVC model for the third dataset.

contribute positively to this class, suggesting elements the model associates with reliability. The red bars indicate features that contribute negatively, suggesting aspects that the model associates with unreliability. Key terms such as "research", "administration", and "world" positively influence the model's decision, whereas words like "development", "foundation", and "gross" seem to push the decision toward unreliability.

Figure 7 shows that specifically, the term "development" equals the 0.08 coefficient, and the word "rank" equals the 0.05 coefficient.

In this LIME visualization, the model predicts a class of "mostly unreliable". The chart illustrates which words increase or decrease that prediction likelihood. Red bars

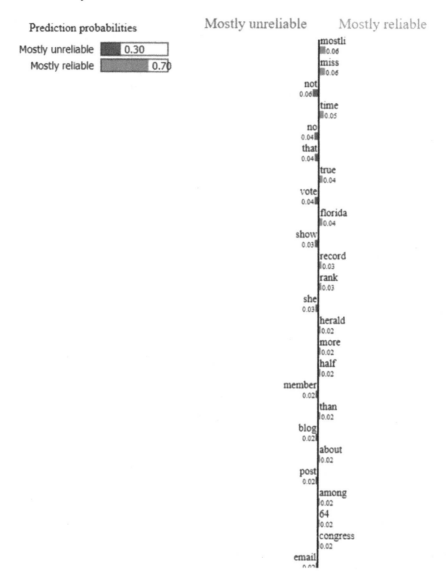

Fig. 6. The text fragment belonging to the "mostly reliable" group on the SVC model for the third dataset is identified.

represent words that push the prediction towards unreliability, while green bars indicate words that contribute to the text being perceived as more reliable. Notable terms like "claim", "article", and "trump" have strong red bars, heavily influencing the model towards an "unreliable" prediction, whereas terms like "senate", "republican", and "vote" have smaller green bars, indicating a lesser influence towards reliability.

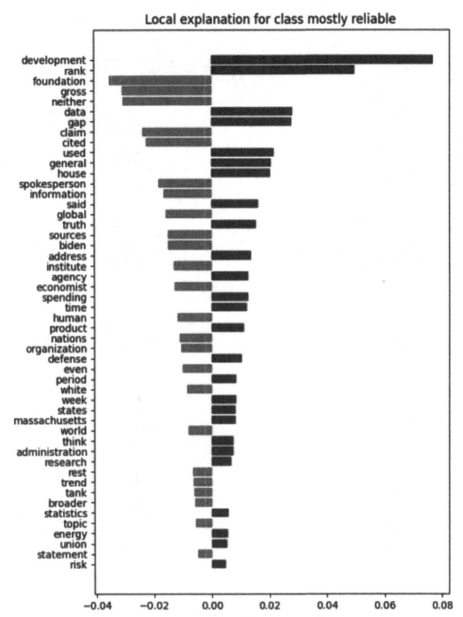

Fig. 7. Local explanation of a class belonging to the "mostly reliable" class in the SVC model. (Color figure online)

Figure 8 explains the "mostly unreliable" interpretation of another LIME explanation. Specifically, the term "claim" equals the -0.125 coefficient, and the word "blue" equals the -0.115 coefficient.

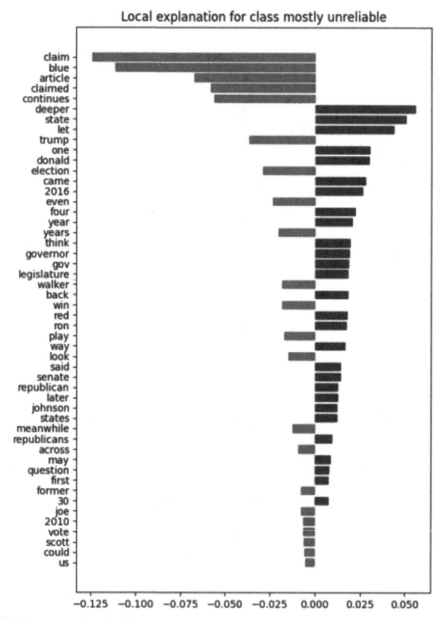

Fig. 8. Local explanation of a class belonging to the "Mostly unreliable" class in the SVC model.

Figure 7 and Fig. 8 demonstrate the results from the third dataset, and the SVC model was used to analyze the text fact-checking.

The illustrations in Fig. 9 are divided into three distinct charts: the learning curve, the model's scalability, and the model's performance:

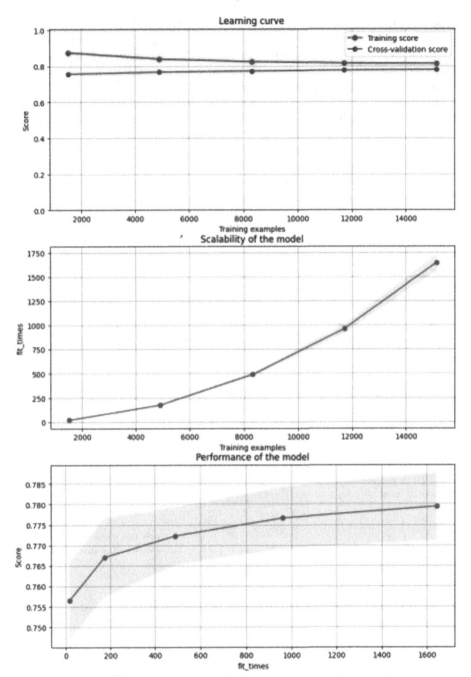

Fig. 9. Chart of SVC model performance with the third dataset. (Color figure online)

- Learning Curve: Shows the number of training examples on the X-axis and the corresponding model score, such as the accuracy and F1 score, on the Y-axis. The red curve represents the score of the training cohort, while the green curve represents the cross-validation score. It provides information about how well the model is learning based on the number of training examples.
- Model scalability: The chart explicates the model's scalability by showing the number of training sets on the X-axis and the time taken to fit the model on the training data on the Y-axis. It visualizes how the fit time changes with fluctuations in the size of the training set, providing a clear representation of the model's scalability.
- Model Performance: This graph shows the relationship between the number of fits (X-axis) and the model performance (Y-axis), allowing us to explore how the model's performance changes with different numbers of fits.

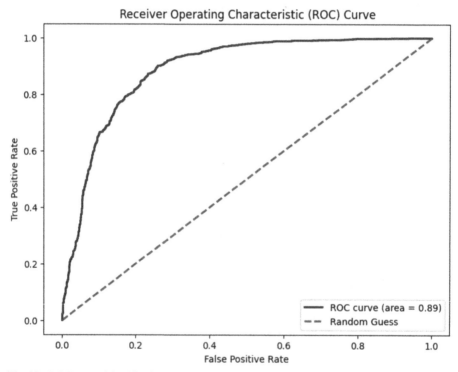

Fig. 10. ROC curve identification for news reliability detection in the SVR Model on the third dataset.

The ROC curve depicted in Fig. 10 is a visual representation of the performance of the news reliability detection model discussed in the provided document. The AUC was 0.89, indicating high accuracy in classifying news articles. This high AUC value indicates that the model has strong discriminative power, far surpassing the baseline performance of 0.5 that would be expected by random chance. As elaborated in this paper, such a model is crucial for identifying misinformation in the media space.

5 Application Infrastructure

The TMining application is a powerful ML tool optimized for processing and analyzing extensive text data. It has a command-line interface that relies on input to execute instructions directly.

The command line commands include.

- -h, --help: Used to provide program usage information and describe available options.
- train -dataset_path./data/factcheck.csv [-x text] [-y target] [-save_to./result] [-model SVC] [-vectorizer TfidfVectorizer] [-kfold 10] [-test_size 0.2]: Primarily used to train ML models on the specified CSV-formatted data.

  dataset_path: path to the dataset.
  x: name of the column containing the input text. Default: "text"
  y: name of the column containing the output labels. Default: "target"
  save_to: the path of saving the trained model file. Default: The path where the program starts. Default model name: "model.mdl".
  model: select a training model. Three models are available: SVC, SVR, and LogisticRegression. Default model: SVC.
  vectorizer: select the text vectorization. Three approaches are available. CountVectorizer, TfidfVectorizer, and HashingVectorizer. Default vectorizer: TfidfVectorizer.
  kfold: number of folds to use for cross-validation. Default 1.
  test_size: size of the test set. Default 0.

- validate -model_path./model.mdl -dataset_path./data/factcheck.csv [-x text] [-y target] [-test_size 0.2]: Validate the model using the provided dataset.

  model_path: path to the trained model.
  dataset_path: path to the dataset.
  x: name of the column containing the input text. Default: "text"
  y: column name containing the output labels. Default: "target"
  test_size: size of the test set. Default: 0.2.

- predict -model_path./model.mdl -text "fake news text": This may be used for prediction via a previously trained model or to extract information from a given text.

  model_path: path to the trained model.
  text: text for prediction.

- visualize -model_path./model.mdl -text "fake news text" [-features 60] [-save_to./result]: Generates an HTML visualization of model predictions for a given text input using LIME (local interpretable model-agnostic explanations).

  model_path: path to the trained model.
  text: next to predict.

features: the maximum number of tokens displayed in the table. Default: 40.
save_to: save the rendered results in HTML. Default: "./results/1.html".

- host -model_path./model.mdl [-address 0.0.0.0] [-port 5000]: Can be used to host the trained model as a service on the specified port, enabling other systems or applications to take advantage of the model for prediction tasks

model_path: path to the trained model.
address: IP address for the API host. Default: 0.0.0.0.
port: port for the API host. Default: 5000.

In the infrastructure sector, this system plays an important role in the extensive text mining process. Its processes include model training to classify text information, providing this capability as a service to other modules in connected systems.

Hosting an application programming interface (API) based on a trained model. There are two endpoints in the API:

- /model/predict?text = here is text (GET method): Gets the text for the prediction and returns the predicted result in JSON format.
- /model/visualize?text = here is text (GET method): Gets the text for the prediction and returns the image with prediction and model explanation.

It clarifies the application functionality needed to address model training, hosting, and API usage for text prediction and visualization.

6 Conclusion

The purpose of this study was to highlight the development of a dynamic ML application that effortlessly detects manipulative context to measure text reliability. Three datasets were used to facilitate this exploration; these texts were segregated into two types of subsets: training and testing. Furthermore, these textual segments have been converted into vectors for easy input into the ML model.

The open-source TMining (Text Mining) application proposed in this study amalgamates diverse methodologies, supplementing the interpretability of models. The application relies on the Natural Language Toolkit (NLTK) for data processing. We ensure the results are clear and reliable when using ML algorithms, such as SVM. The findings of this study confirm the effectiveness of the approach adopted, and the k-fold cross-validation method provides evidence of the model's reliability.

The results showed that the first dataset led to reliable accuracy for all the models used to execute logistic regression, SVC, and SVR. The models' accuracy based on the initial LIAR dataset demonstrated suboptimal results on other papers. As described before, it is not optimal to check whether the statement is true with supervised learning. Moreover, the dataset was rearranged, and the result was much better (~0.6 accuracy) using the TMining application. The models' accuracy based on the second dataset did not show great performance as there was not enough training data. The models' accuracy based on the the third dataset displayed excellent results for all three models and showed the remarkable accuracy of the SVC model in fact-checking (~0.8 accuracy).

Utilizing only supervised ML techniques, without a data source of truth, we cannot accurately ascertain the authenticity of the data, we can only check the manipulative context. The next step of this research will be building a fact-checking tool using LLMs (Large Language Models) to check the source of truth and its related statements.

To recapitulate what was said in this paper, the TMining application represents an optimal solution within its field, exemplifying the pinnacle of current methodologies and techniques advancing the scientific community and can be used for different purposes, from data reliability analysis to authorship and unique text style recognition. The fact that the TMining application is publicly available for the scientific community enriches it even further for broader development and scientific research.

Here is the link to the open-source repository of the TMining (Text Mining) application: https://github.com/MaxLupey/TMining.

References

1. Aïmeur, E., Amri, S., Brassard, G.: Fake news, disinformation and misinformation in social media: a review. Soc. Netw. Anal. Min. **13**(1), 30 (2023)
2. Apejoye, A.: Comparative Study of Social Media, Television and Newspapers' News Credibility (2015)
3. Yuan, L., Jiang, H., Shen, H., Shi, L., Cheng, N.: Sustainable development of in-formation dissemination: a review of current fake news detection research and practice, 458 (2023)
4. Lazer, D.M.J., et al.: The science of fake news. Science **359**, 1094–1096 (2018). https://www.science.org/doi/10.1126/science.aao2998
5. Vlachos, A., Riedel, S.: Fact Checking: Task Definition and Dataset Construction. ACL Workshop (2014)
6. Ribeiro, M. T., Singh, S., Guestrin, C.: "Why Should I Trust You?" Explaining the Predictions of Any Classifier. KDD '16 (2016)
7. Conroy, N.K., Rubin, V.L., Chen, Y.: Automatic deception detection: methods for finding fake news. Proc. Assoc. Inf. Sci. Technol. **52**, 1–4 (2015). https://doi.org/10.1002/pra2.2015.145052010082
8. Castillo, C., Mendoza, M., Poblete, B.: Information credibility on twitter. In: Proceedings of the 20th International Conference on World Wide Web, pp 675–684 (2011)
9. Khanam, Z., Alwasel, B.N., Sirafi, H., Rashid, M.: Fake News Detection Using Machine Learning Approaches
10. Wang, W.Y.: "Liar, Liar Pants on Fire": A New Benchmark Dataset for Fake News Detection. ACL '17 (2017). https://paperswithcode.com/dataset/liar
11. Zhou, X., Zafarani, R.: A Survey of Fake News: Fundamental Theories, Detection Methods, and Opportunities (2020)
12. Tang, D., Qin, B., Liu, T.: Document Modeling with Gated Recurrent Neural Net-work for Sentiment Classification (2015)
13. Zhang, et al.: Simple RNN of Hidden Layer (2019)
14. Chen, Y., Cheng, Q., Cheng, Y., Yang, H.: Applications of Recurrent Neural Net-works in Environmental Factor Forecasting (2019)
15. Shu, K., Wang, S., Liu, H. Beyond News Contents: The Role of Social Context for Fake News Detection (2019)
16. Monti, F., Frasca, F., Eynard, D., Mannion, D., Bronstein, M.M.: Fake News Detec-tion on Social Media Using Geometric Deep Learning (2019)

17. Lupei, M., Mitsa, O., Sharkan, V., Vargha, S., Gorbachuk, V. The identification of mass media by text based on the analysis of vocabulary peculiarities using support vector machines. In: 2022 International Conference on Smart Information Systems and Technologies (SIST) (2022). (https://doi.org/10.1109/sist54437.2022.9945774)
18. Lupei, M., Mitsa, O., Sharkan, V., Vargha, S., Lupei N.: Analyzing Ukrainian media texts by means of support vector machines: aspects of language and copyright. In: International Conference on Computer Science, Engineering and Education Applications, pp. 173–182 (2023)

Advances of Artificial Intelligence/Operational Research Tools in Healthcare

An Advanced Approach to COVID-19 Detection Using Deep Learning and X-ray Imaging

Hela Limam[1(✉)] and Wided Oueslati[2]

[1] Higher Institute of Computer Science, University of Tunis El Manar, Tunisia and BestMod Laboratory, Higher Institute of Management of Tunis, Tunis, Tunisia
hela.limam@isi.utm.tn

[2] Higher School of Business of Tunis, University of La Manouba, Tunisia and BestMod Laboratory, Higher Institute of Management of Tunis, Tunis, Tunisia
wided.oueslati@esct.uma.tn

Abstract. This research paper presents a comprehensive approach for detecting COVID-19 utilizing deep learning algorithms and X-ray imaging. The study proposes a novel methodology based on Convolution Neural Networks (CNN) to accurately classify X-ray images into two distinct classes: "Normal" or "Infected (COVID)". A dataset of X-ray images is employed, and essential preprocessing techniques, data augmentation, transfer learning, and fine-tuning are implemented to construct an effective and efficient model for precise COVID-19 detection. The performance of the developed model is rigorously evaluated using a range of performance metrics. This study underscores the potential and efficacy of employing advanced deep learning methodologies, particularly CNNs, in the automated identification of COVID-19 cases through X-ray images, demonstrating promising results in the realm of medical diagnostics. The proposed approach holds promise for assisting healthcare professionals in swiftly and accurately identifying individuals afflicted with COVID-19, thereby contributing to timely and appropriate medical interventions. The integration of state-of-the-art techniques and the meticulous evaluation of the model's performance positions this research as a valuable contribution to the ongoing efforts in leveraging artificial intelligence for tackling the global COVID-19 pandemic.

Keywords: COVID-19 · Deep Learning · Convolutional Neural Networks · X-ray Imaging · Data Augmentation · Transfer Learning · Fine-Tuning

1 Introduction

The urgency of accurate and timely COVID-19 detection has spurred innovative approaches that leverage cutting-edge technologies. As the global impact of the pandemic continues to unfold, the need for reliable diagnostic methods has become increasingly evident. According to recent statistics from the World Health Organization (WHO), COVID-19 has affected millions of individuals worldwide, leading to significant morbidity and mortality rates. The ability to swiftly and accurately diagnose COVID-19 cases is paramount in containing the spread of the virus, mitigating its impact on healthcare systems, and saving lives. Conventional diagnostic methods, while effective, often face challenges in providing real-time results, especially in the context of a rapidly evolving pandemic. Our research responds to this critical need by proposing a novel methodology that addresses the shortcomings of existing approaches and holds the promise of revolutionizing COVID-19 detection.

The landscape of COVID-19 detection is marked by the imperative for rapid and precise identification. As the pandemic continues to evolve, accurate and timely detection of COVID-19 cases remains essential for implementing effective public health interventions and mitigating transmission risks. However, conventional diagnostic methods such as PCR testing and antigen assays may be time-consuming and resource-intensive, hindering efforts to contain the spread of the virus. Our research aims to overcome these challenges by harnessing the potential of deep learning algorithms, specifically Convolutional Neural Networks (CNN), in conjunction with X-ray imaging.

Proposed methodology has the potential to directly impact COVID-19 detection, public health interventions, and patient outcomes in several ways. Firstly, by leveraging deep learning algorithms and X-ray imaging, our approach enables rapid and accurate identification of COVID-19 cases. This could lead to faster diagnosis and isolation of infected individuals, thereby reducing the risk of transmission within communities. Additionally, the use of deep learning algorithms allows for the detection of subtle patterns in X-ray images that may not be discernible to the human eye, potentially improving diagnostic accuracy and reducing false-negative results.

Furthermore, our methodology has the potential to enhance public health interventions by providing healthcare professionals with a powerful tool for monitoring and managing COVID-19 outbreaks. By accurately identifying and tracking COVID-19 cases, public health authorities can implement targeted control measures, such as quarantine and contact tracing, to limit the spread of the virus. Additionally, the rapid and accurate diagnosis afforded by our approach could facilitate more timely administration of medical interventions, leading to improved patient outcomes and reduced disease severity.

In summary, our proposed approach has the potential to significantly improve COVID-19 detection, public health interventions, and patient outcomes by enabling rapid and accurate identification of COVID-19 cases, enhancing the effectiveness of public health interventions, and facilitating timely medical interventions. By harnessing the power of deep learning algorithms and X-ray imaging, our research strives to make a tangible impact on the ongoing global battle against the COVID-19 pandemic.

This work will first provide an overview of the relevant literature in the field. Following that, we will delve into our research methodology, detailing the study design, data collection, and analysis procedures. The subsequent section will present our findings and data analysis, followed by a comprehensive discussion of the results. Finally, we will conclude with a summary of our key findings and potential avenues for future research.

2 Related Works

In recent years, significant progress has been made in utilizing deep learning techniques, particularly Convolutional Neural Networks (CNNs), for automated COVID-19 detection and diagnosis using medical imaging data. Notably, researchers have developed specialized deep CNN models like COVID-Net, as introduced by Wang et al. [1], tailored for the efficient detection of COVID-19 cases from chest X-ray images. This model demonstrated high accuracy in distinguishing COVID-19 cases from non-COVID-19 pneumonia cases. Additionally, Apostolopoulos and Mpesiana [2] proposed a CNN-based approach that achieved notable classification accuracy in automated COVID-19 detection using X-ray images. These studies underscore the effectiveness of CNNs in enabling rapid screening and diagnosis of COVID-19. Moreover, data augmentation strategies, such as rotation and zooming, have emerged as crucial tools to enhance model generalization, particularly when dealing with limited datasets. Loey et al. [3] investigated various data augmentation techniques applied to chest X-ray images for COVID-19 detection, showcasing improved model robustness and performance. Transfer learning, often coupled with fine-tuning, has proven successful in leveraging pre-trained models for enhanced COVID-19 detection. Aslam et al. [4] effectively utilized this approach by fine-tuning a pre-trained CNN architecture using a COVID-19 dataset, achieving high accuracy in detection. Another advancement involves the integration of multiple imaging modalities, such as X-ray and CT scans, to improve detection accuracy. Jin et al. [5] proposed a hybrid model that fused features from both X-ray and CT images, significantly enhancing overall diagnostic performance. Lastly, addressing the crucial aspect of interpretability in AI models, Ahuja et al. [6] introduced an examinable AI framework for COVID-19 detection, aiding clinicians in understanding model decisions and thereby fostering trust and confidence in AI-assisted diagnoses. These research endeavors collectively represent the evolving landscape of deep learning approaches in COVID-19 detection and diagnosis, encompassing aspects such as specialized model application, data augmentation, transfer learning, integration of imaging modalities, and the importance of explainability in healthcare applications.

Existing approaches in automated COVID-19 detection and diagnosis have made significant strides leveraging deep learning techniques, particularly Convolutional Neural Networks (CNNs). Specialized models like COVID-Net, proposed by Wang et al. [1], have demonstrated high accuracy in distinguishing COVID-19 cases from non-COVID-19 pneumonia cases using chest X-ray images. Similarly, Apostolopoulos and Mpesiana [2] introduced a CNN-based approach achieving notable classification accuracy in automated COVID-19 detection from X-ray images. These studies underscore the effectiveness of CNNs in enabling rapid screening and diagnosis. Data augmentation strategies, such as rotation and zooming, have emerged as crucial tools to enhance model generalization, as demonstrated by Loey et al. [3] in their investigation of various augmentation techniques applied to chest X-ray images for COVID-19 detection. Transfer learning, as explored by Aslam et al. [4], has proven successful in leveraging pre-trained models for enhanced COVID-19 detection. Furthermore, the integration of multiple imaging modalities, as proposed by Jin et al. [5], has shown promise in improving diagnostic performance. However, despite these advancements, limitations persist. Model generalization, data augmentation constraints, transfer learning challenges, effective integration of multiple modalities, and interpretability remain areas requiring further exploration and improvement. These limitations underscore the necessity for continued research to address the evolving challenges in AI-assisted COVID-19 diagnosis and detection.

3 Methodology

In the following, we present in depth the methodology used in our work resulting in eight steps.

1. **Literature Survey:** Our methodology commences with a systematic literature survey. This foundational step involves a comprehensive examination of the current landscape of COVID-19 detection methodologies. By reviewing existing research papers, datasets, and techniques, we gain valuable insights into the challenges and advancements in this domain. This survey not only provides us with a comprehensive understanding of the state-of-the-art but also helps us identify and understand the limitations inherent in existing methodologies.
2. **Conceptualization of Methodology:** Armed with the insights from the literature survey, we pivot towards the conceptualization of our unique methodology. We place particular emphasis on the deployment of Convolutional Neural Networks (CNNs) due to their exceptional capabilities in image classification tasks. The goal is to develop a robust model capable of classifying X-ray images into two distinctive categories: "Normal" and "Infected (COVID)." This conceptualization phase involves designing a methodology that integrates various essential components to ensure the effectiveness and reliability of the model.
3. **CNN Architecture Selection:** The cornerstone of our methodology is the selection of an appropriate CNN architecture. After careful consideration, we opt for the VGG16 model, renowned for its effectiveness in image classification tasks. The VGG16 architecture comprises multiple convolutional layers followed by fully connected layers, allowing for the extraction of intricate features from X-ray images. By leveraging the pre-trained weights of the VGG16 model, we capitalize on learned representations that generalize well to medical imaging tasks, thereby enhancing the model's performance in COVID-19 detection.
4. **Data Collection:** A crucial aspect of our methodology is the acquisition of a high-quality dataset. We meticulously collect X-ray images from reputable sources such as Dr. Cohen's GitHub repository and Kaggle. These images include both COVID-19-positive and normal cases, ensuring a balanced representation of classes. By sourcing diverse and representative samples, we minimize biases in the dataset and facilitate comprehensive model training and evaluation.
5. **Data Preprocessing:** Before feeding the data into the CNN model, we preprocess the X-ray images to enhance their quality and suitability for training. This preprocessing step involves tasks such as resizing, color conversion, and normalization. By standardizing the images and ensuring uniformity in pixel values, we improve the model's ability to learn discriminative features and generalize to unseen data.
6. **Model Training:** With the preprocessed data, we proceed to train the CNN model using the VGG16 architecture. This training phase involves feeding the X-ray images into the model and adjusting its parameters to minimize the loss function. Through iterative optimization techniques such as stochastic gradient descent, we fine-tune the model's weights and biases to improve its performance in COVID-19 detection.
7. **Evaluation:** Once the model is trained, we evaluate its performance using various metrics such as accuracy, sensitivity, specificity, precision, and recall. These metrics provide insights into the model's ability to correctly classify COVID-19-positive and

normal X-ray images. By analyzing multiple evaluation metrics, we gain a comprehensive understanding of the model's strengths and weaknesses, enabling us to refine and optimize the methodology iteratively.
8. **Validation and Testing:** Finally, we validate the model's performance using a separate validation dataset and conduct rigorous testing to assess its robustness and generalization capability. This validation and testing phase ensures that the model performs reliably in real-world scenarios and inspires confidence in its applicability for COVID-19 detection in clinical settings.

In summary, our methodology encompasses a comprehensive and systematic approach to COVID-19 detection using X-ray images. By leveraging state-of-the-art techniques in deep learning, dataset curation, and model evaluation, we aim to develop a robust and reliable model that can assist healthcare professionals in the timely and accurate diagnosis of COVID-19 cases.

4 The Proposed Approach

Within this section, we embark on an in-depth exploration of the methodological framework that forms the bedrock of our innovative approach to COVID-19 detection through the application of deep learning algorithms. Armed with this comprehensive understanding, we then pivot toward the conceptualization of a unique methodology, placing particular emphasis on the deployment of Convolutional Neural Networks (CNNs). Renowned for their exceptional capabilities in image classification tasks, CNNs become the linchpin of our approach, specifically tailored for classifying X-ray images into two distinctive categories: "Normal" and "Infected (COVID)". The comprehensive methodology integrates a myriad of essential components, encompassing data collection, preprocessing, data augmentation, transfer learning, and the subsequent evaluation of the CNN algorithm. Each of these components assumes a critical role in the holistic development of a robust model for COVID-19 detection. To provide a visual representation of the intricate architectural design and the sequential orchestration of these components, Fig. 1 is presented, offering a graphical elucidation of the methodological intricacies that underpin the construction of our proposed model. This figure serves as a visual guide, unraveling the layers of our methodology and showcasing the meticulous steps involved in the creation of our innovative approach to COVID-19 detection through deep learning.

The proposed approach stands out from existing works in several key aspects. Central to our approach is the deployment of CNNs, renowned for their exceptional capabilities in image classification tasks. We specifically tailor CNNs for classifying X-ray images into two distinctive categories: "Normal" and "Infected (COVID)." This tailored approach ensures optimal performance in discerning COVID-19 cases from non-COVID-19 pneumonia cases. It integrates essential components, including data collection, preprocessing, data augmentation, transfer learning, and the subsequent evaluation of the CNN algorithm. Each component plays a critical role in the holistic development of a robust model for COVID-19 detection. Data augmentation techniques enhance model generalization, while transfer learning leverages pre-trained models for improved performance on COVID-19 datasets.

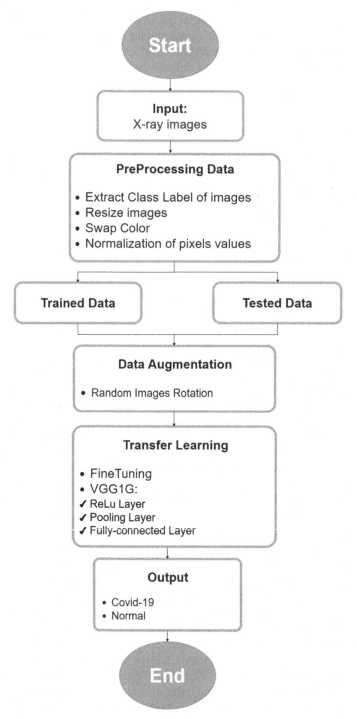

Fig. 1. Proposed Approach for COVID-19 Detection

4.1 Dataset of X-ray Images of Patients

Data collection serves as the foundational phase for constructing machine learning models, offering essential insights to address inquiries and facilitate informed decision-making. In our study focused on COVID-19 detection through X-ray images, the pivotal first step involved sourcing a high-quality dataset. Our quest led us to explore a COVID-19 X-ray image dataset, emphasizing the need for reliable and accessible data. To ensure trustworthiness, we turned to GitHub, a renowned open-source data repository, where we accessed real cases data. The dataset, showcases positive (infected) X-ray images on the left and negative samples on the right. These images served as the training set for a deep learning model implemented with TensorFlow and Keras, enabling automated predictions regarding a patient's COVID-19 status.

The dataset [7] comprises X-ray images of COVID-19-infected patients. Dr. Cohen diligently collected radiological images of COVID-19 cases, making them publicly available on GitHub. To construct the COVID-19 X-ray image dataset, we initially unzipped the metadata file.csv from Dr. Cohen's repository. Subsequently, we selected rows corresponding to positive COVID-19 cases with a posterior-anterior view (PA) of the lungs, the perspective used for "healthy" cases. This meticulous process resulted in a curated set of 25 X-ray images depicting positive COVID-19 cases.

For X-ray images of healthy patients, we leveraged a Kaggle dataset encompassing chest X-rays (pneumonia), seamlessly integrating it with our collection. Consequently, our finalized dataset comprised a balanced set of 50 images—25 depicting X-rays of COVID-19-positive cases and 25 showcasing X-rays of healthy patients. This meticulous curation of data forms the cornerstone of our system's ability to discern and predict COVID-19 status through the utilization of deep learning techniques.

4.2 Initialization of Hyperparameters

There are two types of hyperparameters [8]:

1. **Model Hyperparameters:** These are primarily related to the architecture and size of the neural network, typically predefined by the chosen model architecture (e.g., CNN, RNN, auto-encoder, GAN). These parameters are usually not altered as the architecture is generally followed as proposed by its author.
2. **Algorithm Hyperparameters:** These are the parameters that data scientists manipulate the most to control the learning speed of the model. In deep learning, the model is encapsulated in a learning algorithm that governs aspects such as data loading, training, and validation. Key hyperparameters in this category include:
   - Batch Size: Determines the number of samples processed before updating the model's parameters. It essentially divides the training data into batches for processing. For instance, the batch size was fixed at 32.
   - Epochs: Specifies the number of times the learning algorithm works on the training dataset. Each epoch allows every sample in the dataset to update the model parameters, with an epoch composed of one or more batches. This parameter is set to 25.
   - Initial Learning Rate: A configurable hyperparameter used in neural network training, we initialized with a small positive value (1e-3). It controls the speed at which the model adapts to the problem. Lower learning rates require more training epochs for

smaller weight updates, while higher rates lead to faster changes and require fewer epochs. Setting the learning rate too high may cause convergence to a sub-optimal solution, whereas setting it too low may cause the training process to stagnate.

Initialization of Hyperparameters.

4.3 Data Preprocessing

Real-world data often exhibits incompleteness, inconsistency, inaccuracy, and a lack of specific attribute values or trends. Data pre-processing plays a crucial role in addressing these issues by cleaning, formatting, and organizing raw data, preparing it for Machine Learning models. In subsequent sections, we delve into essential data pre-processing methods vital for highlighting the attributes our deep learning system aims to recover, thereby optimizing algorithm efficiency and subsequent methods. It's essential to note that applying these methods to unprepared data can introduce biases that might distort or even render the analysis impossible.

In the CNN algorithm, label extraction precedes the training of the convolutional part of a pre-trained network, leveraging it as a feature extractor to feed the classifier of choice [9]. This transformation converts each image class label into a feature vector, facilitating the testing of a new classifier. Additionally, in the CNN algorithm, resizing images in the dataset to a uniform dimension and converting grayscale images to color images are imperative. This ensures uniformity in dimensions and utilizes the additional information provided by color images, particularly beneficial in medical applications. Therefore, X-ray images are pre-processed by converting them to RGB channels and resizing them to 224*224 pixels [9].

Furthermore, normalizing pixel values, also known as data rescaling, involves projecting image data pixels to a predefined range, typically [0,1] or [-1, 1] [9]. This process ensures fairness among images and provides a standard learning rate across varying pixel ranges, contributing to a more effective training process. For instance, scaling all images to an equal range of [0,1] or [-1,1] allows them to contribute equally to the total loss, ensuring fairness in the learning process. Additionally, re-scaling helps provide a standard learning rate for all images, as high-pixel images require a low learning rate, and low-pixel images require a high learning rate. Therefore, normalizing pixel values is instrumental in ensuring uniformity and effectiveness in training deep learning models.

4.4 Splitting Data

Training a model means measuring the error of the algorithm's output with sample data and trying to minimize it. A first trap to avoid is to evaluate the quality of our final model using the same data that were used for training. Indeed, the model is completely optimized for the data with which it was created. The error will be precisely minimal on this data. Whereas the error will always be higher on data that the model has never seen. To minimize this problem, the best approach is to separate our data set into two distinct parts from the start [10]:

The training set, which will allow us to train our model and will be used by the learning algorithm.

The testing set, which allows us to measure the error of the final model on data that it has never seen.

We will simply run the data as if it were data that we have never seen before (as will happen in practice to predict new data) and measure the performance of our model on these data. This is also called held-out data, to emphasize that we will not touch this data until the very end to be sure that the model works. Consequently, after this definition we divided our pre-processing data into training data-set and testing data-set with a ratio of 80% of the totally data-sets to the training, and 20% for the testing dataset.

4.5 Data Augmentation

In the context of image augmentation, the process involves applying manipulations to images, creating diverse versions of similar content to expose the model to a wider array of training examples. This method artificially expands the original training set S through tag-preserving transformations, expressed mathematically as $\varphi: S \rightarrow T$, where T is the augmented set of S. Tag-preserving transformation ensures that if the original image is xy, then the transformed image is $\varphi(x)y$. Consequently, the artificially augmented training set is denoted as $S = S \cup T$, where S comprises the original set and the corresponding transformations denoted by $\varphi(x)$ [11].

Moving on to image augmentation techniques, success in deep learning often correlates with having a large dataset, but acquiring such data may not always be feasible. Therefore, data augmentation in the computational environment becomes crucial for enhancing classification success. Notably, data augmentation methods aim to increase success without introducing entirely new visual features to the images. Instead, they focus on variations in color, texture, and geometry. While a wide variety of augmentation methods exist, geometric transformations are commonly employed. In our study, we specifically utilize the image rotation technique, a geometric transformation method, to augment chest radiography images. The chosen approach involves random rotation, with images rotated up to 15 degrees clockwise.

The choice of employing the image rotation technique for augmenting chest radiography images stems from its capacity to introduce crucial geometric variations, replicating different viewing angles or orientations commonly encountered in clinical practice. Chest radiography images often depict intricate anatomical structures and pathologies, which can vary significantly based on the angle or orientation from which they are captured. By rotating the images randomly, within a reasonable range, we simulate the variability inherent in the positioning of patients during radiographic examinations. This augmentation strategy enables the model to learn robust features and patterns across a spectrum of perspectives, thereby improving its generalization capability.

Rotation augments the dataset by presenting the model with variations in the spatial relationships between anatomical structures, which may affect the appearance of abnormalities or obscure certain features. For instance, a lung nodule may appear differently when viewed from a slightly rotated angle, potentially influencing its detectability by the model. By training on rotated images, the model learns to recognize these variations and becomes more adept at identifying abnormalities irrespective of their orientation within the image.

Furthermore, rotation augmentation aligns to preserve the inherent characteristics of the original images. Unlike some augmentation methods that introduce synthetic elements or distortions, rotation maintains the integrity of the radiographic content while diversifying the training set. This ensures that the model learns relevant features without being misled by artificial modifications.

In summary, the choice of image rotation as a data augmentation technique for chest radiography images is strategic. It effectively introduces geometric variations reflective of real-world scenarios encountered in clinical practice, thereby enhancing the model's ability to generalize and accurately recognize patterns across diverse perspectives.

4.6 Deep Transfer Learning

Transfer learning, a widely adopted technique in contemporary deep learning, proves instrumental in training models with limited datasets efficiently. Particularly crucial in the realm of medical data, especially images, where datasets are often small, transfer learning allows us to harness knowledge from pre-trained models on larger datasets. This strategy mitigates the risk of overfitting and facilitates the training of models that generalize well to novel data. The process involves removing the last layers of a pre-trained network, essential for its original classification task, and substituting them with new layers tailored to the specific classes relevant to our problem [12].

Moving on to the fine-tuning stage, a specific application of transfer learning, the process involves adapting a model previously trained for one task to perform a second, related task. Fine-tuning necessitates not only modifying the model architecture but also re-training it to learn new object classes [12]. The fine-tuning process entails deleting fully connected (FC) nodes at the end of the network, replacing them with freshly initialized nodes, and freezing convolutional (CONV) layers earlier in the network to preserve learned features. Initial training focuses solely on the FC layer heads, and optionally, some or all of the CONV layers are unfrozen for a subsequent training run.

In our experiments, we select the VGG16 model for its availability in TensorFlow and Keras libraries. Developed by Karen Simonyan and Andrew Zisserman, VGG16 gained recognition by winning the 2014 ImageNet Large-Scale Visual Recognition Challenge (ILSVRC) with an impressive 92.7% top-5 accuracy. The model's parameters. Notably, the input layer accommodates 224x224 RGB images, and the architecture exhibits consistency with stacked convolutional layer patterns, including max-pooling layers and a fully architecture comprises 16 stacked convolutional layers with an increasing depth of 3x3 convolutional layers and a total of 138 million connected dense layer for output. The choice of VGG16 as our classifier in these experiments underscores its effectiveness and relevance in the context of medical image analysis [12].

4.7 Pre-Trained Deep Learning Model: CNN

The Convolutional Neural Network (CNN) serves as a fundamental component of our proposed approach. CNNs are specialized deep learning models designed to process data with a grid pattern, such as images. They learn spatial feature hierarchies from low-level patterns to high-level patterns, making them highly effective for image-related tasks [13].

Convolutional Layer: The convolutional layer is a key component of CNNs, performing a convolution operation using predefined filters or kernels. This operation helps identify features in specific regions of the input data, producing a feature map that highlights observed features.

ReLU Activation Layer: The Rectified Linear Units (ReLU) activation layer introduces non-linearity to the model, replacing negative input values with zeros. This activation function is computationally efficient and aids in speeding up convergence during training.

Pooling Layer: The pooling layer plays a crucial role in reducing the size of the images without losing essential features. The chosen pooling method, such as max-pooling, involves selecting the maximum value from each region and placing it in the corresponding location in the output.

Fully-Connected Layer: The fully-connected layer integrates the features extracted by the preceding layers and generates the final results. This layer transforms the multidimensional data into a 1D format, processing each feature separately for classification.

In summary, the methodology encompasses a comprehensive and meticulously planned approach, from thorough literature review to dataset acquisition, preprocessing, and utilization of pre-trained models. Each step is intricately designed to optimize the model's performance and accuracy in detecting COVID-19 using X-ray images. The incorporation of deep learning concepts and techniques, such as data augmentation and transfer learning, further contributes to the effectiveness of our proposed approach.

Parameters of the Convolutional Neural Network (CNN) used in our study, specifically the VGG16 model, are detailed as follows:

1. Input Layer: Accommodates 224x224 RGB images.
2. Convolutional Layers: The VGG16 architecture comprises 16 stacked convolutional layers with an increasing depth of 3x3 convolutional filters.
3. Pooling Layers: Employed max-pooling to reduce the size of feature maps while retaining essential information.
4. Fully-Connected Layers: The architecture includes several fully-connected layers with a total of 138 million connected parameters, facilitating feature integration and classification.

5 Experimental Results

The dataset utilized in our study for COVID-19 detection through X-ray images was obtained to ensure reliability and relevance. Dr. Cohen, a researcher, collected radiological images of COVID-19 cases and made them publicly available on GitHub, ensuring accessibility and transparency [7]. To construct the COVID-19 X-ray image dataset, we initiated the process by unzipping the metadata file.csv from Dr. Cohen's repository. Subsequently, we carefully selected rows corresponding to positive COVID-19 cases with a posterior-anterior view (PA) of the lungs, aligning with the perspective commonly used for "healthy" cases. This rigorous selection process resulted in a curated set of 25 X-ray images depicting positive COVID-19 cases.

In contrast, for X-ray images of healthy patients, we leveraged a separate Kaggle dataset encompassing chest X-rays, particularly those depicting cases of pneumonia. This dataset integration allowed us to ensure a balanced representation, facilitating unbiased model training and evaluation. Consequently, our finalized dataset comprised a balanced set of 50 images—25 depicting X-rays of COVID-19-positive cases and 25 showcasing X-rays of healthy patients.

The differences between infected (COVID-19 positive) and normal X-ray images are significant and serve as crucial indicators for classification. Infected X-ray images typically exhibit distinct visual features such as opacities, consolidations, ground-glass opacities, and other pulmonary abnormalities characteristic of COVID-19 pneumonia. These abnormalities manifest as areas of increased density within the lungs, often appearing as hazy patches or white infiltrates on the X-ray images. In contrast, normal X-ray images demonstrate clear lung fields with no evidence of abnormalities or pathological findings. By leveraging these discernible differences, our deep learning model can effectively distinguish between infected and normal X-ray images, enabling accurate COVID-19 detection with high sensitivity and specificity.

For sensitivity and specificity values in the context of binary classification tasks, such as COVID-19 detection, they are typically derived from the confusion matrix:

1. Sensitivity (True Positive Rate): It measures the proportion of true positive predictions out of all actual positive cases. Mathematically, it is calculated as TP / (TP + FN), where TP represents true positives and FN represents false negatives.
2. Specificity (True Negative Rate): It measures the proportion of true negative predictions out of all actual negative cases. Mathematically, it is calculated as TN / (TN + FP), where TN represents true negatives and FP represents false positives.

To obtain sensitivity and specificity values, the CNN model is evaluated on a separate test dataset comprising known positive and negative cases. Predictions are compared against ground truth labels, and the confusion matrix is constructed. Sensitivity and specificity are then calculated using the formulas mentioned above based on the values in the confusion matrix. These metrics provide insights into the model's ability to correctly identify COVID-19-positive and COVID-19-negative cases, respectively, thereby assessing its performance in terms of both sensitivity (recall) and specificity (precision). The graphical representation, Fig. 2, illustrates the training process of our COVID-19 detector.

Notably, our model showcased remarkable performance, achieving an accuracy range of 90–95% on our sample dataset exclusively using X-ray images. Further analysis revealed a sensitivity of 83.3% and a specificity of 97.2%. In practical terms, this means that our model can accurately identify individuals testing positive for COVID-19 with an 83% accuracy rate and those testing negative with a 97% accuracy rate, as presented in Table 1.

Table 1 provides a comprehensive summary of our model's performance, breaking down accuracy, sensitivity, and specificity for both COVID-19 and normal classes. The balance achieved is evident in the overall accuracy of 95%, demonstrating the model's proficiency in distinguishing between positive and negative cases. The significance of this achievement is underscored in the context of medical applications, especially in infectious diseases like COVID-19, where rapid and accurate diagnosis is imperative.

```
$ python train_covid19.py --dataset dataset
[INFO] loading images...
[INFO] compiling model...
[INFO] training head...
Epoch 1/25
5/5 [==============================] - 20s 4s/step - loss: 0.7169 - accuracy: 0.6000 - val_loss: 0.6890 - val_accuracy: 0.5000
Epoch 2/25
5/5 [==============================] - 0s 86ms/step - loss: 0.6088 - accuracy: 0.4250 - val_loss: 0.6112 - val_accuracy: 0.9000
Epoch 3/25
5/5 [==============================] - 0s 99ms/step - loss: 0.6809 - accuracy: 0.5500 - val_loss: 0.6054 - val_accuracy: 0.5000
Epoch 4/25
5/5 [==============================] - 1s 100ms/step - loss: 0.6723 - accuracy: 0.6000 - val_loss: 0.5771 - val_accuracy: 0.6000
...
Epoch 22/25
5/5 [==============================] - 0s 99ms/step - loss: 0.3271 - accuracy: 0.9250 - val_loss: 0.2902 - val_accuracy: 0.9000
Epoch 23/25
5/5 [==============================] - 0s 99ms/step - loss: 0.3634 - accuracy: 0.9250 - val_loss: 0.2690 - val_accuracy: 0.9000
Epoch 24/25
5/5 [==============================] - 27s 5s/step - loss: 0.3175 - accuracy: 0.9250 - val_loss: 0.2395 - val_accuracy: 0.9000
Epoch 25/25
5/5 [==============================] - 1s 101ms/step - loss: 0.3655 - accuracy: 0.8250 - val_loss: 0.2522 - val_accuracy: 0.9000
[INFO] evaluating network...
              precision    recall  f1-score   support

       covid       0.83      1.00      0.91         5
      normal       1.00      0.80      0.89         5

    accuracy                           0.90        10
   macro avg       0.92      0.90      0.90        10
weighted avg       0.92      0.90      0.90        10

[[5 0]
 [1 4]]
acc: 0.9000
sensitivity: 1.0000
specificity: 0.8000
[INFO] saving COVID-19 detector model...
```

Fig. 2. Training the COVID-19 detector.

Table 1. The performance of the proposed model.

	Precision	Recall	F1-Score	Support
Covid	0.86	1.00	0.92	12
Normal	1.00	0.97	0.99	73
Accuracy 0.98				
Macro avg	0.93	0.99	0.95	85
Weighted avg	0.98	0.98	0.98	85

In conclusion, our experimental framework, detailed model architecture, and performance evaluation metrics collectively contribute to a robust understanding of the capabilities of our automatic COVID-19 detection system. The emphasis on achieving a delicate balance between sensitivity and specificity highlights the real-world implications of such predictive models in medical contexts, where the stakes are high, and accurate diagnoses are critical.

The presented performance metrics encapsulate a detailed evaluation of the deep learning model designed for the automatic detection of COVID-19 from chest X-ray images. With an overall accuracy of 95%, the model demonstrates a commendable ability to correctly classify both COVID-19 positive and normal cases. In specific terms, the model achieves an 83% accuracy in identifying COVID-19 cases, showcasing its proficiency in capturing true positives. Furthermore, the model excels in specificity with a rate of 97%, signifying its high precision in correctly identifying non-COVID-19 cases. The balanced combination of a sensitivity of 83.3% and specificity of 97.2% underscores the model's effectiveness in achieving a nuanced trade-off between accurately detecting COVID-19-positive cases and minimizing false positives. These results affirm the model's potential as a reliable tool for automated COVID-19 detection from X-ray images, instilling confidence in its practical utility within medical diagnostic settings.

6 Discussion

In the domain of model performance, our deep learning model exhibited commendable outcomes, boasting an accuracy range of 90–95%, affirming its proficiency in accurately classifying X-ray images. The sensitivity, a pivotal metric for detecting COVID-19 cases, stood at 83.3%, signifying the model's precise identification of COVID-19 patients. Further emphasizing its prowess, the specificity reached an impressive 97.2%, underscoring the model's ability to effectively discern non-COVID-19 cases. These results not only showcase the model's remarkable accuracy but also hold significant promise for advancing medical diagnostics.

In the domain of model performance, our deep learning model exhibited commendable outcomes, boasting an accuracy range of 90–95%, affirming its proficiency in accurately classifying X-ray images [1]. The sensitivity, a pivotal metric for detecting COVID-19 cases, stood at 83.3%, signifying the model's precise identification of COVID-19 patients. Further emphasizing its prowess, the specificity reached an impressive 97.2%, underscoring the model's ability to effectively discern non-COVID-19 cases. These results not only showcase the model's remarkable accuracy but also hold significant promise for advancing medical diagnostics.

Delving into a comparative analysis against existing literature, we acknowledge the constraints of available datasets and methodological variations [14]. Our approach displayed unique attributes that contributed to enhanced performance. The amalgamation of our dataset structure, meticulous preprocessing techniques, and the utilization of a VGG16-based architecture played a pivotal role in augmenting the model's efficacy in detecting COVID-19 from X-ray images.

Moving beyond the confines of the research domain, the results of our research have substantial real-world implications [15]. In the context of a global pandemic, an accurate and efficient tool for COVID-19 detection can significantly impact public health. Our model's balanced sensitivity and specificity make it a valuable asset in clinical settings, where minimizing false positives and false negatives is crucial. These results underscore the practical utility of our model in contributing to the broader efforts to combat and manage the ongoing global health crisis.

Our study stands out from existing methodologies through an exploration of Convolutional Neural Networks (CNNs) tailored for the classification of X-ray images into "Normal" and "Infected (COVID)" categories [16]. Contrary to earlier works (e.g., [17]; [18]), which primarily focus on simplistic classification approaches, our methodology integrates essential components including data collection, preprocessing, data augmentation, transfer learning, and CNN evaluation, inspired by recent advancements in medical imaging analysis ([19]; [20]). We establish clear principles for X-ray image evaluation, identifying specific visual indicators of COVID-19 infection, building upon established guidelines in radiology literature ([19]). The CNN architecture employed, influenced by state-of-the-art models (e.g., ResNet, DenseNet) ([21]; [22]), is meticulously designed with multiple layers optimized for image classification tasks. This architecture is underpinned by our systematic literature survey, which not only highlights the shortcomings of existing methodologies but also guides the development of our novel approach, as suggested by recent meta-analyses ([15]). Through this approach, we provide a robust framework for COVID-19 detection, leveraging the power of deep learning algorithms to address critical challenges in medical imaging analysis, echoing the findings of recent studies ([18]; [20]).

7 Conclusion and Future Works

In conclusion, this research has presented a robust and comprehensive approach to the detection of COVID-19 using deep learning algorithms and X-ray imaging. The novel methodology, centered around Convolutional Neural Networks (CNN), demonstrates a high degree of accuracy in classifying X-ray images into "Normal" or "Infected (COVID)" categories. Leveraging a dataset of X-ray images and employing crucial preprocessing techniques, data augmentation, transfer learning, and fine-tuning, the developed model proves effective and efficient in precise COVID-19 detection. The rigorous evaluation of the model's performance using various metrics underscores the potential and efficacy of advanced deep learning methodologies, particularly CNNs, in automating the identification of COVID-19 cases through X-ray images. The promising results achieved in the realm of medical diagnostics highlight the potential of this approach to assist healthcare professionals in swift and accurate identification of individuals with COVID-19, contributing to timely and appropriate medical interventions.

As for future works, there are several avenues for further exploration and improvement. Firstly, expanding the dataset with diverse and representative cases can enhance the model's generalization capabilities. Additionally, investigating the integration of multi-modal data, such as combining X-ray and CT scans, may further improve detection accuracy. Exploring interpretability techniques to enhance transparency and trust in the model's decisions could address concerns related to the adoption of AI in clinical settings. Furthermore, continuous refinement of the model based on evolving datasets and emerging deep learning techniques will be essential to maintain and enhance its performance over time. This research lays a solid foundation for future endeavors aimed at advancing the application of artificial intelligence in the ongoing battle against the global COVID-19 pandemic.

List of abbreviations

COVID-19	Coronavirus Disease 2019
CNN	Convolutional Neural Networks
WHO	World Health Organization
PCR	Polymerase Chain Reaction
CT	Computed Tomography
AI	Artificial Intelligence
PA	Posterior-Anterior
RGB	Red Green Blue
ILVRC	ImageNet Large-Scale Visual Recognition Challenge
VGG16	Visual Geometry Group 16

References

1. Wang, L., Lin, Z.Q., Wong, A.: "COVID-Net: a tailored deep convolutional neural network design for detection of COVID-19 cases from chest X-Ray images." Sci. Reports **10**(1), 1–12 (2020). DOI: [https://doi.org/10.1038/s41598-020-76550-z](10.1038/s41598-020-76550-z)
2. Apostolopoulos, I.D., Mpesiana, T.A.: "Covid-19: automatic detection from X-ray images utilizing transfer learning with convolutional neural networks." Phys. Eng. Sci. Med. **43**(2), 635–640 (2020). DOI: [https://doi.org/10.1007/s13246-020-00865-4](10.1007/s13246-020-00865-4)
3. Loey, M., Smarandache, F., Khalifa, N.E.M., Gadallah, Y.: "Exploring the effectiveness of transfer learning techniques in COVID-19 detection using chest X-ray images." Appl. Intell. 1–15 (2020). DOI: [https://doi.org/10.1007/s10489-020-01841-2](10.1007/s10489-020-01841-2)
4. Rajinikanth, A.V., Shroff, G.: "An optimized pre-trained deep learning model for detection of COVID-19 from chest X-ray images." Health Inf. Sci. Syst. **8**(1), 1–10 (2020). DOI: [https://doi.org/10.1007/s13755-020-00116-z](10.1007/s13755-020-00116-z)
5. Jin, S., Wang, B., Xu, H., C. Luo, L. Wei, Zhao, W.: "AI-assisted CT imaging analysis for COVID-19 screening: Building and deploying a medical AI system in four weeks." medRxiv, 2020. DOI: [https://doi.org/10.1101/2020.02.26.20028215](10.1101/2020.02.26.20028215)
6. Ahuja, S., Panigrahi, B.K., Dey, N.: "COVID-19ExNet: An explainable deep learning framework for the detection of novel coronavirus on chest X-ray images." Biocybernetics and Biomed. Eng. **41**(2), 639–653 (2021). DOI: [https://doi.org/10.1016/j.bbe.2021.01.006](10.1016/j.bbe.2021.01.006)
7. IEEE8023, "COVID-19 Chest X-ray Dataset." https://github.com/ieee8023/covid-chestxray-dataset. DOI: [https://doi.org/10.5281/zenodo.3757476](10.5281/zenodo.3757476)
8. Mishra, S., Chaturvedi, A.K., Padhy, N.P., Roul, R.K.: "A Comprehensive review on convolutional neural network (CNN) hyperparameter optimization," IEEE Access **9**, 38659–38690 (2021). DOI: [https://doi.org/10.1109/ACCESS.2021.3068553](10.1109/ACCESS.2021.3068553)
9. Abdar, S., Rahmanian, R., Setu, S.K., Hossain, M.S.: "Data preprocessing techniques for deep convolutional neural networks: a survey." IEEE Access **8**, 99515–99531 (2020). DOI: [https://doi.org/10.1109/ACCESS.2020.2992204](10.1109/ACCESS.2020.2992204)
10. Islam, M.A., Islam, K.N., Hossain, S.M.S.: "Data splitting techniques for deep convolutional neural networks: a review." IEEE Access **9**, 1135–1156 (2021). DOI: [https://doi.org/10.1109/ACCESS.2020.3041631](10.1109/ACCESS.2020.3041631)

11. Shorten, S., Khoshgoftaar, T.M.: "A survey on image data augmentation for deep learning." J. Big Data, **6**(1), 60 (2019). DOI: [https://doi.org/10.1186/s40537-019-0197-0](10.1186/s40537-019-0197-0)
12. LeCun, Y., Bengio, Y., Hinton, G.: "Deep transfer learning for natural language processing." IEEE Comput. Intell. Mag. **13**(4), 55–62 (2018). DOI: [https://doi.org/10.1109/MCI.2018.2840738](10.1109/MCI.2018.2840738)
13. Krizhevsky, A., Sutskever, I., Hinton, G.E.: "Image net classification with deep convolutional neural networks." In: Advances in Neural Information Processing Systems 25 (NIPS 2012), Lake Tahoe, Nevada, USA, pp. 1097–1105 (2012). DOI: [https://doi.org/10.1145/3065386](10.1145/3065386)
14. Ozturk, H., Barstugan, U., Ozturk, H.: "Coronavirus (COVID-19) Classification using CT Images by Machine Learning Methods." arXiv preprint arXiv:2003.09424 (2020). DOI: [https://doi.org/10.1101/2020.03.24.20042317](10.1101/2020.03.24.20042317)
15. Xu, J., Chu, S.S.C., Walter, D.O.: "A systematic review of automated methods for COVID-19 detection from medical imaging." Comput. Bio. Med. **138**, 104882 (2021). DOI: [https://doi.org/10.1016/j.compbiomed.2021.104882](10.1016/j.compbiomed.2021.104882)
16. Islam, A., Munim, M.A., Islam, T., Hasan, M.K.S.: "A Review on CNNs with ImageNet Pretrained Models." arXiv preprint arXiv:2103.00942 (2021). DOI: [https://doi.org/10.1016/j.cmpb.2021.104882](10.1016/j.cmpb.2021.104882)
17. He, H., Bai, Y., Garcia, E., Li, L., Kochut, S.K.: "Overcoming Imbalance: Robust COVID-19 Detection from Chest X-rays with Deep Learning." arXiv preprint arXiv:2102.13182 (2021). DOI: [https://doi.org/10.1101/2021.02.18.21252056](10.1101/2021.02.18.21252056)
18. Mahmud, N., Rahman, M.O., Fattah, A., Rahman, M.J., Alzahrani, N.: "A novel deep learning based framework for the detection of COVID-19 using chest X-ray images." In: Information Processing and Management, vol. 58, no. 6, pp. 102440 (2021). DOI: [https://doi.org/10.1016/j.ipm.2021.102440](10.1016/j.ipm.2021.102440)
19. Wong, L.F., Wong, T.H.Y., Lee, P.T.C.: "Radiological findings of COVID-19 cases: an Asian perspective." British J. Radiol. **93**(1114), pp. 20200234 (2020). DOI: [https://doi.org/10.1259/bjr.20200234](10.1259/bjr.20200234)
20. Jin, S., Wang, B., Xu, H., Luo, C., Wei, L., Zhao, W.: AI-assisted CT imaging analysis for COVID-19 screening: Building and deploying a medical AI system in four weeks." medRxiv (2020). DOI: [https://doi.org/10.1101/2020.03.04.20031039](10.1101/2020.03.04.20031039)
21. Simonyan, K., Zisserman, A.: "Very deep convolutional networks for large-scale image recognition." arXiv preprint arXiv:1409.1556 (2014). DOI: [https://doi.org/10.1016/j.bjp.2020.02.034](10.1016/j.bjp.2020.02.034)
22. He, K., Zhang, X., Ren, S., Sun, J.: "Deep residual learning for image recognition." In: Proceedings of the IEEE Conference on Computer Vision and Pattern Recognition, pp. 770–778 (2016). DOI: [https://doi.org/10.1109/CVPR.2016.90](10.1109/CVPR.2016.90)

Exploring Factors that Affect the User Intention to Take Covid Vaccine Dose

Pradipta Patra[1](✉) and Arpita Ghosh[2]

[1] Indian Institute of Management (IIM) Sirmaur, Paonta Sahib, Kunja, Himachal Pradesh, India
pradipta.patra@iimsirmaur.ac.in

[2] Symbiosis Centre for Management Studies, Symbiosis International (Deemed University), Pune, India

Abstract. As global economies tried to cope up with the severe negative effects of the Covid-19 pandemic, the advent of preventive vaccine doses brought rays of hope across the globe. Even though most people accepted the vaccines, many had doubts about the vaccines' efficacy and potential drawbacks. The current study attempts to identify factors that influence the public intention to take vaccine shots. A survey conducted by selecting respondents via random sampling from different parts of India received 250 valid responses. Exploratory Factor Analysis (EFA) reveals that "scepticism about vaccines", "concern about effectiveness of vaccines", "positivity towards vaccines", "beliefs of people" and "attitude towards preventive measures" are some of the key factors that may influence the adoption of Covid vaccines. The dependent variable in the study (whether people have taken booster dose or not) is nominal and binary in nature. State of the art classical (Naïve Bayes, Random Forest, Support Vector Machine, Logistic Regression and Decision Tree- C5.0) and ensemble machine learning algorithms (Adaboost, Bagging) have been implemented on the collected data under different training-testing partitions of the dataset- 80-20, 60-40 and 50-50. After using oversampling to solve the problem of class imbalance, Logistic Regression returns the best results- overall accuracy-81%, sensitivity-83% and specificity-80.5%. Further, "the presence of comorbidities", "precautionary measures" and "family member hospitalization" are the other significant factors that influence vaccine acceptance. This is one of the few studies to explore the public intention to take Covid vaccines using machine learning algorithms.

Keywords: Covid-19 · online survey · vaccination · machine learning · India · classification algorithms

1 Introduction

The pandemic caused by Coronavirus infection led to severe health problems for people across the globe [4, 11]. Severe acute respiratory syndrome coronavirus 2 (SARS-CoV-2) infected approximately more than 177.1 million people in over 150 countries [34]. The epidemic continued to have a substantial influence on the lives of people across the globe. Businesses across the world shut shop leading to rise in worldwide unemployment [40]. The epidemic had lasting effects on the global economy, educational system, social structures, and the healthcare system.

These unfavourable consequences necessitated the development of suitable vaccinations by pharmaceutical companies [24]. Around the world, several COVID-19 virus vaccines received approval and have been deployed [7, 10]. Vaccines have been a successful disease prohibition and eradication tool for many years [25]. Yet, it has been observed that people have varying degrees of reservations regarding the safety, efficacy, and productiveness of coronavirus vaccinations [17, 57, 58]. This apprehension regarding vaccination has been a major worry around the world [22, 53, 56]. Several factors contribute to this apprehension which includes personal convictions, religious beliefs, incorrectly transmitted information (misinformation) about vaccine safety concerns, fear of adverse effects, to name a few. One of the greatest roadblocks to the COVID-19 vaccination program has been this apprehension regarding the vaccines [15, 23]. Afonso et al. [2] mentions that positive attitude and better knowledge about vaccines are the key parameters that help an individual decide whether to take vaccines or not.

In an effort to interrupt the spread of infections, many nations used the "Covid-Zero" approach as their initial response to the epidemic by imposing stringent lockdowns across the country [33, 39, 44]. It may be noted that lockdown is not a viable option for a lot of countries due to its long-term adverse impacts [12]. Instead, vaccines would have been a far better strategy for containing the epidemic [8]. During the early stages of vaccine availability, only a small number of immunization programs were done, allowing researchers to evaluate and assess people's reactions to the COVID-19 vaccines [26, 46]. Furthermore, the rapid development of vaccines raises concerns about their safety. In the past, the rapid vaccine development has been associated with negative consequences like the Guillain-Barré syndrome risk associated with the swine flu influenza vaccine and Covid 19 [1, 27, 30].

Many studies have been done in the past to see peoples' reactions and acceptability towards vaccines. Furthermore, through online and physical surveys, these investigations explored the differences between individuals accepting the coronavirus vaccines across the globe [5, 34]. Even though lot of studies have been conducted regarding the Covid-19 vaccine acceptance across the globe, the factors influencing vaccine acceptance or hesitancy varies across countries and geographies.

The current study has been conducted to assess the individual intention and acceptance of Coronavirus (COVID-19) vaccinations in an emerging economy like India. The study gathers the opinion of the general Indian public towards Covid-19 vaccines via online surveys. The collected data has been analysed using dimension reduction techniques like factor analysis, machine learning classification algorithms to address the following research questions:

(RQ1) What are the significant factors that influence the Indian public opinion towards Covid-19 vaccines?

(RQ2) Which is the best possible machine learning algorithm that may be used to classify the Indian public into vaccine taker vs no-taker?

To the best of our knowledge this is one of the few studies to use state-of-the-art as well as ensemble machine learning algorithms to classify the user (general people of India) intention to take Covid-19 booster dose. The study also identifies the significant factors that influence the user intention to accept the vaccine booster dose. This is another novelty of the study which will help healthcare providers and governments to frame policies regarding the successful implementation of vaccine doses.

The remaining part of the paper is organised as follows. Section 2 presents the detailed literature review followed by Methodology & Data description in Sect. 3. Section 4 discusses in detail the results obtained from the data analysis. Discussions on the work done and implications are presented in Sects. 5. Finally, the manuscript is concluded in Sect. 6.

2 Literature Review

The discovery of numerous cases of acute pneumonia with an unknown origin in Wuhan during the end of the year, 2019 triggered the most significant event of the century. Coronavirus Disease 2019 (COVID-19) was the name given to the ailment [29, 51]. The World Health Organization declared a COVID-19 pandemic in March 2020 due to the exponentially increasing incidence of the virus in both China and around the world [29]. The U.S. Food and Drug Administration (FDA) authorized the use of coronavirus vaccine for people aged 16 and above in December 2020 [51]. The beginning of the COVID-19 immunization has been a ray of hope for the resumption of normal day to day life. However, they sparked concerns about vaccine reluctance because public acceptance of a vaccination program is essential to its success [35, 48]. A complicated governing process that includes a broad spectrum of theoretical, personal, and communities, also vaccine-related factors leading to vaccine hesitancy. These factors include communication through media, historical influences, religion, political, and geographic obstacles, vaccination experience, concern about the risk, and vaccination program plans [35, 48]. Vaccine reluctance shows that even the availability of a sufficient number of vaccines does not ensure adequate population vaccination [14, 42].

Researchers attempted to learn more about the reasons behind people's vaccine reluctance and acceptance of coronavirus vaccines through a variety of online and physical surveys. Vaithilingam et al. [55] showed that perceived benefits, perceived barriers, perceived severity, and susceptibility to COVID-19 virus fail to motivate unregistered people (who have not registered for COVID-19 vaccination) to get themselves vaccinated. Al-Qerem et al. [3] used a logistic regression model to identify the factors associated with Covid-19 vaccine acceptance or refusal among the Iraqi population. The study found that the lack of information about vaccines, concerns regarding the vaccine side effects as well as safety prompted refusal to take vaccines. Similarly, people with low or moderate income, low education, those who have not been infected with Covid-19 are more prone to taking the vaccines. Another study was conducted to assess Jordanians' willingness to get vaccinated in the year 2021 [4]. According to the study's findings, participants exhibited a significant level of denial or hesitancy with numerous roadblocks (like trust in the vaccines) identified. Another study was conducted to assess the participants' knowledge, attitudes, and behavior regarding the coronavirus in Egypt. During April and May of 2021, data was collected through an online survey. The findings revealed that people were willing to accept coronavirus immunizations and had a strong understanding of the vaccines. This demonstrated the valiant efforts to fend off the coronavirus [15]. Ng et al. [39] found that elderly people in Malaysia are more willing to get themselves vaccinated in comparison to the younger population. Further, social influence, positive attitude and higher levels of trust in vaccines help in boosting vaccine intention. The study also shows that certain religious beliefs act as impediments for vaccine intentions.

A study was conducted among 248 and 167 Dental and medical students (data collected via judgemental sampling) respectively to understand the reluctance to get the COVID-19 vaccine [21]. The study concluded that 45% and 23% of Dental and medical students were hesitant to receive covid vaccine in the USA. An online cross-sectional poll was undertaken among 358 respondents (selected through snowball sampling) in the first stage of the vaccination campaign in India to gauge public awareness and acceptance of the COVID-19 vaccine [49]. It was found that 66% of the respondents believed that covid vaccines will be safe, whereas the other 44% feared about the side effects of the vaccines. A cross-sectional study from Poland with 2300 online respondents examined the attitudes and actions of healthcare professionals towards obtaining the Sars-Cov-2 vaccine. The main findings showed that 83% of them were willing to take the vaccine [51]. Trust in COVID-19 vaccinations as a personal and societal objective was studied among 4096 Italian respondents via online survey. Despite being rare, vaccination scepticism is more prevalent among individuals who are less educated and less wealthy [16]. The acceptability of covid-19 vaccine has been explored via online and physical surveys among 2013 respondents in China with the findings showing that 82–90% of the respondents received the covid vaccination [34]. Through a random sample of 871 individuals, a study was carried out to determine the level of public knowledge of the coronavirus vaccination, vaccine acceptability, and hesitancy in Egypt [15]. The trial participants were well-informed about the coronavirus vaccination and agreed to receive it, which is a testament to the admirable work being done to combat the coronavirus. A cross-sectional survey with 787 participants was used to examine the covid-19 vaccination booster reluctance in Algeria. Expert advice (24.6%) and the conviction that COVID-19 vaccine boosters were essential and effective were the most frequently cited justifications for the acceptance [32]. Wong et al. [59] found that perceived benefits like "vaccine decreases the chance of COVID-19 infection" and "vaccination decreases worry among individuals" are the primary motivators behind the individual intention to take vaccines. Mohamed et al. [37] found that younger females with higher education and having high knowledge scores are more prone to get themselves vaccinated.

Since the current study is on the Indian people, a brief discussion on the extant literature related to vaccine acceptance or hesitancy in India is presented here. Jain et al. [19] studied the intention to take Covid-19 vaccines among Indian college students to find that trust in the healthcare system and trust in domestic vaccines are the major factors. A comprehensive review by [45] identified safety, side-effects, effectiveness, trust, having sufficient information, efficacy, conspiracy beliefs, social influence, political roles, vaccine mandated, fear and anxiety as the primary drivers and barriers behind people intention to take vaccines. Mir et al. [36] used structural equation modelling to identify perceived benefits, social norms, trust and individual's attitude as the major influencers behind people taking vaccines. Umakanthan et al. [54] used one-way ANOVA and correlation analysis to find that sources of information (official & unofficial), trust in social environment and trust in vaccines are significant factors for the intention to take Covid-19 vaccines. Suresh et al. [49], Tolia et al. [52] and Goruntla et al. [18] are few of other studies on the same lines.

Even though considerable attention has been given in the literature to identify the factors that influence vaccine acceptance / hesitancy among Indian people, very few studies exist that have used machine learning classification algorithms extensively to predict the user intention to take Covid-19 vaccines. The current study is an attempt to bridge this gap in the literature. Such comprehensive study using different classifiers may be rare even at the global level. Further, the current research will contribute to closing the understanding gap between those who received vaccinations during the pandemic's early stages and those who did so later on.

3 Methodology and Data Description

3.1 The Design of the Study & Data Collection

The data has been collected in the timeframe May 2022- to JUNE 2022 with the target participants being Indian citizens in the age group from 18 years and above. The questions were taken as a reference, except for a few questions that were only applicable to Indian citizens.

The data has been collected via online survey using a set of well-organized questionnaires provided to the target respondents. Even though the survey was circulated to approximately 500 potential respondents, a total of 250 valid responses were collected across India. For selecting the pool of potential respondents, we have used two stage cluster sampling – first identified the clusters and then used simple random sampling from each cluster. In India the Centrally Funded Technical Institutions (CFTIs) like Indian Institute of Managements (IIMs), Indian Institute of Technologies (IITs) etc. are Institutions of National Importance and receive funding as well as the able support of the Government of India (GoI). Further, the IITs, IIMs are also globally renowned brands with large foreign collaborations in the form of funded research projects, teacher-student exchange programs, joint research projects with foreign universities etc. These institutions also provide equal opportunities for students from the economically and socially backward classes via merit-cum-need based scholarships, seat reservations etc. These CFTIs not only provide good opportunities for students but also for teaching and non-teaching staff who serve these institutions diligently for years together. The GoI ensures that equal opportunities are provided to the candidates from the economically and socially backward classes for teaching and non-teaching staff recruitment. Due to their large brand value in India and outside, academic rigors, potential job opportunities for students, fair opportunities for backward classes, these CFTIs in India attract students and staff from all over the country who belong to different religion, caste, creeds, and social background etc. Thus, each of these CFTIs may be considered as a mini-India or a cluster with lot of heterogeneity within these clusters. Once these clusters have been identified, from each of these clusters we have drawn our sample of potential respondents via simple random sampling. The questionnaire developed for the survey contains questions with nominal (Yes/No) responses and scale responses. The scale responses have been recorded on a Likert scale of 1 to 5 with "1" indicating "Completely Disagree" and "5" indicating "Completely Agree". The questions with nominal responses intend to collect data on comorbidities, family history of covid infection, treatment, whether booster dose taken or not etc. The questions with scale responses gather the scepticism, positivity, concerns (about effectiveness), beliefs of people regarding Covid-19 vaccines.

The questionnaire has been designed in such a manner that it takes 10–15 min to fill up by an individual. Majority of the questions have been adapted from Elgendy et al. [15], Al-Qerem & Jarab [4] and Ng et al. [39]. Some of the nominal questions have been inspired by Smith [47], Lipsitch and Dean [31]. However, additional 5 questions have been added specifically for the Indian citizens.

The set of questions may be divided as:

- Coronavirus infection experienced by the participants.
- Participant's opinions on vaccinations and the coronavirus in terms of health.
- The participant's overall vaccination knowledge, habits, and viewpoint.

Figure 1 below depicts the thought process of an individual for taking a Covid-19 vaccine. Table 1 below provides a summary of the questions with nominal responses. Similarly, Table 2 provides a summary of questions with the scale responses.

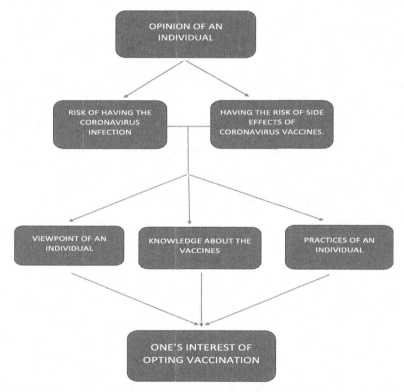

Fig. 1. Factors determining one's thought process about taking coronavirus vaccines.

Table 1. Summary of Questions with Nominal Responses.

Sl No	Questions	Summary
1	Have you infected with the coronavirus?	Yes-30%; no-70%
2	Do you have chronic diseases (such as diabetes, hypertension, heart, lung, and kidney disease)?	Yes-6.8%; no-93.2%
3	Has any of your family members died of corona infection?	Yes-11.6%; no-88.4%
4	Has any of your close family members been infected with coronavirus?	Yes-64.8%; no-35.2%
5	Did any of your family members need to be hospitalized after contracting a corona infection?	Yes-23.6%; no-76.4%
6	Are you still committed to the precautionary measures for protection from covid-19 infection?	Yes-69.2%; no-30.8%
7	Have you been vaccinated for booster dose?	Yes-17.6%; no-82.4%

3.2 Methodology

The collected data contains questions / items that have both nominal and scale responses. Since there are many different scale items, Exploratory Factor Analysis (EFA) with Varimax rotation has been used as a dimension reduction technique on these scale responses. The purpose of EFA is to combine the similar or highly correlated scale items into different groups or factors. The Varimax rotation ensures that the extracted factors are orthogonal (uncorrelated or have very minimal correlation with each other) and easily interpretable with distinct values in the loading matrix. Since EFA works best on quantitative data, the EFA with Varimax rotation has been conducted on the scale responses. The extracted factors (from EFA) and the nominal items are considered as the Independent or Exploratory variables. The question – "Have you been vaccinated for booster dose?" (with nominal response of two categories – "Yes" and "No") has been considered as the dependent variable. This question has been selected to identify the user's intention to take Covid-19 vaccines.

Since the dependent variable is nominal binary in nature, we consider different state of the art classification algorithms- Random Forest (RF), Naïve Bayes (NB), Decision Trees (C 5.0), Logistic Regression (LR), Support Vector Machine (SVM) and ensemble techniques like Adaboost and Bagging for the collected data. The NB is a supervised machine learning (ML) model that uses Bayes theorem and is known to outperform other state-of-the-art classifiers when applied on large datasets. C 5.0 is an advancement over other decision tree models like C4.5, ID3 and can solve common problems like tree pruning, overfitting. The output from C5.0 is easy to interpret and visualize [43]. SVM is another supervised ML model which can process non-linear input data via hyperplane optimization technique [41]. The RF algorithm may be used for both classification and regression. It uses the results from multiple decision trees to arrive at an output with the help of majority voting. The RF algorithm is known to work well even for skewed data [6]. The binary LR model uses maximum likelihood method to find the best possible solution to a classification problem. The right-hand side of the LR model is a linear

Table 2. Summary of Questions with Scale Responses.

Sl No	Questions	Summary
1	A person can be infected with corona more than once	Mean = 4.32; Stdev = 0.994
2	It is necessary to wear masks after taking the coronavirus vaccine	Mean = 4.54; Stdev = 0.812
3	People who have been vaccinated against the coronavirus may get corona infection again	Mean = 4.32; Stdev = 0.919
4	Herd immunity is enough to protect everyone from the coronavirus	Mean = 3.24; Stdev = 1.139
5	It is necessary to take the coronavirus vaccine even if you have already been infected with the coronavirus	Mean = 4.42; Stdev = 1.028
6	I have doubts about vaccines in general without specific reasons	Mean = 2.42; Stdev = 1.321
7	I am concerned about effectiveness of the coronavirus vaccine	Mean = 3.18; Stdev = 1.315
8	I am concerned that there is not enough clinical data	Mean = 3.19; Stdev = 1.293
9	I am concerned about the side effects of the coronavirus vaccine	Mean = 3.02; Stdev = 1.356
10	I do not need the vaccine because I take the preventive measures seriously	Mean = 1.77; Stdev = 1.166
11	I am skeptical about the booster dose	Mean = 2.66; Stdev = 1.267
12	I am still skeptical about the covid vaccines	Mean = 2.24; Stdev = 1.276
13	I think immunity after infection with the virus is better than immunity after taking the vaccine	Mean = 3.08; Stdev = 1.391
14	I think the vaccine itself infects us with the coronavirus	Mean = 2.22; Stdev = 1.401
15	I think a vaccine will eradicate the coronavirus pandemic	Mean = 3.34; Stdev = 1.158
16	I think that the vaccine is the best way to protect against the coronavirus and its complications	Mean = 3.91; Stdev = 1.031
17	The current vaccines are effective against the coronavirus	Mean = 3.95; Stdev = 0.908

combination of the independent variables whereas the left-hand side contains the logit function or log odds ratio. Since the LR is a regression model, apart from classification we also get the statistically significant independent factors [13].

The classification efficiencies of these machine learning algorithms have been evaluated based on measures like: Overall Accuracy; Sensitivity; Specificity; Area Under ROC curve (AUROC) etc. Further, the efficiencies have been cross verified using various partitions of the dataset: 80-20, 60-40 and 50-50. In addition, methodologies like over sampling, under sampling, both over & under sampling and generation of synthetic data via ROSE (Random Over Sampling Examples) have been used to address any inherent class imbalance (for the dependent variable) in the data set. The EFA has been executed in IBM-SPSS software. The classifiers / machine learning algorithms have been implemented using R & R-studio. The objective is to accurately classify the user's intention to get vaccinated with the booster dose and also identify the significant factors that influence this decision. For the purpose of execution of different classification algorithms, all the nominal responses have been recoded as 0–1 with "Yes" labelled as "1" and "No" labelled as "0".

4 Data Analysis and Results

4.1 Exploratory Factor Analysis

The Exploratory Factor Analysis on the scale responses yields five different factors (with absolute eigen-value magnitude greater than 1) and explains about 59% of the total variation in the data. The Bartlett's test of Sphericity is significant indicating that there are significant correlations among the scale items. Kaiser-Meyer-Olkin (KMO) test of sampling adequacy has a value of 0.753 indicating that the collected data is suited for EFA.

Factor 1 loads high on the scale items no - 11, 12, 13 and 14. This factor represents the doubt of people regarding the utility of Covid-19 vaccines. Consequently, it has been named as "Sceptical". Factor 2 loads high on the scale items no - 7, 8, 9 and represents the concerns of the people regarding effectiveness of Covid-19 vaccines. Consequently, it has been named as "Concerned". Factor 3 loads high on the scale items no - 15, 16, 17 and depicts the positive sentiments of the people regarding Covid-19 vaccines. Consequently, it is named as "Positivity". Factor 4 loads high on items no - 2, 4, 5, 10 and shows the beliefs about the preventive measures regarding Covid-19 as well as the vaccines. Hence, it is named as "Preventive". Finally, the last factor 5 (loads high on item no - 1 & 3) speaks about the general beliefs of the people regarding Covid-19 and the vaccines. Hence, it is named as "Beliefs". Table 3 below provides the factor loadings from EFA. A cut-off value of 0.53 has been used to interpret the factors from the rotated factor loadings matrix.

4.2 Classification Algorithms

The five extracted factors from the EFA and the first six items with nominal responses are pooled together to form the explanatory or independent factors. The seventh item among the nominal responses i.e., "Have you been vaccinated for booster dose?" is the dependent variable indicating the user's intention to take vaccines. The dependent variable has a binary nominal response - "Yes" (1) and "No" (0). The results from implementing different machine learning classifiers (NB; RF; SVM; C5.0; LR; Adaboost; Bagging) on different partitions of the collected data is presented in the Table 4. For example, the Naïve Bayes model trained on 80% of the dataset (selected randomly) and tested on the remainder 20% leads to 89.36% overall accuracy, 100% specificity and 16.7% sensitivity scores. The remaining rows of Table 4 are to be read and interpreted similarly.

Table 3. Factor Loadings for EFA.

Scale Items	Sceptical	Concerned	Positivity	Preventive	Beliefs
I am sceptical about the booster dose- Item No 11	0.733				
I am still sceptical about the covid vaccines-Item No 12	0.660				
I think immunity after infection with the virus is better than immunity after taking the vaccine-Item No 13	0.736				
I think the vaccine itself infects us with the coronavirus-Item No 14	0.530				
I am concerned about effectiveness of the coronavirus vaccine- Item No 7		0.816			
I am concerned that there is not enough clinical data- Item No 8		0.743			
I am concerned about the side effects of the coronavirus vaccine- Item No 9		0.749			
I think a vaccine will eradicate the coronavirus pandemic-Item No 15			0.769		
I think that the vaccine is the best way to protect against the coronavirus and its complications-Item No 16			0.841		
The current vaccines are effective against the coronavirus- Item No 17			0.733		
It is necessary to wear masks after taking the coronavirus vaccine- Item No 2				−0.590	
Herd immunity is enough to protect everyone from the coronavirus- Item No 4				0.598	
It is necessary to take the coronavirus vaccine even if you have already been infected with the coronavirus- Item No 5				−0.570	
I do not need the vaccine because I take the preventive measures seriously- Item No 10				0.551	
A person can be infected with corona more than once-Item No 1					0.799
People who have been vaccinated against the coronavirus may get corona infection again- Item No 3					0.812

A careful inspection of Table 3 reveals that even though the overall accuracies and specificity of the classifiers are on the higher side, the sensitivity in all the cases are

Table 4. The classification accuracies from machine learning classifiers.

Sl No	Name of Classifier	Partitions of the dataset	Overall Accuracy (%)	Sensitivity (%)	Specificity (%)
1	NB	80-20	89.36	16.7	100
2	NB	60-40	84.85	7.14	97.65
3	NB	50-50	76.56	26.09	87.62
4	RF	80-20	87.23	0	100
5	RF	60-40	85.86	0	100
6	RF	50-50	81.25	8.7	97.14
7	SVM	80-20	87.23	0	100
8	SVM	60-40	85.86	0	100
9	SVM	50-50	82.03	0	100
10	LR	80-20	89.36	16.7	1
11	LR	60-40	86.87	7.14	100
12	LR	50-50	76.56	21.74	88.57
13	C5.0	80-20	85.11	16.7	95.12
14	C5.0	60-40	77.78	7	89.4
15	C5.0	50-50	78.91	34.8	88.6
16	Adaboost	80-20	85.11	16.7	95.12
17	Adaboost	60-40	76.77	7.14	88.23
18	Adaboost	50-50	75.78	4.35	91.43
19	Bagging	80-20	87.23	16.7	97.56
20	Bagging	60-40	84.85	0	98.82
21	Bagging	50-50	78.91	26.1	90.5

very poor. This is caused to the class imbalance problem in the dependent variable of the collected data. The class imbalance problem is seen through the following ratio.

$$\frac{No\ of\ 0s\ in\ the\ dependent\ variable}{No\ of\ 1s\ in\ the\ dependent\ variable} = 4.68$$

The number of items in the "0" class is nearly 5 times more than the number of items in the "1" class. To overcome the class imbalance problem strategies like over sampling, under sampling, both over & under sampling and Random Over Sampling Examples (ROSE) have been used. All the classification algorithms on all the partitions (80-20; 60-40 and 50-50) have been rerun for each of these strategies. The best results obtained from rerunning the algorithms are presented in the Table 5.

A careful inspection of Table 5 reveals that the Logistic Regression (LR) classifier with over sampling on the 80-20 partition of the dataset performs the best. The mentioned classifier in the given scenario has an Area under the Receiver Operating Characteristic

Table 5. Best Results after removing Class Imbalance.

Sl No	Name of Classifier	Partitions & Methodology	Overall Accuracy (%)	Sensitivity (%)	Specificity (%)
1	Adaboost	80-20 (Under Sampling)	63.83	100	58.54
2	Bagging	80-20 (Under Sampling)	70.21	83.33	68.3
3	SVM	80-20 (Both)	74.47	66.67	75.61
4	LR	80-20 (Both)	76.59	66.67	78.05
5	Adaboost	80-20 (ROSE)	76.6	66.7	78.05
6	Bagging	80-20 (ROSE)	74.47	66.67	75.61
7	NB	80-20 (Over Sampling)	74.47	66.7	70.73
8	SVM	80-20 (Over Sampling)	78.72	66.67	80.49
9	**LR**	**80-20 (Over Sampling)**	**80.85**	**83.33**	**80.49**
10	NB	80-20 (Under Sampling)	70.21	66.7	75.6
11	SVM	80-20 (Under Sampling)	76.6	83.33	75.61
12	LR	80-20 (Under Sampling)	78.72	83.33	78.05

(ROC) curve (AUC) value of 87.4% which is quite high. The ROC curve for the best classifier is presented in the Fig. 2 below. The best classifier also has a Kappa statistics value of 0.43 which is considered to indicate fair reliability for the classifier.

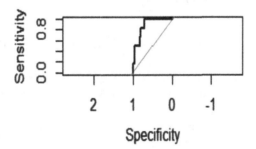

Fig. 2. ROC curve for the best classifier.

The coefficient estimates for the best classifier (LR) is provided in the Table 6.

Table 6. Coefficient Estimates for best classifier (LR).

Coefficients	Estimates	Standard Errors	p-values from Z test
Intercept	−1.3564	0.3398	6.57e-05
Sceptical	−0.4748	0.1357	0.000469
Concerned	−0.3218	0.1339	0.016262
Positivity	0.3606	0.1440	0.012269
PreventiveMeasures	−0.2241	0.1452	0.122561
Beliefs	0.1576	0.1393	0.257834
infected1	0.2605	0.3036	0.390853
comorbid1	2.223	0.4650	1.74e-06
death1	−0.5822	0.4635	0.209130
infect1	0.2563	0.3055	0.401621
hosp1	0.6313	0.3773	0.094333
precaut1	0.7148	0.3107	0.021390

Null deviance: 457.48 on 329 degrees of freedom; Residual deviance: 388.79 on 318 degrees of freedom; AIC: 412.79; Number of Fisher Scoring iterations: 4

Table 6 reveals that independent factor - Sceptical, Concerned and Positivity are significant at a significance level of 5%. Preventive Measures and Beliefs do not significantly influence the log odds ratio of taking Covid-19 vaccines. Among the nominal independent factors, comorbidities (Do you have chronic diseases such as diabetes, hypertension, heart, lung, and kidney disease), hospitalization (Did any of your family members need to be hospitalized after contracting a corona infection) and precaution (Are you still committed to the precautionary measures for protection from covid-19 infection) are statistically significant.

5 Discussion and Implications

The current study attempts to predict the user's intention (among the Indian population) to take Covid-19 vaccines with the help of state-of-the-art classifiers and ensemble techniques. The factors that influence the user's intention to take vaccines have also been identified using Exploratory Factor Analysis (EFA). Further investigation of the techniques used reveals the most important (significant) factors. The study also addresses the problem of class imbalance in the collected dataset. The current work has lot of implications for researchers, practitioners, and policy makers.

From a theoretical perspective, it is true that lot of studies exist that have tried to identify the factors that influence the user's intention to take Covid-19 vaccines across the globe. Important factors like trust in vaccines, trust in healthcare system, positive attitude of individuals, social influence, perceived benefits have been identified as the primary motivators. At the same time factors such as- possible side effects of vaccines, certain religious beliefs, misinformation, rumours have been the main detractors [3, 19, 36, 39, 55, 59]. The current study reinforces the idea that positivity or positive attitude of the Indian people increases their probability of getting vaccinated. Further, scepticism and concerns about the vaccines among the masses decrease their probability of getting vaccinated. The current study also reveals new factors like the presence of co-morbidities among individuals increase their chance of taking the vaccine. This is an important finding since Lipsitch and Dean [31] mention that elderly and people with co-morbidities have the maximum chance of getting severely infected with Covid-19. Further, hospitalization of near and dear ones due to Covid-19 prompts individuals to take vaccines. Finally, people worried about precautionary measures due to Covid-19 are more inclined to get themselves vaccinated. To summarize, the present study not only corroborates some of the explanatory factors (for user's intention to take vaccines) from other studies among the Indian population but also presents some new factors too. This may be considered as a significant contribution to the extant literature.

The results from this study also provide important information for medical practitioners, hospital management, policy makers and governments. These concerned stakeholders must take steps like mass campaigns, door to door campaigns, practical demonstrations to dissuade the negativity and scepticism among the public towards Covid-19 vaccines. There is also lot of confusion among the public regarding the preventive measures surrounding Covid-19 and the covid vaccines. This is another important area for the different stakeholders to work on. There is also lot of positivity among the public regarding the vaccines. These positive points must be emphasized upon via campaigns, advertisements etc. Further, efficient classification algorithms like logistic regression may be used to classify whether a person is likely to get himself/ herself vaccinated or not. After determining the probability of a person getting vaccinated, suitable steps may be taken to convince or facilitate the concerned person. The root-cause of the problem may be studied and ways to eradicate the same may be determined. Even though the current study has been conducted with regards to the Covid-19 vaccination program in India, it has far reached consequences across the globe. United Nations (UN) Sustainability Development Goals (SDGs) 3 emphasizes on Good Health and Well Being for all individuals. Successful vaccination programs may help to attain the mentioned SDG. The explanatory factors identified in the current study may be extended to these other vaccination programs.

6 Conclusion

The current study successfully addresses the research questions given in the first section - identifying the factors the user's intention to take Covid-19 vaccines in India and the best possible model to classify the user's intention. The study has identified "scepticism", "concern towards vaccine effectiveness", "positivity", "presence of co-morbidities", "family history of hospitalization due to Covid-19", and "precautionary measures" as the significant factors that impact user decision to take vaccines in India.

The survey participants were adequately informed about the coronavirus and its preventive vaccines, as part of the survey. Even though the study has been conducted on participants from India, it will be logical to conclude that people (anywhere across the globe) with associated medical conditions (co-morbidities) and who have witnessed the suffering of their family members due to Covid-19 are likely to take the preventive vaccine doses. Further, persons who are careful about Covid-19 precautionary measures will be more than willing to take the vaccines as the final step towards Covid-19 protection. Similarly, people across the globe who are doubtful (sceptical) and have concerns regarding vaccine effectiveness are less likely to opt for the Covid-19 vaccines. Lastly, people with positive mentality are likely to adopt the Covid-19 vaccines irrespective of their nationality. Even though there may be cultural, socio-economic, and healthcare system differences across geographies, the findings of the current study are more general that pervades across geographical boundaries.

It may be noted that the survey conducted as part of this study was not physical. One-to-one interaction (for physical survey) was avoided to maintain social distancing and covid precautions. This was important due to the repeated Covid-19 waves that impacted India between March 2020 to December 2022. Hence, an online survey was the more feasible option. We do not think that the online survey is a restrictive option since we have already argued (in Methodology section) that we have done our random sampling from clusters that represent a mini-India. Consequently, the respondents of our survey come from different age groups, gender, religion, caste, creed, and economic backgrounds. It may be further mentioned that via online surveys, we may be able to reach a wider audience and the give respondents a secure, private setting in which to react honestly and completely [9]. The challenges of a physical survey in a vast country like India are quite daunting with the involvement of lots of resources in the form of labour hours and money. Further, previous experiences have shown that many of the respondents are shy to interact in a physical survey and may not participate.

Given the increasing number of internet users in rural India (331 million in June 2022- as per [50]), we do not think that the online survey conducted as part of this study is a very restrictive option that limits the generalizability of the results. However, we understand that filling an online survey may be challenging for the elderly or for people who are unable to read or write or technologically challenged persons. This is a potential limitation of our study. Since there are not many studies that have used classification algorithms to predict user intention to take Covid-19 vaccines, this study can serve as the primary foundation for future researchers who want to conduct further research by addressing the mentioned limitations.

The study may be improved further by considering a larger sample size of respondents. Future researchers may also consider a wider global population in their samples instead of just India. Further, more advanced algorithms like deep learning and more sophisticated ensemble techniques may be considered.

References

1. Abara, W.E., et al.: Reports of Guillain-Barré syndrome after COVID-19 vaccination in the United States. JAMA Network Open **6**(2), e2253845–e2253845 (2023)
2. Afonso, N.M., Kavanagh, M.J., Swanberg, S.M., Schulte, J.M., Wunderlich, T., Lucia, V.C.: Will they lead by example? Assessment of vaccination rates and attitudes to human papilloma virus in millennial medical students. BMC Public Health **17**(1), 1–8 (2017)

3. Al-Qerem, W., Hammad, A., Alsajri, A.H., Al-Hishma, S. W., Ling, J., Mosleh, R.: COVID-19 vaccination acceptance and its associated factors among the Iraqi population: a cross sectional study. Patient Preference and Adherence **16**, 307–319 (2022)
4. Al-Qerem, W.A., Jarab, A.S.: COVID-19 vaccination acceptance and its associated factors among a Middle Eastern population. Front. Public Health **9**, 632914 (2021)
5. Amani, D., Ismail, I.J.: Investigating the predicting role of COVID-19 preventive measures on building brand legitimacy in the hospitality industry in Tanzania: mediation effect of perceived brand ethicality. Future Bus. J. **8**(1), 13 (2022)
6. Azar, A.T., Elshazly, H.I., Hassanien, A.E., Elkorany, A.M.: A random forest classifier for lymph diseases. Comput. Methods Programs Biomed. **113**(2), 465–473 (2014)
7. Belete, T.M.: Review on up-to-date status of candidate vaccines for COVID-19 disease. Infect. Drug Resist. **14**, 151–161 (2021)
8. Bloom, D.E., Cadarette, D., Ferranna, M.: The societal value of vaccination in the age of COVID-19. Am. J. Public Health **111**(6), 1049–1054 (2021). https://doi.org/10.2105/AJPH.2020.306114, PubMed: 33856880, PubMed Central: PMC8101582
9. Cantrell, M.A., Lupinacci, P.: Methodological issues in online data collection. J. Adv. Nurs. **60**(5), 544–549 (2007). https://doi.org/10.1111/j.1365-2648.2007.04448.x
10. CDC.: http://cdc.gov/coronavirus/2019-ncov/vaccines/different-vaccines.html Retrieved 11 Nov 2023
11. Chen, D., et al.: Recurrence of positive SARS-CoV-2 RNA in COVID-19: a case report. Int. J. Infect. Dis. **93**, 297–299 (2020). https://doi.org/10.1016/j.ijid.2020.03.003
12. Di Domenico, L., Pullano, G., Sabbatini, C.E., Boëlle, P.Y., Colizza, V.: Impact of lockdown on COVID-19 epidemic in Île-de-France and possible exit strategies. BMC Med. **18**(1), 240 (2020). https://doi.org/10.1186/s12916-020-01698-4
13. Dinesh Kumar, U.: Business analytics: the science of data-driven decision making. Wiley (2017)
14. Dror, A.A., et al.: Vaccine hesitancy due to vaccine country of origin, vaccine technology, and certification. Eur. J. Epidemiol. **36**(7), 709–714 (2021)
15. Elgendy, M.O., Abdelrahim, M.E.: Public awareness about coronavirus vaccine, vaccine acceptance, and hesitancy. J. Med. Virol. **93**(12), 6535–6543 (2021)
16. Falcone, R., et al.: Trusting COVID-19 vaccines as individual and social goal. Sci. Rep. **12**(1), 9470 (2022)
17. Geoghegan, S., O'Callaghan, K.P., Offit, P.A.: Vaccine safety: myths and misinformation. Front. Microbiol. **11**, 372 (2020). https://doi.org/10.3389/fmicb.2020.00372
18. Goruntla, N., et al.: Predictors of acceptance and willingness to pay for the COVID-19 vaccine in the general public of India: a health belief model approach. Asian Pac J Trop Med **14**(4), 165–175 (2021)
19. Jain, L., et al.: Factors influencing COVID-19 vaccination intentions among college students: a cross-sectional study in India. Front. Public Health **9**, 735902 (2021)
20. Jordan Times: Jordan to begin COVID-19 vaccination drive by February — Health minister (2020). Accessed 21 Nov 2023
21. Kelekar, A.K., Lucia, V.C., Afonso, N.M., Mascarenhas, A.K.: COVID-19 vaccine acceptance and hesitancy among dental and medical students. J. Am. Dent. Assoc. **152**(8), 596–603 (2021)
22. Kennedy, J.: Populist politics and vaccine hesitancy in Western Europe: An analysis of national-level data. Eur. J. Pub. Health **29**(3), 512–516 (2019). https://doi.org/10.1093/eurpub/ckz004
23. Khan, Y.H., et al.: Threat of COVID-19 vaccine hesitancy in Pakistan: The need for measures to neutralize misleading narratives. Am. J. Trop. Med. Hyg. **103**(2), 603–604 (2020). https://doi.org/10.4269/ajtmh.20-0654

24. Khuroo, M.S., Khuroo, M., Khuroo, M.S., Sofi, A.A., Khuroo, N.S.: COVID-19 vaccines: a race against time in the middle of death and devastation! J. Clin. Exp. Hepatol. **10**(6), 610–621 (2020). https://doi.org/10.1016/j.jceh.2020.06.003. PubMed: 32837093
25. Kwok, K.O., Li, K.K., Wei, W.I., Tang, A., Wong, S.Y.S., Lee, S.S.: Editor's Choice: Influenza vaccine uptake, COVID-19 vaccination intention and vaccine hesitancy among nurses: a survey. Int. J. Nurs. Stud. **114**, 103854 (2021). https://doi.org/10.1016/j.ijnurstu.2020.103854
26. La, V.-P., et al.: Policy response, social media and science journalism for the sustainability of the public health system amid the COVID-19 outbreak: The Vietnam Lessons. Sustainability **12**(7), 2931 (2020). https://doi.org/10.3390/su12072931
27. Langmuir, A.D., Bregman, D.J., Kurland, L.T., Nathanson, N., Victor, M.: An epidemiologic and clinical evaluation of Guillain-Barré syndrome reported in association with the administration of swine influenza vaccines. Am. J. Epidemiol. **119**(6), 841–879 (1984)
28. Larson, H.J., et al.: SAGE Working Group on Vaccine Hesitancy.: Measuring vaccine hesitancy: The development of a survey tool. Vaccine, **33**(34), 4165–4175 (2015). https://doi.org/10.1016/j.vaccine.2015.04.037
29. Lastrucci, V., et al.: SARS-CoV-2 seroprevalence survey in people involved in different essential activities during the General Lock-Down Phase in the province of Prato (Tuscany, Italy). Vaccines **8**(4), 778 (2020). https://doi.org/10.3390/vaccines8040778. GoogleScholar PubMed: 33352743
30. Levison, L.S., Thomsen, R.W., Andersen, H.: Guillain-Barré syndrome following influenza vaccination: a 15-year nationwide population-based case–control study. Eur. J. Neurol. **29**(11), 3389–3394 (2022)
31. Lipsitch, M., Dean, N. E.: Understanding COVID-19 vaccine efficacy. Science **370**(6518), 763–765 (2020). https://doi.org/10.1126/science.abe5938, PubMed: 33087460
32. Lounis, M., Bencherit, D., Rais, M.A., Riad, A.: COVID-19 vaccine booster hesitancy (VBH) and its drivers in Algeria: national cross-sectional survey-based study. Vaccines **10**(4), 621 (2022)
33. Lytras, T., Tsiodras, S.: Lockdowns and the COVID-19 pandemic: what is the endgame? Scandinavian J. Public Health **49**(1), 37–40 (2021). https://doi.org/10.1177/1403494820961293, PubMed: 32981448, PubMed Central: PMC7545298
34. Lyu, Y., et al.: The acceptance of COVID-19 vaccination under different methods of investigation: based on online and on-site surveys in China. Front. Public Health **9**, 760388 (2021)
35. MacDonald, N.E., Eskola, J., Liang, X., Chaudhuri, M., Dube, E., Gellin, B.: Vaccine hesitancy: definition, scope and determinants. Vaccine **33**(34), 4161–4164 (2015). https://doi.org/10.1016/j.vaccine.2015.04.036. GoogleScholar PubMed: 25896383
36. Mir, H.H., Parveen, S., Mullick, N.H., Nabi, S.: Using structural equation modeling to predict Indian people's attitudes and intentions towards COVID-19 vaccination. Diabetes Metab. Syndr. **15**(3), 1017–1022 (2021)
37. Mohamed, N.A., Solehan, H.M., Mohd Rani, M.D., Ithnin, M., Che Isahak, C.I.: Knowledge, acceptance and perception on COVID-19 vaccine among Malaysians: a web-based survey. PLoS ONE **16**(8), e0256110 (2021)
38. Netti, K., Radhika, Y.: A novel method for minimizing loss of accuracy in Naive Bayes classifier. In: 2015 IEEE International Conference on Computational Intelligence and Computing Research (ICCIC) (pp. 1–4). IEEE (2015)
39. Ng, J.W.J., Vaithilingam, S., Nair, M., Hwang, L.A., Musa, K.I.: Key predictors of COVID-19 vaccine hesitancy in Malaysia: an integrated framework. PLoS ONE **17**(5), e0268926 (2022)
40. Nicola, M., et al.: The socio-economic implications of the coronavirus pandemic (COVID19): a review. Int. J. Surg. **78**, 185–193 (2020). https://doi.org/10.1016/j.ijsu.2020.04.018
41. Noble, W.S.: What is a support vector machine? Nat. Biotechnol. **24**(12), 1565–1567 (2006)

42. Omer, S.B., Salmon, D.A., Orenstein, W.A., DeHart, M.P., Halsey, N.: Vaccine refusal, mandatory immunization, and the risks of vaccine-preventable diseases. N. Engl. J. Med. **360**(19), 1981–1988 (2009). https://doi.org/10.1056/NEJMsa0806477
43. Pandya, R., Pandya, J.: C5. 0 algorithm to improved decision tree with feature selection and reduced error pruning. Int. J. Comput. Appl. **117**(16), 18–21 (2015)
44. Pearlman, J., Cheong, D., Huang, C.: Zero-Covid-19 strategy: does the approach still work with rise of Delta variant? The Straits Times Cited March 22, 2022 (2021). http://straitstimes.com/asia/australianz/zero-covid-19-strategy-does-the-approach-still-work-with-rise-ofdelta-variant
45. Roy, D.N., Biswas, M., Islam, E., Azam, M.S.: Potential factors influencing COVID-19 vaccine acceptance and hesitancy: a systematic review. PLoS ONE **17**(3), e0265496 (2022)
46. Sayed, A.M., Khalaf, A.M., Abdelrahim, M.E.A., Elgendy, M.O.: Repurposing of some anti-infective drugs for COVID-19 treatment: a surveillance study supported by an in silico investigation. Int. J. Clin. Pract. **75**(4), e13877 (2021). https://doi.org/10.1111/ijcp.13877
47. Smith (2020). https://www.reuters.com/article/ushealth-coronavirus-southkorea-idUSKCN21V0JQ Retrieved 21 Nov 2020. Reuters
48. Soares, P., et al.: Factors associated with COVID-19 vaccine hesitancy. Vaccines **9**(3), 300 (2021)
49. Suresh, A., Konwarh, R., Singh, A.P., Tiwari, A.K.: Public awareness and acceptance of COVID-19 vaccine: an online cross-sectional survey, conducted in the first phase of vaccination drive in India (2021). https://doi.org/10.21203/rs.3.rs-324238/v1
50. Statista. India: number of internet connections in rural and urban areas 2023 | Statista (2023)
51. Szmyd, B., et al.: Attitude and behaviors towards SARS-CoV-2 vaccination among healthcare workers: a cross-sectional study from Poland. Vaccines **9**(3), 218 (2021)
52. Tolia, V., Renin Singh, R., Deshpande, S., Dave, A., Rathod, R.M.: Understanding factors to COVID-19 vaccine adoption in Gujarat, India. Int. J. Environ. Res. Public Health **19**(5), 2707 (2022)
53. Troiano, G., Nardi, A.: Vaccine hesitancy in the era of COVID-19. Public Health **194**, 245–251 (2021)
54. Umakanthan, S., Bukelo, M.M., Bukelo, M.J., Patil, S., Subramaniam, N., Sharma, R.: Social environmental predictors of COVID-19 vaccine hesitancy in India: a population-based survey. Vaccines **10**(10), 1749 (2022)
55. Vaithilingam, S., Hwang, L.A., Nair, M., Ng, J.W.J., Ahmed, P., Musa, K.I.: COVID-19 vaccine hesitancy and its drivers: an empirical study of the vaccine hesitant group in Malaysia. PLoS ONE **18**(3), e0282520 (2023)
56. Williams, S.E.: What are the factors that contribute to parental vaccine- hesitancy and what can we do about it? Hum. Vaccin. Immunother. **10**(9), 2584–2596 (2014). https://doi.org/10.4161/hv.28596
57. WHO.: Coronavirus Disease (COVID-19) Dashboard. https://covid19.who.int/ Retrieved 20 Jun 2021
58. WHO.: Draft landscape of COVID-19 candidate vaccines. https://www.who.int/publications/m/item/draft-landscape-of-covid-19. Accessed 11 Apr 2021
59. Wong, L.P., Alias, H., Wong, P.F., Lee, H.Y., AbuBakar, S.: The use of the health belief model to assess predictors of intent to receive the COVID-19 vaccine and willingness to pay. Hum. Vaccin. Immunother. **16**(9), 2204–2214 (2020)

Unsupervised Incremental-Decremental Attribute Learning Healthcare Application Based Feature Selection

Siwar Gorrab(✉), Fahmi Ben Rejab, and Kaouther Nouira

Université de Tunis, ISGT, LR99ES04 BESTMOD, Tunis, Tunisie
siwarg9@gmail.com

Abstract. The need for advanced medicare systems that provide realtime decisions and swift outcomes at earlier stages has emerged as a promising candidate, motivated by the intersection of data streams and potent machine learning techniques. These data streams might join the learning process with evolving mixed features, where it is the requirement for handling the incremental and decremental attribute and instance learning tasks. Here we introduce our proposed solution that copes with the weaknesses in healthcare systems: how to learn real-time patient's data to make people's lives advantageous and healthier. Notably, when these new chunks of healthcare data are continuously forthcoming with missing values, redundancies or inconsistencies, data preprocessing is requested. In this paper, a developed incremental and also decremental learning healthcare application is provided based on k-prototypes algorithm and mRMR feature selection technique. It helps to understand new diseases and therapies, and further to advance healthcare monitoring systems and to enhance clinical care based on mRMR feature selection technique. This proposal presents encouraging experimental results compared to the batch k-prototypes method and number of similar incremental healthcare methods. The obtained clusters' inertia and run time results emphasize the scalability and the performance of our proposed real-time healthcare application.

Keywords: Incremental-Decremental Attribute Learning · Mixed Attributes · K-Prototypes · Feature Selection · mRMR · Healthcare Application

1 Introduction

At present, healthcare and medicine, much like various other sectors, are encountering a convergence of two trends: a proliferation of big data and the advancement of machine learning methodologies capable to develop patterns within these data sets. Hence, real-time data mining applications are swiftly evolving to augment healthcare systems by creating tools aimed at enhancing the clinical care process, advancing medical research, and refining efficiency [1]. These tools rely on machine learning algorithms, invented from healthcare data that can make predictions, real-time decisions and even future recommendations. In other words, data mining becomes the only hope for elucidating the patterns that underlie it.

In medicine and healthcare areas, the data comes from multiple sources such as clinical studies, pharmaceutical trials, electronic health records, diagnostic trials, healthcare information entered by patients into their smartphones or recorded on fitness trackers, or even insurance claims data. Every day, a substantial amount of intricate healthcare data, consisting of various types, is continuously generated from the sources mentioned. These multiple data sources mentioned send streams of data records, referred to as data streams, which are sequences of data elements acquired from various origins requiring further analysis and investigation. Hence, it can be acknowledged that one of the primary sources of big data is these data streams [2]. It is received at a very giant speed and volume which makes it a tough and complicated task for batch machine algorithms to create and to maintain the desired results for future significant decisions.

Arising from the fact that healthcare and medical data is of mixed type, means it contains both numerical as well as categorical features, we have directed our research to cluster them using the unsupervised k-prototypes clustering technique. However, these mixed data streams continuously evolve, either by adding new features or by featuring fewer elements. This is due to the unknown full feature space in advance. Indeed, within many practical scenarios, the upcoming chunks of sampling data (data streams) are evolving throughout the learning process. Expressing differently, some data samples are added with continuously arriving data streams, other existing ones are removed (e.g., those that are too noisy or inconsistent). This is exactly the same case for the evolving feature space where data streams may join the learning process with added and/ or deleted attributes. This data describes particulars about patients, their medical tests and analysis, and their treatments. Thereupon, two scenarios might be appealed: either to retraining the model from scratch every time a new data stream takes place which is too time-consuming, or to neglect the old data and focus only on the new emerging ones which leads to lose knowledge and inconsistent models. Thus, it would be more qualified and efficient to update the existing model by including or excluding the evolving data samples with evolving mixed attributes. This is known as incremental and decremental attribute and object learning.

For instance, doctors frequently encounter such scenarios when monitoring a patient's progress or categorizing patients with rare conditions or unique combinations of comorbidities. Indeed, new medical and healthcare data streams are ceaselessly joining. For that reason, there is a huge need to handle such context in healthcare field in order to pull out the most useful and hidden information and to enhance health monitoring systems performance. Accordingly, there is a huge need to manage such dynamic attribute with object spaces in healthcare and medical fields to ensure the greatest real-time patient's follow-up.

Add it to that, the new added features are not all useful for mining and decision making if they come with such forms of anomalies: inconsistency, redundancy [3]. In an effort to identify a subset of the most pertinent features among the newly added ones, we've opted to discard irrelevant and duplicate attributes, which could potentially introduce unwanted correlations in subsequent mining endeavors [4]. While feature selection remains infrequently employed in unsupervised tasks such as clustering, the utilization of dynamic features poses fresh challenges for streamlining the selection process [5].

Such an approach would significantly enhance the subsequent healthcare analysis process. To do so, we have used the minimum-Redundancy-Maximum-Relevance (mRMR) technique [6] as being the minimal-optimal attribute selection algorithm, intended To identify the most concise yet relevant set of features for a specific machine learning task. So, we aim to develop a real-time healthcare monitoring system that well performs when operating in dynamic feature space. This means when new medical parameters are added by doctors, and have to be taken in consideration in order to correctly predict whether the patient's situation is critical and needs doctor's check-up or not. Same thing for intelligent systems that help doctors in decision making, there is a need for incremental and decremental learning at the arrival of new patients' data streams. Hence, the research question of this study is: how can incremental and decremental mixed attribute and object learning lead to enhance healthcare monitoring systems? and whether applying the mRMR feature selection preprocessing technique in unsupervised dynamic context would upgrade the proposals' performance in medical and healthcare fileds? This explains how we have selected proposed method for this problem. In this paper, we present a novel healthcare and medical application to manage such evolving instance and mixed attribute spaces based on incremental and decremental k-prototypes method and using the mRMR feature selection preprocessing technique. Accordingly, our proposal capes with the need to tease out maximum medical and healthcare insights from those number of emerging data points with increasing feature and object domains. This leads to the development of a rich toolkit of a real-time machine learning healthcare method and decision-making system. The primary contributions outlined in this paper include: (i)Incremental and decremental attribute learning healthcare monitoring application… (ii)we extend and upgrade our proposed healthcare application with selecting only relevant features among the new incoming ones with the mRMR feature selection preprocessing technique.

The performance of our proposal is validated based on number of evaluation measures that emphasize its capability compared to the batch k-prototypes method and two other similar healthcare applications. We can cite namely the run time, the clusters' inertia (Sum of Squared Error) and the Silhouette Coefficient score. These used evaluation measures emphasize the quality and the capability of our proposal.

For illustrative purposes, this proposed real-time unsupervised incremental and decremental healthcare application might be certainly useful for advanced medical research. It is also able to enhance the process of clinical care in the following way: a doctor cares about patients' health. He needs to cope with dynamic features and instance domains when all the time new biological parameters have to be monitored, new analysis have to be done, and results should be timely provided to healthcare monitoring systems in order to guarantee best patient's follow-up. Hence, our proposal might deal with this issue and is of great importance for ensuring safe patient's follow-up, and correctly predicted alarms that are crucial for patients' life. This is due to the fact that our proposal deals with these new incoming patients' data streams, joined with evolving features (some added ones and other deleted ones). It also helps in making real-time life decisions. This proposal may be worthwhile in detecting new diseases. For example, corona symptoms are wide and spread. Such dynamic healthcare system can help in diagnosing this pandemic at real-time where patients owing the same symptoms own the same disease.

The rest of the paper is structured in the following way: Sect. 2 is dedicated to theoretically define the standard k-prototypes algorithm and to display many related works to our proposal's context. In the next Sect. 3, we deeply explain our proposed idea. Hereafter, Sect. 4 details the experimental simulations and the performance assessment methodology. In closing, in Sect. 5 we conclude the paper.

2 K-prototypes Method and Related Works

In this section, we firstly present the theoretical definition of the conventional k-prototypes method [7]. Secondly, we illustrate few related works in the context of handling incremental and decremental data mining applications in addition to some related clustering healthcare applications and unsupervised feature selection methods.

2.1 Theoretical Definition of K-prototypes Method

As being proposed by Huang in [7] for clustering large data sets with numerical and categorical values, the k-prototypes clustering algorithm is an extension to the k-means [8] and k-modes [9] algorithms. It is the most popular and outstanding algorithm that capes with mixed data clustering.

Formulation. Considering $X = \{x_1 \ldots x_n\}$, A dataset consisting of n mixed data points, including mr quantitative attributes and mt qualitative attributes, is used. The primary objective of the k-prototype algorithm is to cluster the dataset x into k distinct clusters (groups) while minimizing the cost function j as described in Eq. (1).

$$J = \sum_{i=1}^{n} \sum_{j=1}^{k} u_{ij} d(x_i - c_j) \tag{1}$$

where $u_{ij} \in \{0,1\}$ is an item of the partition matrix $U_n * k$ denoting the membership of a data point i in cluster j;

$c_j \in C = \{c_1 \ldots c_k\}$ is the center of the group j, and $d(x_i - c_j)$ is the dissimilarity metric outlined in the following Eq. (2):

$$d(x_i - c_j) = \sum_{r=1}^{m_r} \sqrt{(x_{ir} - c_{jr})^2} + \sum_{t=1}^{m_t} \delta(x_{it}, c_{jt}) \tag{2}$$

such that x_{ir} and x_{it} depict respectively the values of the numeric attribute r and the categorical attribute t for a data point i and c_{jr} is the mean of the numeric attribute r and the cluster j calculated as follows in Eq. (3):

$$c_{jr} = \frac{\sum_{i=1}^{|c_j|} x_{ir}}{|c_j|} \tag{3}$$

with $|c_j|$ is the number of data points assigned to a cluster j, c_{jt} is mode for categorical attributes t and cluster j, calculated by Eq. (4):

$$c_{jt} = a_t^h \qquad (4)$$

where $f(a_t^h) \geq f(a_t^z), \forall z, 1 \leq z \leq m_c$.

Having $a_t^z \in \{a_t^1 \ldots a_t^{m_c}\}$ is the categorical value z and mc is the number of categories of categorical attribute t, $f(a_t^z) = |\text{ xit = azt } | \text{ pij = 1 }|$ is the frequency count of the attribute value azt; knowing that for categorical features $\delta(p,q) = 0$ when $p = q$ and $\delta(p,q) = 1$ when $p \neq q$.

Algorithm

Algorithm 1 Conventional K-prototypes Algorithm
1: **Input:** X: data set, k: number of clusters
2: **Output:** Cluster centers
3: **Begin**
4:    Choose at random k initial cluster centres from the data set X.
5:    Assign each data point in X to its nearest cluster center using Equation (2).
6:    Update the cluster centres after each allocation using Equation (3) and Equation (4).
7:    If the updated cluster centers are identical to the previous ones then terminate, otherwise, go back to step 5.
8: **End.**

2.2 Related Works

Incremental and Decremental Attribute and Instance Learning Algorithms

In the context of online learning, several incremental and decremental learning algorithms have been proposed:

- In [10], Dong et al. introduced an enhanced metric learning method for managing both incremental and decremental attributes. This method simultaneously handles instance and attribute evolutions by integrating a smoothed Wasserstein metric distance.
- In [11], a new online incremental and decremental learning algorithm was developed based on a variable support vector machine. This approach addresses the long execution times and low efficiency of traditional Support Vector Machine algorithms with large-scale training samples.
- Additionally, the study in [12] presents a one-pass learning approach for incremental and decremental features. This method scans each instance only once and does not require storing the entire dataset, effectively accommodating the nature of evolving streaming data.

Meanwhile, lee et al. have developed a novel incremental and decremental learning method for the least-squares support vector machine in [13], which can adapt a pre-trained model to changes in the training data set, without retraining the model on all the data from scratch, and where the changes may include addition and deletion of data samples. Recenty, Gorrab et al. have proposed an incremental and decremental attribute learning algorithm based on k-prototypes clustering method for mixed data streams that can accommodate a large number of updates, owing to the training samples which become available one after another over time and with evolving feature space [14]. Recently, authors in [15] developed a dynamic unsupervised k-prototypes algorithm that incrementally and decrementally clusters continuously arriving data streams accompanied by new mixed attributes at real-time and within memory and time restrictions based on a developed merge and split methods.

Unsupervised Feature Selection Techniques
Feature selection preprocessing techniques are widely used and implemented within multiple supervised machine learning algorithm, and still rarely used in unsupervised tasks. Thereby, Almusallam et al. have extended the k-means algorithm in [5] to produce an efficient unsupervised attributes selection method for streaming features applications where the number of attributes increases while the number of observations remains fixed. Also, another study in [3] proposes a real-time feature selection k-prototypes method for incremental features and object learning. Likewise, in [16] Shi et al. have newly produced an unsupervised adaptive attributes selection method with binary hashing model, aiming to exploit the discriminative information under the unsupervised scenarios. Another study in [17] developed an unsupervised spectral feature selection algorithm for high dimensional data in which features are clustered using advanced Self-Tuning spectral clustering method based on local standard deviation, so as to detect the global optimal feature clusters. Lastly, the study in [18] proposed an unsupervised feature selection method based on a variance threshold technique in a context of handling incremental attribute learning task for medical and healthcare applications. Namely in [19], authors proposed a a novel attribute selection method, namely: low-redundant unsupervised attribute selection based on data structure learning and feature orthogonalization which selects low-redundant features by using the orthogonal representation idea to obtain uncorrelated features. While Unsupervised Feature Selection (UFS) methods have attracted considerable interest in different research domains due to their wide application in problems where unlabeled data appears, a filter unsupervised feature selection method for mixed data is presented in [20] that addresses the unsupervised feature selection problem in the following two stages: combination of spectral feature selection to identify relevant features and a pair-wise redundancy analysis to remove those features with a high correlation with others. As importantly, study in [21] proposes a technique for grouping breast cancer from breast ultrasound images utilizing deep learning and feature-selection algorithms based on meta-heuristic methods: firstly, they extract features from ultrasound breast images using a standard deep learning model, then they choose an optimal set of features from this deep feature set employing an unsupervised version of the Whale Optimization Algorithm (U-WOA) in order to reduce the number of redundant and noninformative features initially extracted from the deep learner and to build a more robust framework.

Unsupervised clustering Healthcare Applications
As long as clustering methods help in discovering and recognizing hidden patterns in healthcare domain, Numerous clustering algorithms are presently suitable for clustering healthcare data. In that deal, study in [22] has developed an application of unsupervised

clustering methods to Alzheimer's disease (AD) to offer insights into which clustering technique best suits the partitioning of patients of AD based on their similarity and further discuss the implications of the use of clustering algorithms in the treatment of AD. Authors in [23] have also proposed an unsupervised machine learning approach for the discovery of latent disease clusters and patient subgroups employing electronic health records; they utilized Latent Dirichlet Allocation (LDA), a generative probabilistic model, and proposed a new model named Poisson Dirichlet Model (PDM) to model patients' disease diagnoses and to alleviate age and sex factors. Papers such as [24] has proposed a visual topic model for clustering healthcare data, which aims to properly extract health topics for analyzing discriminative and coherent latent features of tweet documents in healthcare applications. As well, study in [2] provides a better insight into a real-time healthcare and medical application encompassing various medical case studies such as patient monitoring, disease management, and clinical support systems to enhance disease prediction. Besides, Gorrab et al. propose in [18] an innovative incremental k-prototypes based feature selection method for medical and healthcare applications that improves the process of clinical care, advances medical research, and improves efficiency based on feature selection. Finally, the study discussed in [25] addresses the challenge of diagnosing and analyzing the prevalence of COVID-19 cases, as well as dealing with a large volume of patient records lacking labels. It introduces an enhanced version of a recent direct method that jointly estimates non-negative cluster indices and spectral embeddings. The proposed model brings about improvements and demonstrates promising results in the categorization of radiographs.

3 Proposed Healthcare Approach

This section pulls out the practical scenario of our proposed approach along with its objectives and hidden particularities. In addition, the feature selection module is subsequently detailed.

3.1 Definition and Approach Presentation

To begin, we present an online incremental and decremental attribute and object learning healthcare application designed for mixed data stream clustering, utilizing a feature selection preprocessing technique. Online feature learning is often favored for classifying large-scale data thanks to its mean computational cost. But in practical scripts, features evolve such as when some features disappear and new ones emerge, many metric learning models fail to adapt effectively, even though they handle evolving instances well. This situation is commonly encountered by doctors when analyzing patient data, health monitoring records, and various medical situations. This is owing to the fact that healthcare features are continuously increasing and disappearing in a streaming data environment. Thus, our proposed Incremental and Decremental Healthcare K-prototypes approach (IDHKprototypes) based mRMR feature selection technique can deal with many medical and healthcare real tasks where the features are evolving, with some of them being vanished and some others being augmented. Add it to that, our proposal capes with the weaknesses of the previously proposed IHK-prototypes healthcare application [2], in terms of learning and further analysing medical data streams with evolving attribute and instance spaces. As well, our proposed IDHK-prototypes based mRMR approach is capable of handling continuously emerging mixed data streams with evolving feature and object spaces within memory and time restrictions. Furthermore, it makes a pre-selection of new joined attributes using the mRMR technique in order to choose the most relevant and significant ones.

So, our proposal clusters the continuous incoming medical data streams, with evolving mixed feature and instances (newly joined ones and / or vanished ones), within constraints of time and memory consumption. Keeping these constraints in one's head, our proposed real-time healthcare solution fulfills the following requirements entirely:

Provides real-time results by performing fast, incremental and decremental feature and object learning of incoming healthcare mixed data streams without relearning from the beginning.

Adapt dynamically to changing of the data and selects a group of relevant incoming features among all new emerging ones.

Offer a dynamic model representation that remains compact and does not expand with the number of observations and attributes processed. This is, specifically, without requiring that the whole input data being loaded in the memory from the beginning of the learning procedure.

The following Algorithm 2 well describes our proposal's steps of functioning.

Algorithm 2 Incremental and Decremental Healthcare K-prototypes based mRMR Algorithm

---
1: **Input:** X: data set, k: number of clusters
2: **Output:** k cluster centers
3: **Begin**
4:   Choose at random k initial cluster centres from the data set X.
5:   Assign each data point in X to its nearest cluster center using Equation (2).
6:   Update the cluster centres after each allocation using Equation (3) and Equation (4).
7:   If the updated cluster centers are identical to the previous ones **then** terminate, otherwise, go back to step 5.
8:   If a new healthcare mixed data stream emerges with vanished attributes and data objects and/ or escorted with new ones **then** Select a subset of relevant features amongst new ones based on mRMR feature selection technique,
    Return to step 5,
    Merge the resulting clusters from initial and data stream models
    Impute missing values for all resulted merged clusters, otherwise, terminate.
9: **End.**

Figure 1 enlightens an overview of the developed healthcare and medical approach, delineated in three main steps which will be elaborated upon later. The overview of our proposal, presented in Fig. 1, is detailed so:

1. Several healthcare and medical sources send continuously mixed data streams. These latters may arrive with departed features as well as they might be escorted with new added ones, in a context of evolving feature and object spaces.

2. As a mandatory step, we highly prioritize selecting the most pertinent patient's features amongst the new emerging ones. To do so, we have opted to use the minimum-Redundancy-Maximum-Relevance (mRMR) technique [6] as being the minimal-optimal feature selection algorithm.
3. Subsequently, the selected features would be injected to the incremental and decremental learning module, based on the recently proposed IDKprototypes method [14] founded on the merge technique.

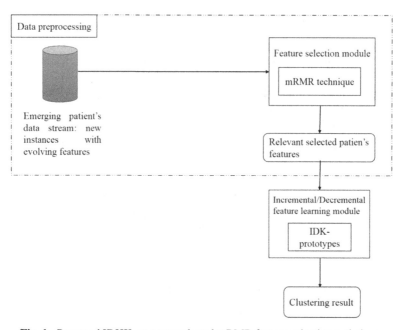

Fig. 1. Proposed IDHK-prototypes based mRMR feature selection technique.

To provide a clearer understanding and define the merge technique that underpins the incremental-decremental feature learning module, it is essential to detail the merge algorithm proposed in [14]. The merge procedure, as incorporated into our proposed IDHK-prototypes, is outlined in the following Algorithm 3:

Algorithm 3 Merge algorithm
1: **Input:** clusters = $\{c_1, c_2, c_3, c_4, c_5, c_6\}$
2: **Output:** clusters = $\{c'_1, c'_2, c'_3\}$
3: **Begin**
4:   For each cluster A_i in $\{c_1, c_2, c_3\}$ do
5:     For each cluster B_j in $\{c_4, c_5, c_6\}$ do
6:       $DB \leftarrow computeDaviesBouldinIndex(A_i, B_j)$
7:       $CH \leftarrow computeCalinskiHarabaszIndex(A_i, B_j)$
8:     end For
9:   end For
10:   For each cluster A_i in $\{c_1, c_2, c_3\}$ do
11:     For each cluster B_j in $\{c_4, c_5, c_6\}$ do
12:       If $computeDaviesBouldinIndex(A_i, B_j) = \text{Max}(DB)$ then
13:         $Cluster1 \leftarrow A_i$
14:         $Cluster2 \leftarrow B_j$
15:       end If
16:       If $computeCalinskiHarabaszIndex(A_i, B_j) = \text{Min}(CH)$ then
17:         $Cluster3 \leftarrow A_i$
18:         $Cluster4 \leftarrow B_j$
19:       end If
20:     end For
21:   end For
22:   If (Cluster1=Cluster3) and (Cluster2=Cluster4) then
23:     NewCluster $\leftarrow$ Merge (Cluster1,Cluster2)
24:     Delete (Cluster1)
25:     Delete (Cluster2)
26:   end If
27:     $SSE1 \leftarrow SSE(Merge(Cluster1, Cluster2))$
28:     $SSE2 \leftarrow SSE(Merge(Cluster3, Cluster4))$
29:   If SSE1 < SSE2 then
30:     NewCluster $\leftarrow$ Merge (Cluster1,Cluster2)
31:     Delete (Cluster1, Cluster2)
32:   Else
33:     NewCluster $\leftarrow$ Merge (Cluster3,Cluster4)
34:     Delete (Cluster1, Cluster2)
35:   end If
36: **Return** clusters = $\{c'_1, c'_2, c'_3\}$
37: **End.**

According to the merge algorithm, presented in Algorithm 3, we go forward to detail and explain some used functions.

- Compute (index):

  - **Input:** Clusters.
  - **Output:** Davies-Bouldin index (DB) and Calinski-Harabasz (CH) score matrices.

Compute (index) function is the basic function in the merge algorithm since it determines the best couple of clusters to be merged as long as it calculates the Davies-Bouldin index and the Calinski-Harabasz score.

- Min (CH/DB), Max (CH/DB):

- **Input:** Matrix of indexes.
- **Output:** lowest or highest index.

Max (CH/DB) and Min (CH/DB) functions are used in our proposed merge algorithm to search respectively about the lowest and the highest indexes from the input matrices.

– Merge (clusters):

- **Input:** clusters (i and j).
- **Output:** cluster ij.

The Merge(clusters) function combines the two most similar clusters into a single cluster, incorporating elements from both original clusters. This merge function effectively integrates the extended number of attributes and instances, as well as those that are eliminated, with the newly joined data stream, ensuring a smooth incorporation into the learning process.

– Delete(cluster):

- **Input:** clusters.
- **Output:** removed clusters.

After fusing the most appropriate clusters into one single cluster, we need to delete each single one of them so that to remove redundancy.

NewCluster:

As indicated by its name, NewCluster refers to the cluster result obtained from the merge procedure, which aligns with the main objective of our approach: attributes and objects are incrementally and decrementally learned using our proposed IDHK-prototypes method.

3.2 Feature Selection Module

Undoubtedly, various feature selection preprocessing techniques can be employed to extract a subset of the most relevant features. However, feature selection is rarely utilized in unsupervised tasks, such as clustering, making this study particularly challenging. To identify the smallest relevant subset of features for our unsupervised clustering task in the medical and healthcare fields, we have chosen to use the mRMR technique, as proposed in [6]. This technique is designed to determine the smallest subset of features that ensures minimum redundancy and maximum relevance. In other words, it intends to select the minimum number of useful features for small memory consumption, less time require, better performance, and enhanced results eligibility. Into the bargain, the mRMR feature selection technique requires to calculate the relevance of each attribute in the present data stream so that to be able to arrange them for further selection. Therewith, the mRMR technique operates iteratively as follows:

– At each iteration, it identifies the prime feature to select, according to the rule identifying its relevance, and then appends it to the basket of selected features.

- It is essential to mention that features added into the bucket can never come out. Here, one may ask what is the selective rule?
- The answer is as follows: At each iteration, we aim to select the most relevant attribute, which has the maximum relevance to the target variable. Simultaneously, we strive for minimum redundancy by selecting features that have the least correlation with those chosen in earlier iterations.
- Specifically and when using mRMR feature selection technique, you are basically required to make a prime choice; that is to choose the number of features k you aim to select.

The following Fig. 2 elucidates an overview of the mRMR technique.

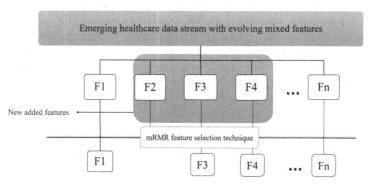

Fig. 2. Overview of minimum Redundancy Maximum Relevance feature selection technique.

To evaluate the joined healthcare features at each iteration i, a score f is computed as follows in Eq. (5):

$$score_i(f) = \frac{relevance(f \mid target)}{redundancy(f \mid features\ selected\ until\ i - 1)} \quad (5)$$

The key question now is how to determine the numerator and the denominator of the calculated score in Eq. (5):

- Numerator: This represents the relevance of a feature f at the i-th iteration. It is measured by the F-statistic between the attribute f and the target variable.
- Denominator: This represents the redundancy. It is calculated as the average correlation between the feature f and all previously selected features at earlier iterations.

Here, we have opted to use the mRMR feature selection technique as being an effective and straightforward baseline approach to feature selection. Systematically, new emerging mixed features are firstly sorted according to a predetermined score, then selected accordingly. Besides the mRMR feature selection technique is more efficient than the variance feature selection technique, proposed in a similar study in [18] owing to the fact that we are not required to fix a variance threshold in order to selection a subset of relevant features.

4 Simulations and Performance Assessment

Here comes the time to analyze number of experiments performed on real-mixed medical and healthcare databases. This is in the intention of testing the performance of our proposed healthcare application for feature selection with streaming objects and features, towards the srandard k-prototypes, the IHK-prototypes, and the FSIHK-prototypes algorithm.

4.1 Framework

To evaluate the performance of our proposed IDHK-prototypes based mRMR method, we have opted to compare it with other existing streaming feature healthcare applications such as IHK-prototypes method [2], FSIHK-prototypes method [18], and the standard k-prototypes method. We used the latest version of python language and executed them on an i5 processor with 4 GB memory. In fact, the simulations were conducted on 4 real-life healthcare and medical databases, containing from 7 to 32 mixed features. These benchmarking data sets are described in Table 1.

Table 1. Description of the used healthcare databases.

Dataset	Objects	Attributes Acronym
Stroke prediction	5110	12 SP
Pharmaceutical drug spending	1036	7 PDS
Breast cancer wisconsin	569	32 BCW
Personality scale analysis	315	8 PSA

Aiming to provide the motivation of the proposed healthcare study and the advantages that our real-time model would propose, we present its implementation in a real-world scenario as follows: to model the evolving streaming features/ objects situation, we take the BCW example as a medical data set, described in Table 1. At first, we introduced an input data having 200 data points and 29 numeric and categorical attributes. Afterwards, a new data stream joins the learning procedure with 369 instances (170 old ones, 99 new ones, and 30 old ones are vanished) and 32 features (25 old ones, 7 new ones, and 4 old ones vanished) joined attributes. In this situation, our real-time healthcare application is supposed to apply its proposed feature selection module based on mRMR feature selection technique so that to gain a the most significant incoming mixed features (see Fig. 1). Consequently, the selected features would be injected in the incremental-decremental learning module based on IDK-prototypes algorithm. to get the final clustering results without retrain of the model from scratch, each time a recent healthcare data stream comes out with evolving attributes. Contrarily to the batch k-prototypes algorithm, that is supposed to learn the initial data points and retrain the model from beginning at the arrival of a new data stream.

4.2 Evaluation Measures

Since evaluating the efficiency of an unsupervised clustering method is more complex than calculating recall and precision for a supervised machine learning algorithm, we chose to assess our proposal using indices that measure cluster cohesion and separation. Each evaluation measure is assigned an up or down arrow to indicate whether a higher or lower value is recommended, respectively.

1. The run time in seconds (RT $\downarrow$) which considers all the period needed to cluster the input data and gain the last clustering result.
2. The Sum of Squared Error (SSE $\downarrow$): determined using the following Eq. (6), SSE estimates the clusters' dispersion in accordance to their centroids. Consequently, SSE values are closer to zero reflects more dissimilar clusters.

$$SSE_{x \in C} = \sum_{i=1}^{n} \sum_{j=1}^{m} (x_{kj} - c_j) \quad (6)$$

where c_j is the centre of the cluster j.

1. The Silhouette Coefficient [26] (SC $\uparrow$): a high SC score consider a well-defined model. SC values are bounded between [+1 for clusters with high density, -1 for incorrect clustering]. It is composed of two scores a: the average distance from a data point to all other data objects within one particular cluster; b: the average distance between a data point and all other points in the next nearest cluster. SC is determined using Eq. (7):

$$s = \frac{b - a}{\max(a, b)} \quad (7)$$

Next, we will present and elaborate the simulations' results shown in Table 2 and provide detailed discussion of these findings.

Table 2. Compared RT, SSE and SC results of K-prototypes vs. IHK-prototypes vs. FSIHK-prototypes vs. IDHK-prototypes based mRMR for each data set.

Data sets	K-prototypes			IHK-prototypes			FSIHK-prototypes			IDHK-prototypes		
	RT	SSE	SC	RT	SSE	SC	RT	SSE	SC	RT	SSE	SC
SP	87.788	1.920	0.363	56.229	1.651	0.388	55.605	1.814	0.423	53.642	1.602	0.591
PDS	11.090	1.956	0.469	6.772	1.577	0.523	5.704	1.432	0.619	4.231	1.347	0.745
BCW	7.188	2.130	0.457	4.543	1.547	0.467	4.491	1.997	0.512	3.891	1.512	0.609
PSA	2.956	2.587	0.487	1.794	1.682	0.576	1.559	1.412	0.766	1.246	1.366	0.816

In Table 2, we compare the capability of our new proposal with conventional k-prototypes method, that requires the loading of the entire input data into memory from the start of the learning process. This results in high memory usage and increased processing time; (2) the IHK-prototypes method [2] which consists in an innovative incremental attribute learning healthcare application for data stream clustering that does not select relevant features from new incoming ones; (3) the FSIHK-prototypes [18] that deals only with incremental attribute and object learning tasks and selects only pertinent streaming features before starting the incremental learning process using the variance threshold feature selection technique.

Run Time Results Analysis

Analysis of Run Time Results. As shown in Table 2, our proposed IDHK-prototypes-based mRMR method requires nearly half the time wanted by the batch k-prototypes method across all datasets. It also performs better than both IHK-prototypes and FSIHK-prototypes in terms of execution time. For instance, the PDS dataset requires 11.09, 6.772, 5.704, and 4.231 s for the learning process when using the k-prototypes, IHK-prototypes, FSIHK-prototypes, and our new IDHK-prototypes method, respectively. Simply put, by learning from the input data on pharmaceutical drug spending and the accompanying evolving data streams—featuring both new and removed features that better capture the situation as learning progresses—we can achieve predictive data mining analysis in nearly half the time. This enables more coherent real-time decision-making.

Sum of Squared Error Results Analysis

This new developed method has achieved the most reduces SSE scores across the four utilized datasets. Compared to the other methods mentioned, it produces a more coherent model with better-defined clusters. This indicates a model with maximum similarity within clusters and minimal similarity between clusters, as a lower SSE value corresponds to a better-defined model. For example, the calculated SSE values for the PDS dataset are 1.956, 1.577, 1.432, and 1.347 for, respectively, the mentioned methods. This was depicted in the following Fig. 3.

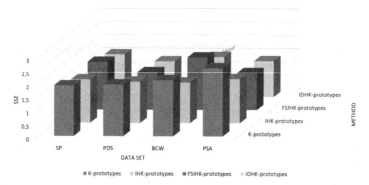

Fig. 3. SSE scores comparing Batch K-prototypes, IHK-prototypes, FSIHK-prototypes, and IDHK-prototypes for each dataset.

Silhouette Coefficient Results Analysis

As for inspection of clusters, the highest SC scores (adjacent to + 1) are those of our proposed IDHKprototypes based mRMR method for all used data sets, compared to all other mentioned methods. The SC scores for the PDS data set are respectively 0.469, 0.523, 0.619, and 0.745 (the closest to 1) for accordingly batch k-prototypes, IHK-prototypes, FSIHK-prototypes and our new proposed IDHK-prototypes. Hence, we can admit that our proposal is the optimum real-time attribute and object learning method in an evolving healthcare features context that can incrementally and decrementally learn new joined medical data streams and define a more consistent model with better defined clusters. Besides, our proposed IDHK-prototypes method is the presents the best results for all used healthcare data sets and with all evaluation measures. This explains the optimality and the better quality of our proposed dynamic method against the other compared methods. Numeric results are better illustrated in Fig. 4.

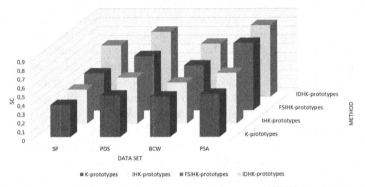

Fig. 4. The SC scores of K-prototypes vs. IHK-prototypes vs. FSIHK-prototypes vs. IDHK-prototypes for each dataset.

Conclusively, the results obtained below confirm that incremental and decremental feature and object learning tasks present a compelling alternative. Our proposed IDHK-prototypes-based mRMR approach addresses an essecial matter in big data within the medical and healthcare sectors. Additionally, conducting feature selection before modeling patient streaming data, which involves added or removed features or instances, decreases execution time, as less data leads to faster training. This may accelerate the learning process and reduces overfitting. This is assured by minimizing the chances of making wrong medical decisions in relation to noisy data. Consequently, the SSE and SC criteria highlight the effectiveness and capability of our proposal. Furthermore, we strongly advocate for the application of our proposed IDHK-prototypes method to demonstrate its ability to overcome the limitations of the conventional k-prototypes technique, the IHK-prototypes, and the FSIHK-prototypes methods, by achieving a coherent clustering model in half-time reduced than the usual learning time (with the time savings proportional to the dataset size). In a few words, our proposed IDHK-prototypes based mRMR healthcare application can help patients make the right real-time decisions. For descriptive purposes, it is effective in identifying patients who need proactive care or lifestyle changes to prevent the deterioration of their health conditions. It may help, as well, in applying number of analysis and synthesis on these healthcare big data so as to generate online predictive models and values for early interventions. This value translates into the ability to comprehend and predict diseases and outcomes at earlier stages, thereby promoting patient health, advancing the fields of medicine and healthcare, reducing costs, and improving life span with bolstering medicare quality.

5 Conclusion

In this study, we present a new incremental-decremental attribute-object learning method based k-prototypes algorithm and using mRMR feature ordering technique for medical and healthcare applications. The problematic remains in dealing with incoming mixed medical data streams that ceaselessly join the learning process to make patients life's easier to manage. This real-time medical-aid system performs well in a mixed data streaming context where features are evolving (some of them are added and others are removed as time proceeds). The main goal of this study is to adjust a pre-trained model to variations in the training dataset without the need to completely retrain the model from the beginning. Into the bargain, the changes may include adding and/ or deleting mixed attributes of continuously joined data samples. Furthermore, the mRMR feature selection technique is highly effective in such context to select only relevant features from the new added ones so that to be trained. Experiments conducted on real medical data sets validate the scalability and effectiveness of our proposal. However, other feature selection techniques might be more effective than the mRMR technique in such context. Inspired by these promising results, we want to improve this application by the incremental class learning level in the medical field. Moreover, we suggest to apply more feature selection techniques to enhance our proposal's potency.

References

1. Price, I.I., Nicholson, W.: Artificial intelligence in health care: applications and legal issues (2017)
2. Gorrab, S., Rejab, F.B., Nouira, K.: Advanced incremental attribute learning clustering algorithm for medical and healthcare applications. In: Rojas, I., Valenzuela, O., Rojas, F., Herrera, L.J., Ortuño, F. (eds) International Work-Conference on Bioinformatics and Biomedical Engineering, pp. 171– 183. Springer, Cham (2022). https://doi.org/10.1007/978-3-031-07704-3_14
3. Gorrab, S., Rejab, F.B., Nouira, K.: Real-Time K-prototypes for incremental attribute learning using feature selection. In: Alyoubi, B., Ben Ncir, CE., Alharbi, I., Jarboui, A. (eds) Machine Learning and Data Analytics for Solving Business Problems, pp. 165–187. Springer, Cham (2022). https://doi.org/10.1007/978-3-031-18483-3_9
4. Hancer, E.: A new multi-objective differential evolution approach for simultaneousclustering and feature selection. Eng. Appl. Artif. Intell. **87**, 103307 (2020)
5. Almusallam, N., Tari, Z., Chan, J., Fahad, A., Alabdulatif, A., Al-Naeem, M.: Towards an unsupervised feature selection method for effective dynamic features. IEEE Access **9**, 77149–77163 (2021)
6. Peng, H., Long, F., Ding, C.: Feature selection based on mutual information criteria of max-dependency, max-relevance, and min-redundancy. IEEE Trans. Pattern Anal. Mach. Intell. **27**(8), 1226–1238 (2005)

7. Huang, Z.: Extensions to the k-means algorithm for clustering large data sets with categorical values. Data Min. Knowl. Discov. **2**(3), 283–304. Springer (1998). https://doi.org/10.1023/A:1009769707641
8. MacQueen, J.: Some methods for classification and analysis of multivariate observations. In: Proceedings of the Fifth Berkeley Symposium on Mathematical Statistics and Probability, vol. 1, issue 14, pp. 281–297 (1967)
9. Huang, Z.: A fast clustering algorithm to cluster very large categorical data sets in data mining. In: DMKD, vol. 3, issue 8, pp. 34–39. Citeseer, (1997)
10. Dong, J., Cong, Y., Sun, G., Zhang, T., Tang, X., & Xu, X. Evolving metric learning for incremental and decremental features. IEEE Trans. Circ. Syst. Video Technol. **32**(4), 2290–2302 (2021)
11. Chen, Y., Xiong, J., Xu, W., Zuo, J.: A novel online incremental and decremental learning algorithm based on variable support vector machine. Cluster Comput. **22**, 7435–7445 (2019)
12. Hou, C., Zhou, Z.H.: One-pass learning with incremental and decremental features. IEEE Trans. Pattern Anal. Mach. Intell. **40**(11), 2776–2792 (2017)
13. Lee, W.H., Ko, B.J., Wang, S., Liu, C., Leung, K.K.: Exact incremental and decremental learning for LS-SVM. In: 2019 IEEE International Conference on Image Processing (ICIP), pp. 2334–2338. IEEE (2019)
14. Gorrab, S., Rejab, F.B.: Incremental-decremental attribute learning algorithm based on K-prototypes for mixed data stream clustering. Int. J. Comput. Inform. Syst. Ind. Manag. Appl. **13**, 149–159 (2021)
15. Gorrab, S., Ben Rejab, F., Nouira, K.: Split incremental clustering algorithm of mixed data stream. Progress in Artificial Intelligence **13**(1), 51–64 (2024)
16. Shi, D., Zhu, L., Li, J., Zhang, Z., Chang, X.: Unsupervised adaptive feature selection with binary hashing.: IEEE Trans. Image Process. (2023)
17. Wang, M., Han, H., Huang, Z., Xie, J.: Unsupervised spectral feature selection algorithms for high dimensional data. Front. Comput. Sci. **17**(5), 1–14 (2023)
18. Gorrab, S., Rejab, F.B., Nouira, K.: Innovative incremental K-prototypes based feature selection for medicine and healthcare applications. In: International KES Conference on Innovation in Medicine and Healthcare, pp. 282–291. Springer Nature Singapore, Singapore (2023). https://doi.org/10.1007/978-981-99-3311-2_25
19. Samareh-Jahani, M., Saberi-Movahed, F., Eftekhari, M., Aghamollaei, G., Tiwari, P.: Low-redundant unsupervised feature selection based on data structure learning and feature orthogonalization. Expert Syst. Appl. **240**, 122556 (2024)
20. Solorio-Fernaíndez, S., Carrasco-Ochoa, J.A., Martínez-Trinidad, J.F.: Filter unsupervised spectral feature selection method for mixed data based on a new feature correlation measure. Neurocomputing **571**, 127111 (2024)
21. Pramanik, P., Pramanik, R., Naskar, A., Mirjalili, S., Sarkar, R.: U-WOA: an unsupervised whale optimization algorithm based deep feature selection method for cancer detection in breast ultrasound images. In: Handbook of Whale Optimization Algorithm (pp. 179–191). Academic Press (2024)
22. Alashwal, H., El Halaby, M., Crouse, J.J., Abdalla, A., Moustafa, A.A.: The application of unsupervised clustering methods to Alzheimer's disease. Front. Comput. Neuroscience **13**, 31 (2019)
23. Wang, Y., et al.: Unsupervised machine learning for the discovery of latent disease clusters and patient subgroups using electronic health records. J. Biomed. Inform. **102**, 103364 (2020)
24. Rajendra Prasad, K., Mohammed, M., Noorullah, R.M.: Visual topic models for healthcare data clustering. Evol. Intell. **14**(2), 545–562 (2021)
25. Dornaika, F., El Hajjar, S., Charafeddine, J.: Towards unsupervised radio graph clustering for COVID-19: the use of graph-based multi-view clustering. Eng. Appl. Artif. Intell. **133**, 108336 (2024)

26. Rousseeuw, P. J.: Silhouettes: a graphical aid to the interpretation and validation of cluster analysis. J. Comput. Appl. Math. **20**, 53–65 (1987)
27. Kwedlo, W.: A clustering method combining differential evolution with the K-means algorithm. Pattern Recogn. Lett. **32**(12), 1613–1621 (2011)

On Solving the Physicians Scheduling Problem at an Emergency Department: A Case Study from Canada

Ghada Yakoubi[1,2], Chahid Ahabchane[2], and Safa Bhar Layeb[1,3](✉)

[1] LR-OASIS, National Engineering School of Tunis, University of Tunis El Manar, Tunis, Tunisia
safa.layeb@enit.utm.tn
[2] University of Quebec in Abitibi-Temiscamingue, Rouyn-Noranda, QC, Canada
[3] Centre Génie Industriel, Université Toulouse, IMT Mines Albi, Albi, France

Abstract. Emergency departments play a pivotal role in healthcare delivery since they represent the first line in hospitals to face emergency patients. This department is where medical care is provided to patients whose arrival rate is uncertain and a high fluctuation in the daily requirements is presented. Yet, managing its human resources efficiently is imperative to improve the quality of its services and provide the right treatments at the right time. This work aims to establish a schedule for physicians in emergency departments which represents a challenging task as it requires respecting several rules incorporating diverse aspects. The objective of this study is to allocate workdays and shifts to emergency department physicians to align physicians' productivity with patients' demands, without decreasing physicians' preferences, to maximize the number of patients treated at the emergency department. Taking into consideration the stochastic and time-varying patients' demand, we presented a mathematical programming approach that determines feasible physicians' schedules respecting the emergency department's hard requirements, minimizing patients' coverage violations, along with the violation of physicians' preferences. A real-life case study was carried out in a public Hospital in Quebec, Canada. Different scenarios were then created to study the effect of each goal on the problem. Afterward, we applied a meta-heuristic solution approach belonging to the iterative evolutionary algorithms with several feature configurations to solve the physicians scheduling problem. The results obtained from the computational study of the several scenarios were discussed as a function of the degree of satisfaction of the goals under which the system operates allowing to conclude the best scenario that generates a one-month schedule respecting all goals without deterioration.

Keywords: physicians scheduling problem · emergency department · genetic algorithm · evolutionary algorithm

1 Introduction

Access to sufficient healthcare is a vital requirement for every citizen and the emergency department stands as an imperative component within hospital facilities and services. The optimization of these services has been extensively studied due to its significance and a large attention was directed towards optimizing emergency departments' human resources [1]. Physicians are part of these resources and their scheduling plays a crucial role in planning the emergency department. This department is where medical care is provided to patients whose arrival rate is uncertain and a high fluctuation in the daily requirements is presented. Yet, immediate attention for these patients is needed. Indeed, the lack of high-quality services may lead to undesirable consequences such as patients leaving the emergency department before getting the treatment [2], an increase in suffering among patients, and a rise in both physicians and patients' dissatisfaction [3]. On the contrary, having the right members of physicians with the right skills and performances can affect outstandingly the quality of services and the emergency department performance. That's why improving the emergency department physicians' planning and creating physicians' schedules is a high priority and a major challenge to be tackled at emergency departments. The emergency department is open 24 h per day, 7 days a week, which requires physicians to work different shifts and to work as much on nights and on weekends as during the day. Different research works were done previously on physicians scheduling problems and adverse aspects, such as work conditions, preferences, fairness, resident specifics, flexible shifts, and stochastic demand, were considered. However, these aspects combination is exceptionally distinctive which creates a diversity and a complexity for physicians scheduling problems that makes it more challenging [4]. This study aims to optimize the efficiency of an emergency department by creating physicians' schedules that efficiently allocate them to shifts, considering specific requirements and constraints. Delving into the complexity of the problem, the study meticulously scrutinizes various characteristics, encompassing diverse instances, constraints, and the dynamic interplay between demand and supply within the emergency department. This was followed by a mathematical programming approach that models a real case study of an emergency department located in the province of Quebec, Canada. We have tailored a genetic algorithm to solve this complex real-world problem. The proposed approach was implemented to effectively address the identified challenges of the problem.

This study consists of four sections, each serving a well-defined purpose. In Sect. 2, an overview of healthcare personnel scheduling problems is presented with an extensive view of physicians' scheduling problems highlighting their instances, characteristics, and optimization methods. In Sect. 3, a comprehensive understanding of the problem is established spotlighting the proposed solving approach. Computational experiments are conducted in Sect. 4 showcasing the different scenarios applied to the studied optimization problem and comparing their performances. Concluding remarks and future work extensions follow in Sect. 5.

2 Literature Review

In the literature, many research works were done on the healthcare personnel scheduling problems, as in the works of [4–6] and [7]. This problem represents a real challenge and has received significant attention over the last decades aiming to provide the appropriate service quality to patients while enhancing, planning, and scheduling the hospital resources [8]. In Sect. 2.1, the diverse healthcare personnel scheduling problems that have been discussed in previous literature studies are presented with special emphasis on emergency department physicians accentuating the different problem instances and characteristics. Studies on physicians' productivity and performances within the emergency department are discussed in Sect. 2.2. The various literature works that consider the diverse nature of patient demands are outlined in Sect. 2.3.

2.1 Personnel Scheduling Problems in Emergency Departments

The nurse and physician scheduling problems fall under the umbrella of personnel scheduling challenges. The major personnel rostering problems focus on nurse scheduling, with 56 articles published between 2010 and 2020 [8], taking into consideration the quality-of-care services, the hospital's financial aspect, patients' demand, and work regulations [5]. However, the physicians scheduling problem was under-emphasis although it represents a real problem that arises in hospitals [8]. These two problems may seem to have the same objective of performing a schedule for a time period. However, they differ in several ways. The nurse scheduling was stated as a two-step process aiming to create a schedule satisfying the collective union agreements. On the other hand, the physicians' scheduling problem was described as a one-step process striving to create a schedule satisfying physicians' demand and their preferences with no formal scheduling rules [9]. In addition, [10] affirmed that nurse and physician scheduling are both multi-objective optimization problems where nurse schedules adhere to some written rules, but physician schedules are more driven by individual preferences. Despite their differences, their mathematical formulation is not quite different and the same mathematical approach to nurse scheduling problems can successfully be applied to the physicians scheduling problems [11]. Although, the nurse scheduling problems may help in formalizing and solving the physicians' scheduling problems. Over the last decades, this topic has received appreciable attention from researchers. Exploring physicians' scheduling problems in Canada dates back to the late 1990s [4]. This increase of attention on healthcare personnel scheduling problems was based on an increased interest in hospital organizations in terms of optimized scheduling and allocation of medical staff members (i.e., physicians and nurses) [12]. Different classifications of physicians scheduling problems were discussed in the existing literature whether it is a staffing problem (addressing strategic decision making such as determining the appropriate size of the workforce and they generally concern long-term planning periods), a rostering problem (focusing on tactical or offline operational planning and are related to mid-term planning periods that span typically from one week to a few months) or a replanning problem (considering working schedule adjustments for short-term periods such as re-scheduling physicians in case of when an unpredictable event happens) [4]. Establishing offline operational planning for physicians of an emergency department and creating a schedule that covers a given time horizon is the primary focus of this study. The objective is to create timetables where physicians are assigned to workloads aiming to meet the emergency department restrictions and legislation to have feasible schedules [13], incorporating adverse soft constraints. Hard constraints settings can vary based on the country, state, or even from one hospital to another of the same region [4]. For example, [14] cited that exactly one physician is required to perform a shift during the planning period. However [15] specified a range of the overall number of residents that need to be assigned to a day or a night shift. When requirements are preferable but not compulsory, they are called soft constraints and usually, in physicians' scheduling problems, they include the physician staff members' preferences [13]. In that context, some hard constraints of one hospital can be seen as soft constraints in another hospital. Giving the constraint of assigning the staff member to his/her required number of day and night shifts, [15] classified it among the hard constraints. On the other hand, this constraint was seen as a non-compulsory constraint hoping to perform a schedule that satisfies it the most [6]. Several studies on physicians scheduling and rostering problems have appeared in different research including different instances namely the time horizon and the number of different working shifts since the working days are typically divided into planning intervals, each interval with a specific length. The most common planning horizon of the addressed problem is four weeks or one month. [16] established an optimal one-month schedule, with three shifts per day, for doctors in a pediatric department that meets working contracts and shift requirements. [17] generates a 4-week timetable for staff members in a medical department taking into consideration the working regulations constraints. Fairness is also a key aspect in physician scheduling problems. Scheduling staff members in a hospital medical department, [17] aims to balance the working hours and the preferred working shifts between medical staff members to increase the equity between them. [18] assigned shifts to resident physicians over a one-month planning horizon to generate a schedule that balances the working hours and the workload of each shift to enlarge the righteousness between resident physicians. Another discussed physicians' scheduling problems key aspect in the literature is physician preference satisfaction since it affects the delivered quality of the care service. [4] indicates that generally, more than 60% of the considered publications involve staff preferences and this holds the physician scheduling problems. [19] describes the employee preferences as a set of undesirable shifts in month days for each employee. [18] established a timetable where employees preferred availability days and shifts were considered. [17] established a 28-day planning schedule where employees' preferences were based on ranking the working shifts with three preference ranks: good, normal, or bad, and employees were asked to give their two preferred days off of each week. [7] created a one-month planning schedule where employee preferences were presented as a list for each staff member the day, its preference if it is a day off or on, and the preferred working shift. 1,4/on/night and 5/off are two examples of preferences. [20] developed a personalized staff scheduling method where preferences were based on flexible starting times and number of hours to work each week chosen by staff members.

2.2 Productivity of Physicians in Emergency Departments

Physicians' productivity is very common in healthcare. Many studies have discussed the impact of physicians' productivity on emergency department services. [21] suggested the Poisson regression model as a stochastic optimization model to forecast physicians' productivity, represented by the number of new patients attended to by a physician within a one-hour time frame defined as the patient-per-hour rate (PPH), based on some factors that affect physicians' productivity levels. Using the PPH rate, they applied the k-means clustering method aiming to group physicians having similarities in their productivity levels into clusters. The number of different clusters was determined by the Elbow method. A Two-stage stochastic integer program was established to minimize the mismatch between the number of patients at the emergency department and the total physicians' productivity. [3] suggested a productivity index of physicians based on patients' consultation length and the number of patients, forecasted hourly using Facebook Prophet, each physician can serve. They proposed a mixed-integer programming model aiming to balance patients' demands with physicians' capacities to satisfy patients' demands and guarantee physicians' satisfaction and fairness. Other studies generalized one physician's productivity to all physicians. In [22], historical data from April 1, 2011, to March 31, 2012, was used to conclude that physicians' productivity is equal to 4.1, 5.9, and 2 patients per hour in the acute care area by acute care physicians, fast track clinic by fast track physicians and fast track clinic by acute care physicians respectively. Other studies did not forecast physicians' productivity. However, they considered the assignment of individual providers or pairs of providers based on their skill levels. When an individual provider possesses a skill level denoted as "l", they are only able to independently treat patients with acuity levels of "l" or below. For providers whose skill levels are below "l", a team comprising two providers is formed to address the needs of patients with acuity levels up to "l" [23].

2.3 Patients Demand in Emergency Departments

Due to the overcrowding and the increase in patient demand that hospital departments have faced in recent years, patients' pain and suffering are prolonged, and wait times and patient dissatisfaction rates are highly increasing. Different studies tackled this problem and some of them took into consideration patients' acuity levels that vary from 1 (being most urgent) to 5 (being least urgent) based on the Canadian Triage and Acuity Scale, aiming to respond efficiently to patient demand by modeling their arrival rates [21]. An examination of patient demand distributions for each period using historical data and patient acuity level was conducted to estimate the duration that patients with a given acuity level would require to get the treatment done [23]. Another method was suggested by applying a two-step methodology to predict the demand. In the first stage, the total number of patients arriving every day of the scheduling period for every type of emergency room was forecast using a decomposable time series model. Then, a distribution of the demand among the periods of each day was carried out using an intra-day distribution model and using historical data [3]. In [22], an estimation of patients' arrival rates was applied by fitting Poisson generalized additive models to the historical arrival data which allows concluding the patterns in the arrival rates if they vary by day or the variation is between weekdays and weekends and if holiday weekends affect the patient demand. In other studies, patient demand was modeled as an uncertain/stochastic nature of arrival volumes per hour [21], and multiple other methods were applied in the literature to forecast the demand at emergency departments such as autoregressive integrated moving average models, multivariate vector autoregressive models, generalized autoregressive conditional heteroskedasticity models, Linear regression models and nonlinear regression models, Poisson regression models and artificial neural networks [3].

2.4 Solving Approaches

In the literature addressing the healthcare personnel scheduling problems (including the physicians scheduling problems), various solving approaches have been explored, showcasing a diverse range of methodologies and strategies. Some studies have emphasized the utilization of exact methods that search for the optimal solutions. These methods guarantee finding at least one optimal solution for the studied problem after an unspecified amount of time [24]. Accordingly, exact methods encompass fundamental mathematical techniques, including but not limited to linear programming, non-linear programming, mixed integer programming, goal programming, and dynamic programming [25]. Different techniques were applied to solve the physicians' scheduling problem in emergency rooms such as the Column Generation technique and the Constraint Programming approach that were applied as exact approaches aiming to create physicians' schedules satisfying both resources constraints and their requirements [14]. Conversely, the physicians scheduling problem was proven to be NP-hard in many literature works such as [26] and [16]. Due to the in-feasibility of the exact methods approaches in NP-Hard problems that have exponential or worse growth as the scale increases, approximate method approaches can provide near-optimal solutions in a reasonable amount of time. These approximate approaches are divided into two categories. The first category addresses the heuristic methods that offer a trade-off of time for solution quality. They provide near-optimal solutions much faster while dealing with complex objectives, yet without ensuring the optimal solution [24]. Some examples of heuristics that were applied in previous literature studies are the Hill-Climbing Algorithm [10] and Local Search [11]. The second category pertains to metaheuristic algorithms which are specialized methods aimed at addressing problems classified as NP-hard. Generally, these approaches are used when traditional heuristics fail to solve the given problem [5]. They optimize the problem by improving a candidate solution iteratively based on a given measure of quality and are divided into two categories [27]: Population Based Meta-heuristics such as Genetic Algorithm [28–30], Keshtel algorithm [19], and Trajectory Based Metaheuristics namely Tabu Search [9, 14, 19] and Simulated Annealing [13, 28]. This work has uniquely addressed the simultaneous improvement of various aspects within the emergency department, including patient coverage demand and physicians' preferences, while also introducing the concept of balancing physicians' performance, all while adhering to emergency department regulations. This holistic approach, which integrates multiple characteristics that have not been collectively studied before, aims to enhance the number of patients treated within the emergency department as a primary objective. By considering the optimization of different facets alongside the balance of physicians' workload and efficiency, this study contributes to a more comprehensive understanding of emergency department management, ultimately striving to enhance patient care and operational effectiveness.

3 An Optimization Model of the Physicians Scheduling Problem

In this section, the Physicians Scheduling Problem of a real-case study of an emergency department is formulated. This emergency department belongs to Rouyn-Noranda Hospital which is located in the Abitibi-Temiscamingue region of Quebec, Canada. It is composed of different healthcare providers offering 24–7 services for patients with different acuity levels and receiving an astounding influx of over 22,000 visits annually which presents a testament to its critical role as a primary healthcare provider in the community. The primary objective is to create physicians' schedules maximizing the number of patients treated at the emergency department taking into consideration some criteria. This can be seen as a minimization of the mismatch between the total patient demand of the emergency department which is calculated by the number of patients arriving at the emergency department and the supply provided by the emergency care calculated by the number of patients treated by all emergency department physicians. In the healthcare literature, physician productivity is commonly assessed by measuring the number of patients attended to by a physician, often referred to as the patient-per-hour rate (PPH). Physicians exhibiting a higher performance or productivity tend to experience shorter average patient length of stay and service times [21]. However, generating physicians' schedules does not solely depend on physicians' productivity but also from patient arrival rates. Matching physicians' productivity with uncertain demands helps to decrease patient waiting times and reduce the number of patients who leave the emergency department without receiving medical attention. Yet, patient arrival rates hold a dynamic aspect that remains subject to change at any given moment, influenced by a multitude of factors spanning the day-to-day and week-to-week fluctuations [3]. This fluctuated nature of patient arrival rates necessitates an adaptable and responsive approach to physicians' scheduling to ensure optimal physicians' scheduling. Different mathematical programming models established in prior research incorporating physicians' performances and the stochastic nature of patient demand were proposed in the literature attempting to produce optimal emergency department physicians' schedules such as in [22] and [23]. These studies indicate that aligning physicians' productivity with patient arrivals helps in reducing the unmet patient demand within the emergency department and potentially enhancing its overall efficiency and responsiveness.

3.1 Problem Description

This paper focuses on physicians' scheduling problem at the emergency department of Rouyn-Noranda Hospital located in Quebec, Canada. The emergency department of the studied problem is composed of different healthcare providers. However, this study focuses only on physicians and one substitute physician and their affectation on the shifts taking into account adverse constraints. The emergency department is described as a 24–7 service where physicians' services should be available at any time of the day and every day. The planning horizon is one month, each day with three shifts (night, day, and evening) and each shift with an 8-h duration. Night, day, and evening shifts start from 11 pm, 7 am, and 3 pm to 7 am, 3 pm, and 11 pm respectively. Patients with different acuity levels arrive at the emergency department following the Poisson distribution with an average of 4 patients per hour. The patients' arrival rate is stochastic and depends from one day to another and from the hour of the day. Each patient needs a consultation length with a minimum of 15 min and a maximum of 30 min depending on her/his acuity level which means that physicians' productivity varies from 2 patients to 4 patients per hour depending also on patients' demand. In this study, physicians and substitute productivity vary also based on their performances. Each individual has a deterministic performance that does not vary during the scheduling period but depends on the individual physician or substitute. We assume also that a fixed number of physicians is initially given and that each physician is characterized by some preferences and availability. The objective of the problem under study is to allocate feasible schedules to physicians while satisfying the emergency department working regulations, minimizing the covering violations of the scheduling period, and balancing physicians' daily performances while also taking into consideration physicians' preferences. We assume that the system is initially empty and that at the end of the day, 90% of patients not treated will leave the system without being seen. A comprehensive description of the problem's symbols can be found in Table 1.

Table 1. Symbols and definitions.

Symbol	Definition
$w1$	Weight for hard constraints
$w2$	Weight for demand coverage
$w3$	Weight for physicians' performances imbalance
$w4$	Weight for physicians' preferences
HCv	Hard constraints violations
$Covv$	Demand coverage violations
Bv	Physicians' performances imbalance
Pv	Physicians' preferences
D	Set of the number of days of the scheduling period
Sd	The supply (physicians' productivity) of a given day d
Dd	The demand (patients' demand) of a given day d
$PhysicianProductivityD$	The total number of patients that can be treated by the physicians affected to the schedule during the time horizon D
$PatientTotalDemandD$	The total number of patients looking for treatment at the emergency department during the scheduling horizon D

Based on the annotations presented in the previous Table, a mathematical model was developed to minimize the hard constraints violations, patients covering violations during the scheduling period, physicians' performances balancing deviation, and physicians' dissatisfaction (i.e., physicians' preferences violations).

Minimize:

$$w1 * HCv + w2 * Covv + w3 * Bv + w4 * Pv \quad (1)$$

Hard constraints violations (HCv) are calculated based on the emergency department compulsory constraints that intervene in the feasibility of the schedule. These constraints depend on each hospital based on its general agreement. For the problem under study, the generated schedules should satisfy the constraints as follows:

- Constraint 1: All shifts must be filled
- Constraint 2: One physician should be affected per shift
- Constraint 3: A physician is allocated a maximum of one shift per day
- Constraint 4: A physician who is assigned to a night shift must not be assigned to a day and evening shift on the same day
- Constraint 5: A physician who is assigned to a day shift must not be assigned to the evening shift of the same day and to the night shift of the next day
- Constraint 6: A physician who is assigned to an evening shift must not be assigned to a night and day shift on the next day
- Constraint 7: Physicians should be affected based on their availability

The second component of the objective function (1) is used to calculate the covering violations of a given schedule that need to be minimized (Covv). It represents the total under-covering and over-covering of patients' demand for the emergency department during the planning period considering the deterministic nature of physicians' performances and patients' stochastic arrivals (2). That reads as:

$$PhysicianProductivity_D - Covv+ + Covv- = PatientTotalDemand_D, \forall D. \quad (2)$$

The third component deals with minimizing physicians' daily performance imbalance (Bv) which represents a prerequisite sought after by the emergency department. To the best of our knowledge, considering this criterion while generating physicians' schedules within the emergency department has not been studied yet. To model this component, we propose the following expression:

$$Bv = max(\frac{sd}{Dd}) - min(\frac{sd}{Dd}), \forall d \in D. \quad (3)$$

In addition, the physicians' scheduling problem under study proves to be more intricate than it appears, given the adverse requirements that physicians seek in an optimal schedule. These demands are represented as soft constraints for the problem (aiming to minimize their violations Pv) and commonly include guaranteeing for a given physician one or more preferences among those described as follows:

- Not working more than a maximum desired number of night shifts per month
- Not working more than a desired number of night shifts, day shifts, and evening shifts in a row
- Not working less than a given number of night shifts in a row
- Not work more than a given number of night series, day series and evening series
- Having a minimal given number of days off during the scheduling period

To calculate preference violations, a penalty is added to the variable Pv any time a physician preference is not respected.

3.2 Illustrative Example

Let's consider an illustrative example of a feasible schedule created for 6 physicians (Physician 1 to Physician 6), for four days scheduling period (Day 1 to Day 4) with three shifts per day: morning shift (M), evening shift (E) and night shift (N). Physicians' availability for each shift can be represented as shown in Table 2 where 1 indicates that the physician is available for the corresponding shift on that day and 0 indicates that the physician is not available for the corresponding shift on that day.

Table 2. Example of Physicians Availability.

	Day1			Day2			Day3			Day4		
Physician	M	E	N	M	E	N	M	E	N	M	E	N
1	1	1	1	1	0	0	1	1	1	0	0	1
2	0	1	1	1	1	1	1	0	0	0	0	0
3	0	0	0	0	0	1	1	1	1	1	1	1
4	1	1	1	1	0	0	0	0	1	1	1	1
5	0	0	0	0	1	1	1	1	0	0	1	1
6	1	1	1	1	0	0	1	1	1	1	1	0

Based on this availability, a feasible schedule can be created while assigning physicians to morning, evening, and night shifts for each day. An example of two schedules is given in Fig. 1.

(a)

	Day1	Day2	Day3	Day4
M	1	1	2	3
E	2	5	5	6
N	6	3	4	1

(b)

	Day1	Day2	Day3	Day4
M	4	6	6	3
E	2	2	5	6
N	1	3	4	4

Fig. 1. Example of physicians' schedules.

Both schedules adhere to the hard requirements of the emergency department described in the previous section and accommodate physicians' availability (Table 2). However, their efficacy differs in addressing patients' demands due to variations in individual physician performance, as well as in satisfying various components of the objective value.

Moreover, if a schedule fails to satisfy at least one hard constraint or assigns a physician to a shift for which they are not available, the schedule cannot be accepted. An example of a rejected schedule is depicted in Fig. 2, where physician 1 is assigned to more than one shift per day (M and N). Additionally, this physician is not available for day 2, shift N. Therefore, the schedule cannot be accepted.

	Day1	Day2	Day3	Day4
M	1	1	2	3
E	2	5	5	6
N	6	1	4	1

Fig. 2. Example of a rejected schedule.

3.3 Solving the Physicians Scheduling Problem

Genetic Algorithm. Reaching optimal or near optimal solutions for large-scale optimization problems, within an effective period of time, is continuously seeking more attention from researchers. Therefore, various approaches were developed, over the last decades, combining intelligently different concepts for exploring and exploiting the search space. These methods are defined as meta-heuristics. Inspired by phenomena arising in nature, they aim to find near–optimal solutions for optimization problems. The physicians' scheduling problem was proven to be NP-hard since the schedule planner cannot manually enumerate all possible cases, especially with the increase of the number of constraints, to efficiently solve large-scale physicians scheduling problems [26]. This section deals with the exploitation of the genetic algorithm that was established by John Holland in 1975. It is defined as an adaptive search procedure that is based on the process of natural genetics and Charles Darwinian survival of the fittest to find useful solutions to complex problems. As such, the genetic algorithm represents intelligent exploitation of a random search within a defined search space to solve the NP-hard problem under study [31]. It is worth noting that the Genetic Algorithm is a population-based algorithm where, in each iteration, it searches for a population of points in the solution space rather than only for a single-point solution. Also, the next population selected by the genetic algorithm is based on a random number of generators and probabilistic transition rules [32]. The genetic algorithm is composed of four main elements: the initial population composed by some individuals, called also chromosomes, the selection method to select individuals from the population, the crossover method where two selected parents will exchange one part or some parts of the chromosome based on the crossover method chosen and finally the mutation method where one or more genes of a chromosome are converted to different ones. In genetic algorithms, crossover and mutation represent the two most central methods since they allow for to diversification of individuals and generate a population with a variety of chromosomes [32]. An essential component of the genetic algorithm is the fitness function that defines the criterion for scoring chromosomes of the population and for probabilistically selecting them to pass to the next generation [33]. The fitness function is calculated based on the objective of the problem. As an illustration, for the problem under study, the fitness function is characterized by the negative value of the objective function described in Sect. 3.1 endeavoring to achieve its minimization as presented in (1). The more the fitness score is greater, the more the solution is getting closer to the optimal one.

To solve the physicians' scheduling problem using a genetic algorithm, every component of the solver should be chosen carefully. Hence, getting inspired by the different encoding methods and genetic algorithm operators used in previous studies will help in deciding how to encode the solution of the problem under study and which operators to apply. Two different methods for the initialization of a population of chromosomes were described previously, whether each chromosome is represented by the traditional bit-string structure or by using two-dimensional array chromosome structures encoded in a data structure [31]. Many encoding methods were introduced in the literature such as with characters, integer values, binary values, or strings. A chromosome encoding for the doctor scheduling problem was presented in [16] as a matrix with N*D dimension, as a doctor-day view, (N: number of doctors and D: number of days within the planning period). Each cell of the matrix denotes a character of the working shift as M, E, N, or o representing the morning shift, evening shift, night shift or a day off respectively. A chromosome encryption as a two-dimensional array was implemented in [19], where each element of the array is an integer value equal to 0,1,2 or 3 referring to day off, morning shift, evening shift, or night shift respectively. However, a one-dimensional array chromosome was performed in [34], where each day is split into three shifts, apiece characterized by an integer number referring to the nurse number affected by that shift. After initializing the initial population with N chromosomes randomly and affecting a fitness score for each chromosome, these values are used as a benchmark to judge individuals' strength, and depending on that, a selection phase will take place. Different selection methods were implemented in literature works namely roulette wheel selection, Tournament selection, and elitism selection. The principle of roulette wheel selection is based on giving each individual from the population a probability of being selected proportional to his fitness [35]. Chromosomes with higher fitness will have more chances to be selected since their probabilities will be higher than those with weaker fitness [33]. Tournament selection is the most popular selection technique in genetic algorithms since it is easy to implement. It is based on selecting "n" (referred to as tournament size) random individuals from the population and returning the fittest one between them for further operations in the genetic algorithm process [36]. Elitism is a selection method that can be applied solely or in addition to other selection operators. It ensures that the best individuals from the current population will pass to the next generation which allows for the reduction of genetic drift [37]. As described in [38], a percentage equal to 10 percent from the worst new individuals obtained are replaced by the same number of better old individuals from the previous population. The crossover operator in the genetic algorithm is critical since it intervenes in the diversity of the population. Choosing wisely this operator can affect the genetic algorithm's performance. In healthcare personnel scheduling problems, multiple methods are applied such as one single point crossover that is divided into one column point crossover [17] and one-row point crossover [39] (in the case of a two-dimensional array chromosome), multi-point crossover [38] and partial mapped crossover [40]. This operator generates offspring by melding parts of two parent chromosomes. The next operator that generates offspring is the mutation operator. Through this component, random changes on chromosome bits take place. The most used mutation methods are: Resorting the genes of one column [17], Random swapping mutation [41], Cost Bit Matrix [42], Dynamic mutation [40] and Random mutation [28]. For the implementation of the genetic algorithm to solve the physicians scheduling problem under study, a two-dimensional array chromosome with integer values representing the number of each physician was implemented to generate the initial population randomly. This encoding method helped to verify the first two hard constraints of the emergency department. Added to that, the roulette wheel selection as well as the elitism were used as selection operators, the one single point crossover was used as a crossover operator, and finally, the random swapping mutation also as reciprocal exchange mutation was applied as a mutation operator. The selection of these operators was determined by prioritizing the most frequently utilized ones, ensuring an improvement in schedules across a wide array of problems similar to the problem under analysis.

Recovery Scheme. Applying the genetic algorithm operators conduct a high probability to generate infeasible schedules that do not respect the hard requirements of the emergency department. Accordingly, a recovery scheme was performed in addition to the genetic algorithm to repair infeasible schedules allowing to respect the hard requirements of the emergency department when dealing with physicians' affectation to more than one shift in three consecutive shifts and physicians availability taking into account their performances. This procedure was inspired by previous studies since they achieved a repair process to promote feasible schedules [17, 19, 38, 43]. The flow chart of the optimization approach is provided in Fig. 3.

The solving approach for the problem under study starts with a population of chromosomes initialized randomly based on emergency department constraints to finally generate offspring with enhanced results. With the inspiration of the different encoding methods in previous studies, an integer two-dimensional array chromosome structure was chosen since it promotes the validation of the first two compulsory constraints which helps to minimize the verification of hard constraints in the recovery scheme. Two different structures of the individual chromosome (physicians' schedule) were defined whether it is a schedule per day or a schedule per shift as presented in Fig. 4.

Different procedures to create the individual chromosome were performed following different methods and taking into account physicians' availability as well as the emergency department's hard requirements. The first method is based on affecting physicians randomly starting with the first day and first shift until reaching the last day and last shift of the scheduling period.

A verification step took place during the implementation of this method which includes verifying physicians' availability as well as verifying the physicians affected by the previous two shifts and considering them for the affectation of the new physician to satisfy the problem constraints. Method 2 is based on calculating the number of physicians available for each day and each shift, then giving a percentage of availability for each tuple (day, shift) and ordering them in ascending order. Physicians' affectation will take place then starting from (day, shift) having the smallest availability. The third method is built based on choosing a physician randomly among all physicians and affecting it to a day and a shift randomly until fulfilling all shifts of the scheduling period.

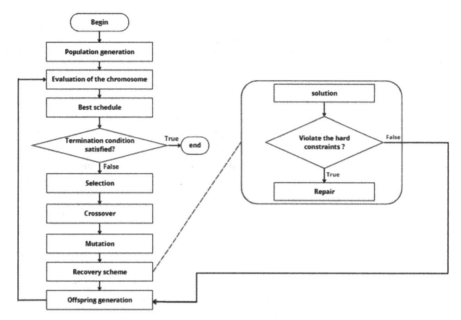

Fig. 3. The flowchart of the genetic algorithm with a recovery scheme. (a) Genetic Algorithm. (b) Recovery scheme.

(a)

	M	E	N
Day1	4	9	2
Day2	7	6	4
...			
Day n	2	7	9

(b)

	Day1	Day2	...	Day n
M	4	7		2
E	9	6		7
N	2	4		9

Fig. 4. The Chromosome Encoding Method. (a) Schedule per day. (b) Schedule per shift.

The last method does not differ much from method 3. The only difference between them is physicians' availability. In method 3, the (day, shift) for which the chosen physician will be affected does not take into consideration its availability. However, this condition was incorporated in method 4. All methods produced feasible schedules after execution. However, the computational time for the first method was much less compared with the other proposed methods. As a result, the choice of the chromosome creation method went to the first method.

For the crossover operator, one single-point crossover was employed in this study. The one column (Fig. 5) as well as the one-row point crossover (Fig. 6), indicating crossover per day and crossover per shift respectively, were executed to conclude that the one-row point crossover does not always guarantee the diversity of schedules. As a result, the one-column point crossover was chosen (Fig. 5).

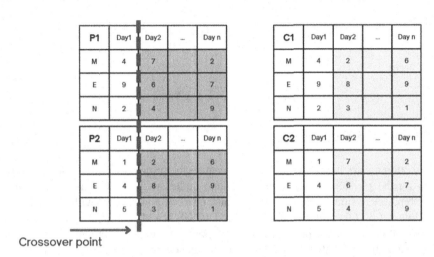

Fig. 5. One Column Point Crossover.

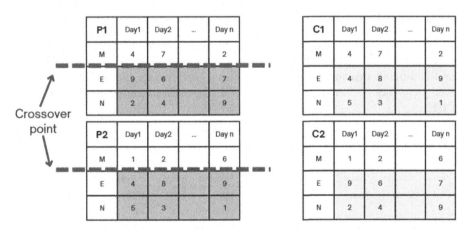

Fig. 6. One Row Point Crossover.

To finish with the mutation operator, the random swapping mutation, where two genes of the same chromosome are chosen randomly and an exchange of information between them takes place to produce a diversified offspring (Fig. 7).

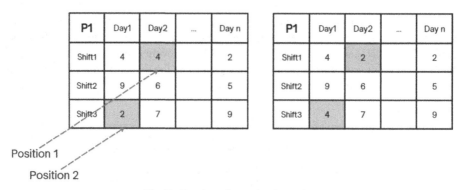

Fig. 7. Random Swapping Mutation.

4 Numerical Experiments

This section covers the accomplishments to meet the project goals. The programming codes for the physicians' scheduling problem under study, as well as the resolution algorithm, were written in Python and executed on a computer system with specifications including a 1.30 GHz Intel i7 processor and 12 GB RAM. The computational results obtained after testing the proposed solving approach aiming to maximize, at the first place, the number of patients treated at the emergency department considering adverse hard and soft constraints, are detailed in this section.

4.1 Instances Description

To evaluate the performance of the proposed model, eight scenarios with different weights (w1, w2, w3, w4) linked with each component of the objective function (hard constraint, demand coverage, physicians' performance imbalance, and physicians' preference respectively) were conducted. The different scenarios are represented in Table 3.

Table 3. Scenarios.

Scenarios	Scenario1	Scenario2	Scenario3	Scenario4	Scenario5	Scenario6	Scenario7	Scenario8	Scenario9
w1	10000	10000	10000	10000	10000	10000	10000	10000	10000
w2	10	0	0	10	10	10	10	10	10
w3	0	10	0	10	1	10	10	1	1
w4	0	0	10	0	0	10	1	1	0.5

For all scenarios, the minimization of hard constraints is the most important objective for the problem under study since they intervene in the schedule feasibility. Hence, a large weight was associated with this objective. Scenario 1 describes the case where physicians' preferences and performance balancing are not taken into consideration.

Only over-covering and under-covering were optimized. Scenario 2 takes into account the minimization of physicians' preferences violations only and Scenario 3 characterizes the situation where only performance balancing is optimized. As a priority was established for each objective component, the remaining scenarios were elaborated to test the effectiveness of the combinations between the different objectives. Scenario 4 describes the case where objectives 2 and 3 are fairly taken into account. Scenario 5 represents the same previous situation; the only difference is that goal 2 is more prioritized than goal 3. All the goals of the objective function are fairly taken into account in Scenario 6. Scenarios 7, 8, and 9 characterize the same condition as Scenario 6, however, in the first scenario, goal 3 is more prioritized than goal 4, and in the second situation, goals 3 and 4 are taken fairly with a prioritization in goal 2 and in the last scenario, all priority orders were considered.

Table 4 details the parameter values of the proposed and implemented genetic algorithm.

Table 4. The default parameters adopted in genetic algorithm.

Parameters	Value
Population size	20
Crossover rate	0.6
Mutation rate	0.1
Number of elite members	10
Stopping criteria	Execution time = 3 min

4.2 Results

In this section, numerical experiments are conducted to evaluate the performance of each scenario using the proposed solving algorithm for the concerned physicians' scheduling problem. Seven categories of key performance indicators are used and results of each KPI for each scenario are presented in Table 5.

Results of the first three scenarios show that enhancing only one objective among the given four objectives can affect the performance of the other goals such as the balancing violations, physicians' preferences, as well as the total covering violations, were deteriorated by 0.001, 6, and 0.11 in Scenario 1, Scenario 2 and Scenario 3 respectively.

Hard Constraints. For the hard requirements, since the recovery scheme was implemented within the genetic algorithm, the different scenarios tackled produced optimized schedules respecting these constraints. As a result, all hard constraints were satisfying.

Total Covering Violations. After establishing ten simulations for the proposed scenarios, the maximum variation of the average total covering violations is provided by the first and fifth scenarios with values equal to 14.3% and 16% respectively.

Under-Covering and Over-Covering of Patients' Demand. The minimum of under-covered patients' demand was provided by Scenario 5 then Scenario 1 with almost similar results. However, results were deteriorated in Scenarios 2,3 and 9. Close results regarding the over-covering of patients' demand were presented in Scenarios 1 and 5 with the smallest addressed to the fifth Scenario. Yet, degradation results were noted for Scenarios 3, 4, and 8.

Performances Balancing. The balancing violations indicator gives an idea of the physicians' performance balancing per day. Computational results show that physicians' performances are more balanced when Scenarios 2, 4, 5, and 9 are used to solve the problem, rather than using Scenarios 1, 3, 6, 7, and 8 where results got deteriorated.

Physicians' Preferences. The sixth indicator denotes the respect for physicians' preferences aiming to be maximized. Scenario 6 appears to be the greatest scenario with the minimum preference violations and the maximum preference variation between the initial and the final schedule. Scenarios 1, 3, 4, 7, 8, and 9 are also acceptable since they produce violations slightly bigger than the best one. On the contrary, Scenarios 2 and 5 got worse when it came to physicians' preferences.

Patients Left Without Being Seen. The emergency department aims to decrease the number of patients untreated. In other words, decreasing the number of patients left without being seen by a physician. However, schedules produced by Scenarios 3 and 9 expand the number of these patients which made them unacceptable. Conversely, Scenario 5 is the best concerning this indicator by diminishing the number of patients left without being seen with approximately 57 patients during the scheduling period.

Best Scenario. Choosing the best scenario that respects all emergency department goals respecting their priority orders is critical. As discussed previously, the solving approach, namely the genetic algorithm, does not always guarantee the validation of all objectives since deterioration can appear for some of them. Considering the priority orders while achieving the different scenarios, Scenario 5 seems to be the scenario that responds best to the goals' priorities and maximizes the most the number of patients treated at the emergency department (best results for UC demand and number of patients left without being seen) which represents the priority among the objectives while assuring an improvement in performance balancing with 5%. Yet, a slight deterioration in physicians' preferences took place.

The results from the manually generated schedule by the emergency department manager, which prioritizes the satisfaction of hard requirements within the emergency department, are presented in Scenario 10 (Table 5). A comparison is made with the optimal schedule produced by the proposed solving approach. Both schedules adhere to the hard constraints, as they are imperative for ensuring feasibility. However, in terms of minimizing the covering violations, Scenario 5 achieves a score of 0.85, while Scenario 10 scores 1.29. This translates to 57.93 and 68.96 of under-covered demand for Scenario 5 and Scenario 10 respectively, alongside 16.10 and 54.26 of over-covered demand. The higher level of over-covered demand in Scenario 10 suggests a potential concern regarding physicians' over-productivity. This could indicate a situation where highly efficient physicians are being allocated to shifts with lower demand. Regarding balancing

Table 5. Computational results of each scenario.

	Scenario1	Scenario2	Scenario3	Scenario4	Scenario5	Scenario6	Scenario7	Scenario8	Scenario9	Scenario10
	10.63	2.99	320	13.29	10.47	314.86	38.57	38.84	17.89	x
	9.20	2.54	150	11.90	8.84	133.77	25.54	27.79	14.95	0
	13.45	17.71	53.12	10.45	15.56	57.51	33.78	28.45	16.43	x
	3	3	3	3	3	3	3	3	3	x
1. Hard constraints										
	0	0	0	0	0	0	0	0	0	0
2. Total covering violations										
	1.063	1.16	1.17	1.04	1.01	1.15	1.02	1.053	1.13	x
	0.92	1.014	1.28	1.01	0.85	1.09	0.96	1.048	1.06	1.29
	0.143	0.02	0.11	0.03	0.16	0.06	0.06	0.005	0.07	x
3. Under covering of patients' demand										
	65.94	65.59	68.4	64.9	64.94	67.92	63.17	67.33	68.23	x
	2.97	2.67	3.11	2.91	2.91	3.08	2.81	3.04	3.10	x
	61.38	67.73	73.49	63.37	57.93	97.55	62.87	63.44	68.81	68.96
	2.71	3.07	3.41	2.82	2.52	3.06	2.79	2.83	3.13	3.14
	4.10	3.84	4.58	3.53	3.72	4.36	4.75	4.42	4.16	4.11

(continued)

Table 5. (continued)

	Scenario1	Scenario2	Scenario3	Scenario4	Scenario5	Scenario6	Scenario7	Scenario8	Scenario9	Scenario10	Scenarios
4.	Over covering of patients' demand										
	27.17	42.56	38.08	25.34	22.96	35.98	28.74	23.64	33.34	x	Initial objective value
	17.47	34.82	40.83	26.95	16.10	28.61	19.35	30.95	20.39	54.26	Final objective value
	7.55	14.10	8.53	6.52	4.90	9.55	4.35	11.97	6.73	8.20	Gap (%)
											CPU time (min)
5.	Balancing violations										
	0.305	0.29	0.39	0.289	0.37	0.285	0.33	0.30	0.35	x	Number of violations
	0.306	0.25	0.41	0.181	0.32	0.335	0.39	0.31	0.29	0.38	Initial covering violations
	-0.001	0.04	-0.02	0.108	0.05	-0.05	-0.06	-0.01	0.06	x	Final covering violations
6.	Physicians' preferences										
	38	28	32	52	46	30	25	28	31	x	Variation
	30	34	15	40	50	12	12	17	20	46	Initial under covered demand
	8	6	17	12	4	18	13	11	11	x	% initial under covered demand
7.	Patients left without being seen						2815				Final under covered demand
	594	590	616	584	566	611	569	606	614	x	% final under covered demand
	552	610	662	570	509	608	566	571	619	621	Max under covered demand / day
	41.1	19.3	45.84	13.76	57.42	3.27	2.66	34.93.	5.22	x	

(continued)

Table 5. (*continued*)

Scenarios
Initial over covered demand
Final over covered demand
Max over covered demand
Initial balancing violations
Final balancing violations
Variation
Initial Preferences violations
Final preferences violations
Variation
Total demand
Initial number of patients
Final number of patients
Variation

deviation, Scenario 5 exhibits superior balance in physicians' performance compared to the manually generated schedule. In terms of physicians' preferences, Scenario 10 marginally outperforms Scenario 5. The final key performance indicator revolves around the number of patients left without being seen, amounting to 509 for Scenario 5 and 621 for Scenario 10. In conclusion, while Scenario 5 shows a minor decline in physicians' satisfaction violations, it emerges as the more robust option across most KPIs.

5 Conclusion

In this study, the physicians scheduling problem at an emergency department was studied. This study captures the stochastic nature of patients' demand and the deterministic nature of physicians' performances that depend on the individual physician and the number of patients aiming to get treatment at any given moment. The purpose of this study is to implement the genetic algorithm as a population-based metaheuristic to solve the physicians' scheduling problem at the emergency department under multiple hard and soft constraints and take into consideration three other criteria known as covering violations, physicians' performances balancing, and physicians' preferences satisfaction. The first goal is to maximize the number of patients treated at the emergency department while aiming to minimize some other violations.

A comprehensive literature review was conducted on large-scale constrained problems, similar to the one under study, to extract various characteristics and instances of this problem. Then, a mathematical model was developed highlighting the different parts of the cost function to address the objectives outlined in this study. Different experiments on nine problem scenarios were executed using the solving approach, the genetic algorithm, to finally, conclude that Scenario 5 improved the best coordination between physicians' productivity and patient arrivals with a slight deterioration in physicians' preferences. This study tailored the genetic algorithm to address a real case study, which, at first sight, its application might seem not complicated. Yet, adapting it to suit the specific characteristics of the problem under investigation is not trivial where selecting an appropriate chromosome representation and genetic algorithm methods and operators represent crucial choices and can significantly impact the quality of the generated results.

For future work extensions, it is recommended to refine the weights of the objective function in the problem under study, along with fine-tuning the parameters of the genetic algorithm aiming to improve the results. Additionally, this study suggests several avenues for future research, including forecasting patients' demand at the emergency department, considering the stochastic nature of physicians' performances to better reflect reality, and exploring the efficacy of other recent evolutionary algorithms. These investigations aim to determine if alternative algorithms outperform the proposed state-of-the-art genetic algorithm for the physicians' scheduling problem studied.

References

1. Ahsan, K.B., Alam, M.R., Morel, D.G., Karim, M.A.: Emergency department resource optimisation for improved performance: a review. J. Ind. Eng. Int. **15**, 253–266 (2019). https://doi.org/10.1007/s40092-019-00335-x

2. Derlet, R.W., Richards, J.R.: Overcrowding in the nation's emergency departments: complex causes and disturbing effects. Ann. Emergency Med. **35**(1), 63–68 (2000). https://doi.org/10.1016/S0196-0644(00)70105-3
3. Camiat, F., Restrepo, M.I., Chauny, J.-M., Lahrichi, N., Rousseau, L.-M.: Productivity-driven physician scheduling in emergency departments. Health Syst. **10**(2), 104–117 (2021). https://doi.org/10.1080/20476965.2019.1666036
4. Erhard, M., Schoenfelder, J., Fügener, A., Brunner, J.O.: State of the art in physician scheduling. Eur. J. Oper. Res. **265**(1), 1–18 (2018). https://doi.org/10.1016/j.ejor.2017.06.037
5. Ernst, A.T., Jiang, H., Krishnamoorthy, M., Sier, D.: Staff scheduling and rostering: a review of applications, methods and models. Eur. J. Oper. Res. **153**(1), 3–27 (2004). https://doi.org/10.1016/S0377-2217(03)00095-X
6. Bruni, R., Detti, P.: A flexible discrete optimization approach to the physician scheduling problem. Oper. Res. Health Care **3**(4), 191–199 (2014). https://doi.org/10.1016/j.orhc.2014.08.003
7. Pasandideh, M., Behmanesh, R.: A new mathematical model for staff scheduling in the operating room based on their preferences. Econ. Comput. Econ. Cybern. Stud. Res. **56**(4) (2022). https://doi.org/10.24818/18423264/56.4.22.06
8. Abdalkareem, Z.A., Amir, A., Al-Betar, M.A., Ekhan, P., Hammouri, A.I.: Healthcare scheduling in optimization context: a review. Health Technol. **11**, 445–469 (2021). https://doi.org/10.1007/s12553-021-00547-5
9. Carter, M.W., Lapierre, S.D.: Scheduling emergency room physicians. Health Care Manage. Sci. **4**, 347–360 (2001). https://doi.org/10.1023/A:1011802630656
10. Gunawan, A., Lau, H.C.: The bi-objective master physician scheduling problem. PATAT (2010)
11. Rousseau, L.-M., Pesant, G., Gendreau, M.: A general approach to the physician rostering problem. Ann. Oper. Res. **115**(1–4), 193–205 (2002). https://doi.org/10.1023/A:1021153305410
12. Gunawan, A., Lau, H.C.: Master physician scheduling problem. J. Oper. Res. Soc. **64**(3), 410–425 (2013). https://doi.org/10.1057/jors.2012.48
13. Burke, E.K., Li, J., Qu, R.: A Pareto-based search methodology for multi-objective nurse scheduling. Ann. Oper. Res. **196**, 91–109 (2012). https://doi.org/10.1007/s10479-009-0590-8
14. Gendreau, M., et al.: Physician scheduling in emergency rooms. In: Practice and Theory of Automated Timetabling VI: 6th International Conference, PATAT 2006 Brno, Czech Republic, August 30–September 1, 2006 Revised Selected Papers 6, pp. 53–66. Springer (2007). https://doi.org/10.1007/978-3-540-77345-0_4
15. Topaloglu, S.: A multi-objective programming model for scheduling emergency medicine residents. Comput. Ind. Eng. **51**(3), 375–388 (2006). https://doi.org/10.1016/j.cie.2006.08.003
16. Alharbi, A., AlQahtani, K.: A genetic algorithm solution for the doctor scheduling problem. In: The Tenth International Conference on Advanced Engineering Computing and Applications in Sciences, pp. 91–7 (2016)
17. Lin, C.-C., Kang, J.-R., Hsu, T.-H.: A memetic algorithm with recovery scheme for nurse preference scheduling. J. Ind. Prod. Eng. **32**(2), 83–95 (2015). https://doi.org/10.1080/21681015.2014.997815
18. Wang, C.-W. et al.: A genetic algorithm for resident physician scheduling problem. In Proceedings of the 9th Annual Conference on Genetic and Evolutionary Computation, pp. 2203–2210 (2007). https://doi.org/10.1145/1276958.1277380
19. Hamid, M., Tavakkoli-Moghaddam, R., Golpaygani, F., Vahedi-Nouri, B.: A multi-objective model for a nurse scheduling problem by emphasizing human factors. Proc. Inst. Mech. Eng. **234**(2), 179–199 (2020). https://doi.org/10.1177/0954411919898895

20. Koruca, H.I., Emek, M.S., Gulmez, E.: Development of a new personalized staff-scheduling method with a work-life balance perspective: case of a hospital. Ann. Oper. Res., pp. 1–28 (2023). https://doi.org/10.1007/s10479-023-05244-2
21. Zaerpour, F., Bijvank, M., Ouyang, H., Sun, Z.: Scheduling of physicians with time-varying productivity levels in emergency departments. Prod. Oper. Manag. 31(2), 645–667 (2022). https://doi.org/10.1111/poms.13571
22. Savage, D.W., Woolford, D.G., Weaver, B., Wood, D.: Developing emergency department physician shift schedules optimized to meet patient demand. Can. J. Emergency Med. 17(1), 3–12 (2015). https://doi.org/10.2310/8000.2013.131224
23. Ganguly, S., Lawrence, S., Prather, M.: Emergency department staff planning to improve patient care and reduce costs. Decis. Sci. 45(1), 115–145 (2014). https://doi.org/10.1111/deci.12060
24. Little, C., Choudhury, S.: A review of the scheduling problem within Canadian healthcare centres. Appl. Sci. 12(21), 11146 (2022). https://doi.org/10.3390/app122111146
25. Mansini, R., Zanotti, R.: Optimizing the physician scheduling problem in a large hospital ward. J. Sched. 23(3), 337–361 (2020). https://doi.org/10.1007/s10951-019-00614-w
26. Hidayati, M., Wibowo, A., Abdulrahman, S.: Preliminary review on population based approaches for physician scheduling. In: 2018 Indonesian Association for Pattern Recognition International Conference (INAPR), pp. 90–94. IEEE 2018). https://doi.org/10.1109/INAPR.2018.8627005
27. Boussaïd, I., Lepagnot, J., Siarry, P.: A survey on optimization metaheuristics. Inf. Sci. 237, 82–117 (2013). https://doi.org/10.1016/j.ins.2013.02.041
28. Kundu, S., Mahato, M., Mahanty, B., Acharyya, S.: Comparative performance of simulated annealing and genetic algorithm in solving nurse scheduling problem. Proc. Int. Multi Conf. Eng. Comput. Sci. 1, 96–100 (2008)
29. Rochman, E.M.S., Rachmad, A., Santosa, I., et al.: The application of genetic algorithms as an optimization step in the case of nurse scheduling at the bringkoning community health center. J. Phys. Conf. Ser. 1477, 022026. IOP Publishing (2020). https://doi.org/10.1088/1742-6596/1477/2/022026
30. Ohki, M.: Many-objective nurse scheduling using NSGA-II based on Pareto partial dominance with linear subset-size scheduling. IJCCI, 118–125 (2018). https://doi.org/10.5220/0006894501180125
31. Maenhout, B., Vanhoucke, M.: Comparison and hybridization of crossover operators for the nurse scheduling problem. Ann. Oper. Res. 159, 333–353 (2008). https://doi.org/10.1007/s10479-007-0268-z
32. Ghaheri, A., Shoar, S., Naderan, M., Hoseini, S.S.: The applications of genetic algorithms in medicine. Oman Med. J. 30(6), 406 (2015). https://doi.org/10.5001/omj.2015.82
33. Haldurai, L., Madhubala, T., Rajalakshmi, R.: A study on genetic algorithm and its applications. Int. J. Comput. Sci. Eng. 4(10), 139–143 (2016)
34. Ilmi, R.R., Mahmudy, W.F., Ratnawati, D.E.: Optimasi penjadwalan perawat menggunakan algoritma genetika. S1. DORO: Repository Jurnal Mahasiswa PTIIK Universitas Brawijaya 5(13) (2015)
35. Jebari, K., Madiafi, M., et al.: Selection methods for genetic algorithms. Int. J. Emerging Sci. 3(4), 333–344 (2013)
36. Shukla, A., Pandey, H.M., Mehrotra, D.: Comparative review of selection techniques in genetic algorithm. In: 2015 International Conference on Futuristic Trends on Computational Analysis and Knowledge Management (ABLAZE), pp. 515–519. IEEE (2015). https://doi.org/10.1109/ABLAZE.2015.7154916
37. Ahn, C.W., Ramakrishna, R.S.: Elitism-based compact genetic algorithms. IEEE Trans. Evol. Comput. 7(4):367–385 (2003). https://doi.org/10.1109/TEVC.2003.814633

38. Puente, J., Gómez, A., Fernández, I., Priore, P.: Medical doctor rostering problem in a hospital emergency department by means of genetic algorithms. Comput. Ind. Eng. **56**(4):1232–1242 (2009). https://doi.org/10.1016/j.cie.2008.07.016
39. Lin, C.-C., Kang, J.-R., Chiang, D.-J., Chen, C.-L.: Nurse scheduling with joint normalized shift and day-off preference satisfaction using a genetic algorithm with immigrant scheme. Int. J. Distrib. Sens. Netw. **11**(7), 595419 (2015). https://doi.org/10.1155/2015/59541
40. Leksakul, K., Phetsawat, S., et al.: Nurse scheduling using genetic algorithm. Math. Prob. Eng. (2014). https://doi.org/10.1155/2014/246543
41. Alfadilla, N., Sentia, P.D., Asmadi, D., et al.: Optimization of nurse scheduling problem using genetic algorithm: a case study. In: IOP Conference Series: Materials Science and Engineering, vol. 536, p. 012131. IOP Publishing (2019). https://doi.org/10.1088/1757-899X/536/1/012131
42. Kim, S.-J., Ko, Y.-W., Uhmn, S., Kim, J.: A strategy to improve performance of genetic algorithm for nurse scheduling problem. Int. J. Softw. Eng. Appl. **8**(1), 53–62 (2014). https://doi.org/10.14257/ijseia.2014.8.1.05
43. Saraswati, N.W.S., Artakusuma, I.D.M.D., Indradewi, I.G.A.A.D.: Modified genetic algorithm for employee work shifts scheduling optimization. J. Phys. Conf. Ser. **1810**, 012014 (2021). https://doi.org/10.1088/1742-6596/1810/1/012014

Temporal Emotional and Thematic Progression (TETP): A Novel Analysis of Mental Health Discussions on Social Platforms

Sharath Kumar Jagannathan[✉] and Gulhan Bizel

Data Science Institute, Saint Peter's University, 2641 JFK Boulevard, Jersey City, NJ 07306, USA
{sjagannathan,gbizel}@saintpeters.edu

Abstract. In an era where online platforms increasingly host critical conversations on mental health, understanding the evolution of these discussions is vital. Traditional analysis methods, while insightful, often overlook the sequential interplay of emotions and themes, a gap our research aims to fill. This research paper introduces the Temporal Emotional and Thematic Progression (TETP) methodology, a novel approach that tracks the temporal dynamics of mental health discourse on social media by charting discussions within a two-dimensional emotional-thematic space. Utilizing sentiment analysis and Latent Dirichlet Allocation (LDA) for thematic extraction, we convert discussion sequences into trajectories, revealing patterns in how users navigate through emotional and thematic phases over time. A preliminary analysis of mental health-related posts on Reddit not only confirms the viability of TETP but also uncovers distinct pathways of discourse evolution, offering insights into the collective journey of online communities grappling with mental health issues. By elucidating the temporal aspect of online discussions, TETP aims to enhance understanding of digital mental health landscapes, improve platform moderation strategies, and inform mental health practitioners about prevalent online discourse patterns, ultimately contributing to better online support ecosystems for mental health.

Keywords: Temporal Analysis · Sentiment Analysis · Latent Dirichlet Allocation (LDA) · Mental Health Discussions Online Platforms

1 Introduction

In the digital age, there has been a significant growth in the usage of online platforms as a key route for individuals to engage in debates, exchange experiences, and seek help on a wide range of issues, including mental health. While these platforms offer solace to many, they also create a large amount of unstructured data, which contains significant insights into the dynamics of mental health talks. Traditional approaches like sentiment analysis and topic modeling have proved essential in understanding these debates. Their shortcoming is that they ignore the sequential and temporal components of these interactions. This research offers the Temporal Emotional and Thematic Flow (TETP) technique, a unique approach meant to capture the chronological flow of online mental health debates while accounting for both their emotional and thematic components.

1.1 Online Platforms and Mental Health

In the digital age, there has been a significant growth in the usage of online platforms as a key route for individuals to engage in debates, exchange experiences, and seek help on a wide range of issues, including mental health. While these platforms offer solace to many, they also create a large amount of unstructured data, which contains significant insights into the dynamics of mental health talks. Traditional approaches like sentiment analysis and topic modeling have proved essential in understanding these debates. Their shortcoming is that they ignore the sequential and temporal components of these interactions. This research offers the Temporal Emotional and Thematic Flow (TETP) technique, a unique approach meant to capture the chronological flow of online mental health debates while accounting for both their emotional and thematic components.

1.2 Traditional Sentiment Analysis

Sentiment analysis, often referred to as opinion mining, involves the use of natural language processing and text analysis to identify and extract subjective information from source materials. Traditional sentiment analysis tools categorize text as positive, negative, or neutral. While these tools have been instrumental in gauging the overall sentiment of a text, they often treat each digital interaction as a discrete entity, neglecting the sequential and interconnected nature of online discussions.

1.3 Topic Modeling and LDA

Latent Dirichlet Allocation (LDA) is a popular method for topic modeling, which allows sets of observations to be explained by unobserved groups. In the context of text analysis, these unobserved groups or topics help in understanding the underlying thematic structure of a large corpus of text. LDA has been widely used in analyzing online content, from news articles to social media posts.

1.4 Temporal Analysis in Social Media Data

Temporal analysis involves studying the sequence and timing of events. In the realm of social media, temporal analysis can provide insights into how discussions evolve over time. These analyses are crucial in understanding the progression and evolution of online discussions, especially in platforms where discussions are dynamic and rapidly changing.

The Temporal Emotional and Thematic Progression (TETP) Analysis Method employs a structured research methodology, starting with the 'Data Collection' phase, during which data is harvested from various online platforms through APIs, with a focus on mental health discussions. Following data collection, the methodology divides into two parallel streams: 'Emotional Analysis' and 'Thematic Analysis.' In Emotional Analysis, the process involves extracting sentiments and classifying them into positive, negative, or neutral categories. Both emotional and thematic data converge in the '2D Mapping' phase, where each piece of data is plotted in a two-dimensional space. The x-coordinate represents thematic attributes, while the y-coordinate captures emotional attributes. This spatial representation is critical for visualizing the relationship between the two dimensions of the data.

The 'Trajectory Formation' stage involves connecting these data points to form trajectories, which illustrate the flow and transition of discussions over time. This step is essential for observing patterns and shifts in the narrative of mental health discourse.

In the 'Analysis' phase, the trajectories are examined to identify trends and sentiments fluctuations across timeframes. The final stage is the 'Conclusion', where the synthesized data informs practical applications in mental health support and provides insights for predictive modeling. This concluding part interprets the results to enhance understanding and intervention strategies within the context of mental health on social media platforms.

2 Background and Related Work

Tadesse et al. (2020) explores the use of deep learning techniques to identify signs of suicide ideation in social media forums. Essentially, it uses advanced computer algorithms to analyze online posts and detect whether individuals are exhibiting thoughts or intentions related to suicide. This study applies complex computer models to sift through social media posts, aiming to find signs that a person might be thinking about suicide. It's a cutting-edge approach to potentially save lives by catching early warning signs online [1]. Cao, Zhang, & Feng (2020) work focuses on constructing personal knowledge graphs from social media data to enhance the detection of suicidal ideation. In simpler terms, it creates a detailed map of an individual's social media activity to better understand and predict their mental health risks. Here, the focus is on building detailed profiles of social media users to better predict suicidal thoughts. It's like creating a map of a person's online behavior to understand their mental health risks more accurately [2].

Almeida, Briand, & Meurs (2017) deals with identifying early signs of depression from user-generated content on social media. It aims to detect depression risks by analyzing the things people write online, before these risks manifest more severely. This research looks at social media posts to find early indicators of depression, helping to identify mental health issues before they escalate [3]. Amini et al. (2016) compares different statistical and machine learning methods, like logistic regression and neural networks, to identify high-risk groups for suicide. It's about using various computer-based approaches to find patterns that might indicate a higher risk of suicide in certain groups of people. This paper compares various computerized methods to identify groups at a higher risk of suicide. It's about finding patterns in data that can alert us to who might need help the most [4]. Roy et al. (2020): This study uses machine learning to predict the future risk of suicidal ideation based on social media data. It means that by analyzing what individuals post online, the study attempts to foresee who might experience suicidal thoughts in the future. This study uses advanced data analysis to predict who might experience suicidal thoughts in the future based on their social media activity. It's an approach that could provide crucial early interventions [5].

Eichstaedt et al. (2018) demonstrates how language used on Facebook can predict depression, highlighting a potential tool for early diagnosis by analyzing social media language patterns [6]. Sawhney et al. (2021) introduces a novel method for detecting phases of emotional states related to suicide ideation on social media, offering a more nuanced understanding of mental health states [7]. Zogan et al. (2020) explores a hybrid deep learning model that combines various types of data to detect depression on social media, aiming for an explainable and effective approach [8]. Ma & Cao (2020) focuses on identifying suicide risks on social media using a dual-attention mechanism, which considers the context of user posts more deeply for accurate risk assessment [9]. Cao et al. (2019) develops a method for detecting latent suicide risk on microblogs using specialized word embeddings and layered attention to analyze text, offering a sophisticated tool for early intervention [10].

The Table 1 showcases a variety of approaches and contributions to the field of mental health analysis through social media, highlighting the diversity and depth of research in this area.

The studies encompass innovative approaches to mental health analysis, leveraging social media, dynamic Bayesian networks, and NLP techniques. Usman Naseem et al.'s work introduces a novel architecture for mental health identification on social media by integrating users' emotional histories and posting patterns, demonstrating improved detection of at-risk individuals [41]. Frank Iorfino et al. employ dynamic Bayesian networks to unravel the complex interrelations among various health factors in young individuals receiving mental healthcare, pinpointing key pathways from childhood disorders to adolescent depression and beyond, which underscores the necessity for early intervention [42]. SocialNLP 2023 explores cutting-edge NLP applications on social media to analyze textual data for insights into human emotions or behaviors, aligning with the overarching theme of employing advanced computational methods to enhance mental health diagnostics and understanding. Together, these studies highlight the significance of interdisciplinary approaches in tackling the multifaceted challenges of mental health surveillance and intervention in the digital age [43].

The studies reviewed align closely with the overarching theme of your paper on Temporal Emotional and Thematic Progression (TETP) in analyzing mental health discussions on social media. These works collectively illustrate advancements in sentiment analysis, topic modeling, and temporal dynamics understanding in online mental health discourse. They offer insights into the evolution of these discussions, contributing to the development of your novel TETP approach. By mapping emotional and thematic shifts over time, your TETP technique stands to benefit from these studies' advancements in detecting nuanced emotional states, thematic changes, and risk factors in online mental health conversations. This collective body of research provides a solid foundation for your paper's innovative approach to understanding and managing mental health in the digital sphere.

The research surveyed complements the TETP approach by offering diverse perspectives on analyzing online mental health discussions. These studies underscore the importance of nuanced sentiment analysis and advanced topic modeling in understanding the complex dynamics of mental health discourse on social platforms. This body of work enriches your research, providing valuable context and methodologies that could enhance the TETP's capability to map the emotional and thematic evolution of online mental health conversations, thereby offering deeper insights into the patterns and trends in digital mental health narratives.

Table 1. Approaches and contributions by other Authors.

Author(s) and Year	Focus and Contribution
Song et al. (2018)	Detecting depression from social media text, offering an interpretable approach [11].
Ophir et al. (2020)	Analyzing Facebook posts for suicide risk detection using deep neural networks [12].
Shah et al. (2020)	Hybrid method for suicidal ideation detection combining various techniques [13].
McHugh et al. (2019)	Examines the relationship between suicidal ideation and actual suicide [14].
Stone (2021)	Analyzes changes in suicide rates in the United States [15].
B. RN., Weiland, & Peterson (2009)	Comprehensive overview of various illnesses for a general audience [16].
De Choudhury (2013)	Role of social media in addressing mental health challenges [17].
Shing, Resnik, & Oard (2020)	Model to prioritize suicide risk assessment based on social media language [18].
Niederkrotenthaler (2017)	'Papageno effect' and its positive influence on suicide prevention [19].
Chancellor et al. (2019)	Ethical concerns in using social media data for mental health inference [20].
Coppersmith et al. (2015)	Identifying depression and PTSD from Twitter data [21].
Milne et al. (2016)	Triaging content in online peer-support forums [22].
Preotiuc-Pietro et al. (2015)	Detecting mental illness through linguistic analysis in large-scale studies [23].
Resnik et al. (2015)	Identifying depression-related language on Twitter with supervised topic modeling [24].
Tsugawa et al. (2015)	Recognizing depression through patterns of Twitter activity [25].
De Choudhury et al. (2016)	Identifying shifts towards suicidal ideation from social media content [26].
Sawhney et al. (2021)	Detecting varying degrees of suicide ideation on social media [27].
Yang et al. (2016)	Hierarchical attention networks for document classification [28].
Demszky et al. (2020)	GoEmotions dataset for emotion analysis in computational linguistics [29].

(*continued*)

Table 1. (*continued*)

Author(s) and Year	Focus and Contribution
Kumar (2021)	Real-time mental health analytics using IoMT and social media datasets [30].
Turcan & McKeown (2019)	'Dreaddit,' a Reddit dataset for stress analysis in social media [31].
Matero et al. (2019)	Assessing suicide risk using advanced language processing techniques and BERT [32].
Lin et al. (2020)	Detecting depression from social media interactions with 'SenseMood' [33].
Gui et al. (2019)	Cooperative multimodal approach for depression detection on Twitter [34].
Shen et al. (2017)	Depression detection using a multimodal dictionary learning approach [35].
Tadesse et al. (2019)	Identifying depression-related posts in Reddit forums [36].
Zogan et al. (2021)	'DepressionNet,' enhancing depression detection on social media [37].
Haque, Reddi, & Giallanza (2021)	Using deep learning for suicide and depression identification on social media [38].
Zogan et al. (2020)	Hybrid deep learning model for detecting depression from social media [39].
Santos, C.M., Passos, A.M., and Uitdewilligen, S., 2016	Explored the impact of temporal mental model accuracy on team learning and performance, finding that high accuracy did not significantly affect learning, but low accuracy reduced learning behaviors and indicated that inaccurate shared models can result in closed minds, with team adaptation mediating between learning and performance [40].

3 Methodology

3.1 Data Collection

To examine the development of emotional and thematic elements in mental health discussions online, our research utilized data from Reddit, known for its wide-ranging discussions and varied user base. We specifically focused on subreddits related to mental health, such as r/mentalhealth, r/depression, and r/anxiety, over a selected period. This approach ensured a thorough collection of conversations. For each Reddit post and comment, key data like the text, timestamps, user IDs, and relevant metadata were gathered, forming the basis of our analysis. An API, or Application Programming Interface, is a set of protocols and tools for building software applications. It specifies how software components should interact, allowing different programs to communicate with each other. In the context of Reddit, the Reddit API enables developers to programmatically access and interact with Reddit data. This includes retrieving posts, comments, and user information from various subreddits. Utilizing the Reddit API allows for the collection and analysis of vast amounts of data from Reddit, including posts related to specific topics like mental health, thereby providing valuable insights into public discussions and trends.

3.2 Sentiment Analysis

To analyze the emotional tone of each online interaction, sentiment analysis was employed. This involved using a mix of pre-trained models and custom classifiers to categorize posts and comments into positive, negative, or neutral sentiments.

The process began with pre-processing, where text data was cleaned to remove any URLs, special characters, and irrelevant words. This step was crucial to ensure consistency and accuracy in analysis. The text was then tokenized and lemmatized, breaking it down into basic elements for easier processing by the sentiment analysis models.

In the model application phase, advanced sentiment classifiers, like BERT-based models, were used to assign sentiment scores to each text entry. These scores were then normalized and classified into one of the three sentiment categories, providing a structured understanding of the emotional undertones in the online discussions. This approach allowed for a detailed and nuanced analysis of the sentiments expressed in online mental health discussions.

3.3 Latent Dirichlet Allocation (LDA)

To extract the central themes in mental health discussions, we employed Latent Dirichlet Allocation (LDA), a sophisticated method for topic modeling. Initially, the text data underwent meticulous pre-processing, involving cleaning, tokenization, and lemmatization, like sentiment analysis. This step was essential to ensure the data was primed for accurate thematic extraction and then constructing a dictionary and corpus. A unique dictionary comprising individual words was created, and a corresponding corpus was formulated using the bag-of-words model. This foundation was pivotal for the LDA model, as it relies on word frequency and distribution to discern topics.

The LDA model training and topic assignment as described in algorithm 1 were executed. The model was trained on the prepared corpus, with a predetermined number of topics to be identified, such as 20. Each post and comment in the dataset were then assigned to the most relevant topic based on the highest probability. This step enabled us to systematically categorize and understand the prevalent themes in the online mental health discourse, thereby providing deep insights into the nature of these conversations.

Algorithm 1. **Topic Modeling using Latent Dirichlet Allocation (LDA).**

1. Data Preparation:

    - Import essential libraries: pandas, sklearn's CountVectorizer and LatentDirichletAllocation, matplotlib.pyplot.
    - Load the dataset into a pandas DataFrame.

2. Text Vectorization:
   - Initialize CountVectorizer with parameters: max_df = 0.95, min_df = 2, and stop_words = 'english'.
   - Apply CountVectorizer to the text data to create a Document-Term Matrix (DTM).

3. LDA Model Initialization and Training:
   - Set the number of topics (e.g., 5).
   - Create an LDA model instance with the defined number of topics and a fixed random state.
   - Fit the LDA model to the Document-Term Matrix.

4. Topic Exploration and Visualization:
   - Implement a function to print the top words in each topic identified by the LDA model.
   - Display the top words for each topic using the feature names from the vectorizer.

5. Topic Assignment to Documents:
   - Assign the most likely topic to each document in the dataset, based on the LDA model output, and add this as a new column in the DataFrame.

3.4 Mapping Discussions to 2D Space

Upon identifying emotional and thematic attributes, the discussions were visually represented in a 2D space. This involved organizing the conversations chronologically using timestamps, a process called temporal sequencing. Each discussion was then converted into a vector, with its thematic attribute, derived from.

Latent Dirichlet Allocation, defining the X-coordinate and its emotional attribute, determined by sentiment analysis, marking the Y-coordinate. These vectors were sequentially connected, forming trajectories that illustrate the progression and evolution of the discussions over time. This innovative approach provides a dynamic and insightful visual representation of online mental health discourse.

4 Validity Considerations in TETP Analysis

4.1 Internal Validity

To ensure the internal validity of our Temporal Emotional and Thematic Progression (TETP) analysis, we carefully considered the relationship between the independent variables (e.g., thematic, and emotional dimensions of social media posts) and the dependent variable (e.g., the evolution of mental health discussions). One key aspect of our approach involved the use of pre-trained models and custom classifiers for sentiment analysis, which were validated against established benchmarks to ensure their reliability in classifying emotional tones accurately.

To assess the reliability of our thematic analysis, we employed Latent Dirichlet Allocation (LDA) with parameters fine-tuned based on extensive cross-validation to ensure that the topics extracted were representative of the underlying discussions. Although Cronbach's Alpha is traditionally used for survey data reliability, our analytical methods are based on computational techniques where internal consistency is ensured through algorithmic validation against known outcomes and through the reproducibility of results across different data sets.

External Validity

External validity, or the generalizability of our findings, was addressed by selecting a diverse and representative sample of mental health discussions from Reddit. By analyzing data from a range of mental health-related subreddits over a specific period, we aimed to capture a wide spectrum of mental health narratives. Additionally, the scalability and flexibility of the TETP method to various social media platforms suggest that our findings have relevance beyond the specific context of Reddit, offering insights into the broader dynamics of online mental health discourse.

To further enhance the external validity of our study, future research directions will involve applying the TETP analysis framework to different social media platforms and languages, thus assessing the universality of the identified patterns across cultural and linguistic boundaries. This will help in understanding the extent to which our findings can be generalized to the global online discourse on mental health.

4.2 Theoretical Contributions

This study introduces a novel Temporal Emotional and Thematic Progression (TETP) analysis method, marking a significant advancement in the field of computational linguistics and mental health discourse analysis. Theoretically, the TETP framework contributes to our understanding of online mental health narratives by:

Mapping Emotional and Thematic Evolution. Demonstrating how mental health discussions evolve over time within digital communities, offering a dynamic view that traditional static analysis methods do not capture.

Integrating Multidimensional Data Analysis. By analyzing emotional and thematic dimensions concurrently, TETP provides a more holistic understanding of online discourse, challenging existing theories that often consider these aspects in isolation.

Enhancing Theories of Digital Communication. The findings support and extend theories of online communication by illustrating how digital platforms serve as critical spaces for mental health support and community building.

These contributions underscore the importance of temporal analysis in understanding the complex nature of online mental health discourse, providing a foundation for future research in this rapidly evolving field.

4.3 Managerial Implications

The implications of the TETP analysis method extend beyond theoretical advancements, offering practical insights for various stakeholders:

For Social Media Platform Moderators. TETP can inform the development of more nuanced content moderation strategies, enabling platforms to identify and support mental health discussions more effectively, potentially leading to the creation of safer online spaces for vulnerable users.

For Mental Health Practitioners. By identifying prevalent themes and emotional trajectories in online discussions, practitioners can gain insights into the concerns and mood states prevalent among online communities, informing the development of targeted interventions.

For Policymakers and Public Health Officials. The methodological approach can guide the development of evidence-based policies and initiatives aimed at supporting mental health in digital environments, emphasizing the need for collaboration between tech companies, healthcare providers, and regulatory bodies.

5 Analysis

The analysis involves a method using the PRAW library, enabling the retrieval of posts in various categories, such as 'hot', 'new', 'top', 'controversial', and 'rising,' from subreddits dedicated to mental health topics, including depression, anxiety, bipolar disorder, ADHD, OCD, and PTSD. By implementing this approach, we can gather valuable insights into user interactions, sentiment analysis, and trending discussions within these online communities, contributing to a deeper understanding of mental health-related discussions on Reddit.

In Fig. 1, we analyze the prevalence of the topics with 'Anxiety and Physical Symptoms', 'Seeking Help and Expressing Difficulties', 'Family Dynamics and Relationships', 'Mental Health Over Time', 'Social Interactions and Self-perception' across seven distinct categories using LDA. This method was applied to examine the frequency of mentions under each of these categories.

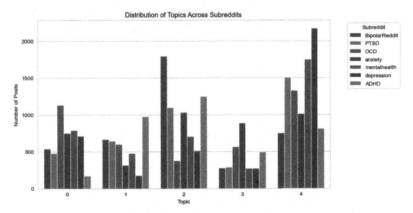

Fig. 1. Exploration of Five Distinct Topics with seven categories of subreddits.

Figure 2 depicts the sentiment polarity distribution and Fig. 3 on the sentiment's distribution using box plot in posts across various subreddits discussing mental health topics. 'Mental Health' skews predominantly positive, 'Anxiety' tends towards the negative, and 'Depression' shows a distribution in both positive and negative directions with outlier data.

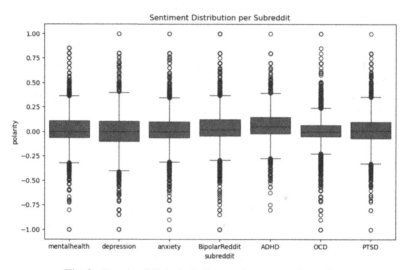

Fig. 2. Emotional Polarity in Discussions across the topics.

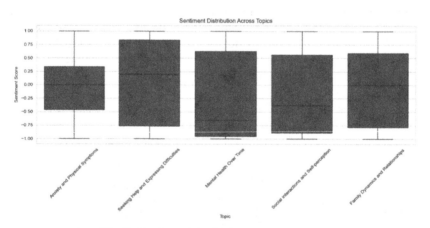

Fig. 3. Sentiment Distribution across the topics.

In Fig. 4, a marked decline in sentiment polarity is observed from September 2023 to October 2023, indicating a notable shift in emotional expressions within the dataset during this specific timeframe. This observation underscores the dynamic nature of the data, as highlighted by the comprehensive analysis employing Latent Dirichlet Allocation.

The five topics identified through a Latent Dirichlet Allocation (LDA) approach reflect common themes in discussions related to mental health. Topic #1 centers on the experiences of individuals with ADHD and PTSD, highlighting the complexity of their emotions and struggles in daily life. Topic #2 delves into the nuances of living with OCD, underscoring the persistent thoughts and uncertainties that individuals face. Topic #3 focuses on bipolar disorder, emphasizing the challenges of managing medication, depression, and manic episodes. Topic #4 explores a broader spectrum of mental health issues, capturing the feelings of uncertainty and desire for a better understanding of one's own mental state. Lastly, Topic #5 addresses the realities of dealing with anxiety, including the impact on sleep patterns, the use of medication, and the experience of panic attacks. These topics collectively offer a multifaceted view of mental health, underscoring the diverse experiences and challenges faced by individuals dealing with various mental health conditions.

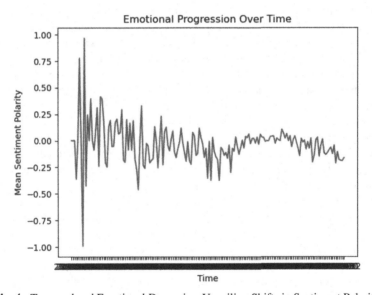

Fig. 4. Temporal and Emotional Dynamics: Unveiling Shifts in Sentiment Polarity.

Figure 5 presents a word cloud capturing the most frequent words in positively toned posts. This visual representation offers insights into prevalent positive themes within the dataset.

Fig. 5. Positive Sentiment Word Cloud.

6 Results

In Fig. 6, "Sentiment Over Time per Subreddit", shows fluctuating sentiment polarity scores from 2010 to 2024, with each subreddit color-coded. Sentiment appears to vary significantly, with no clear trend toward positive or negative sentiment over time. The Fig. 7 on "Distribution of Sentiment Polarity Scores", presents a histogram of sentiment scores, revealing a bimodal distribution with peaks at the extremes of sentiment (-1 and 1). This suggests that posts tend to be highly polarized, with many strongly negative and positive expressions, and fewer neutral sentiments.

The chart in Fig. 8 illustrates the distribution of discussions on various mental health topics across different subreddits. It suggests certain subreddits tend to focus on specific aspects of mental health. For example, discussions in the anxiety subreddit are heavily weighted towards topics involving daily symptoms, whereas the bipolar subreddit has a notable concentration on medication and treatment strategies. This data is valuable for understanding community engagement and prevalent concerns within each subreddit, which could inform targeted support and information dissemination strategies for individuals seeking help in these online communities.

The temporal sentiment analysis across various mental health subreddits has unveiled distinct patterns in user interactions and emotional discourse. The pronounced bimodal distribution of sentiment polarity underscores the critical role these forums play in facilitating both outpourings of distress and sources of encouragement. These findings reflect the journal's focus on the intricate dynamics of online mental health communities, emphasizing the importance of supportive digital ecosystems for individuals navigating the complexities of mental wellness. The analysis presented herein offers valuable insights into the ebb and flow of mental health narratives in digital spaces, as captured through the sentiment of shared experiences over time. The stark polarization in sentiment underscores the dual nature of these platforms—spaces where individuals seek and provide solace amidst their struggles. These patterns not only enrich our understanding of online mental health discourse but also highlight the imperative for nuanced moderation and support systems within these virtual communities.

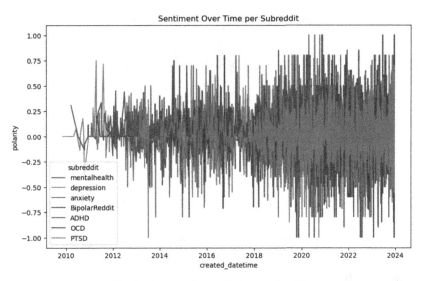

Fig. 6. Sentiment over time per subreddit.

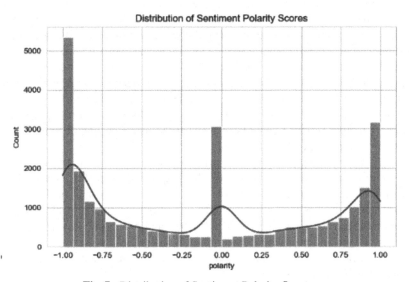

Fig. 7. Distribution of Sentiment Polarity Scores.

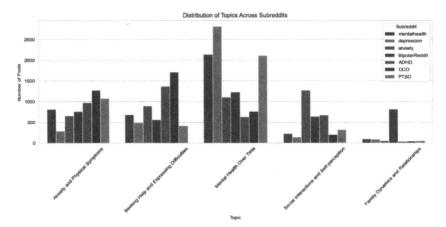

Fig. 8. Distribution of topics across subreddits.

7 Conclusion

The Temporal Emotional and Thematic Progression (TETP) analysis has provided a groundbreaking perspective on the dynamics of mental health discussions on social media. By mapping the intricate interplay between sentiment and themes over time, this research uncovers the nuanced journeys individuals embark upon in these digital communities. The TETP approach offers a compelling framework for future studies, promising to enhance the support systems for mental health management online and providing a base for practitioners to align interventions with the identified patterns. This study not only advances our understanding of digital discourse but also paves the way for more effective, timely, and context-aware responses to mental health needs in the ever-evolving landscape of social media.

TETP method marks a significant stride in the realm of mental health discourse analysis on social platforms. It delineates the temporal shifts and thematic currents in online conversations, offering a holistic view of the digital mental health landscape. This research enhances predictive capabilities and intervention strategies, fostering a deeper engagement with mental health narratives and potentially revolutionizing support frameworks within online communities. The implications of this study extend beyond academia, signaling a new direction for mental health practitioners and platform moderators in the digital age.

Our study emphasizes the importance of rigorous methodologies to minimize biases, particularly in data collection from diverse social media platforms. Ethical considerations, especially concerning privacy and the sensitive nature of mental health discussions, were paramount, guiding our approach to anonymize data and ensure respectful analysis. The validity of our suicidal ideation detection models was continually assessed through iterative testing and validation against established psychological frameworks, ensuring accuracy and reliability. Recognizing the dynamic nature of language, especially in the context of mental health, our models incorporate adaptive learning mechanisms to account for evolving expressions and terminologies. We also acknowledge the importance of differentiating genuine suicidal ideation from expressions of distress, employing nuanced linguistic analysis and contextual evaluation to mitigate the risk of false positives. Efforts to ensure cultural sensitivity and inclusivity in our language models were made, with plans for ongoing refinement to better reflect the diverse voices within digital communities. Additionally, our study's methodologies include provisions for periodic updates to algorithms, ensuring they remain responsive to emerging trends and patterns in mental health discourse online. Collaborations with mental health professionals are actively sought to ground our computational findings in practical, real-world applicability, aiming to enhance the support structures for individuals navigating mental health challenges in digital spaces.Top of Form.

References

1. Tadesse, M.M., Lin, H., Xu, B., Yang, L.: Detection of suicide ideation in social media forums using deep learning. Algorithms **13**(1), 7 (2020)
2. Cao, L., Zhang, H., Feng, L.: Building and using personal knowledge graph to improve suicidal ideation detection on social media. IEEE Trans. Multimedia (2020). https://doi.org/10.1109/tmm.2020.3046867
3. Almeida, H., Briand, A., Meurs, M.-J.: Detecting early risk of depression from social media user-generated content. In: CLEF (Working Notes) (2017)
4. Amini, P., Ahmadinia, H., Poorolajal, J., Amiri, M.M.: Evaluating the high risk groups for suicide: a comparison of logistic regression, support vector machine, decision tree and artificial neural network. Iran. J. Public Health **45**(9), 1179 (2016)
5. Roy, A., Nikolitch, K., McGinn, R., Jinah, S., Klement, W., Kaminsky, Z.A.: A machine learning approach predicts future risk to suicidal ideation from social media data. NPJ Digit. Med. **3**(1), 1–12 (2020)
6. Eichstaedt, J.C., et al.: Facebook language predicts depression in medical records. Proc. Natl. Acad. Sci. **115**(44), 11203–11208 (2018)
7. Sawhney, R., Joshi, H., Flek, L., Shah, R.: Phase: learning emotional phase-aware representations for suicide ideation detection on social media. In: Proceedings of the 16th Conference of the European Chapter of the Association for Computational Linguistics: Main Volume, pp. 2415–2428 (2021)
8. Zogan, H., Razzak, I., Wang, X., Jameel, S., Xu, G.: Explainable depression detection with multi-modalities using a hybrid deep learning model on social media (2020). arXiv preprint arXiv:2007.02847
9. Ma, Y., Cao, Y.: Dual attention based suicide risk detection on social media. In: 2020 IEEE International Conference on Artificial Intelligence and Computer Applications (ICAICA), pp. 637–640 (2020)
10. Cao, L., et al.: Latent suicide risk detection on microblog via suicide-oriented word embeddings and layered attention (2019). arXiv preprint arXiv:1910.12038
11. Song, H., You, J., Chung, J.-W., Park, J.C.: Feature attention network: interpretable depression detection from social media. In: PACLIC (2018)
12. Ophir, Y., Tikochinski, R., Asterhan, C.S., Sisso, I., Reichart, R.: Deep neural networks detect suicide risk from textual Facebook posts. Sci. Rep. **10**(1), 1–10 (2020)
13. Shah, F.M., Haque, F., Nur, R.U., Al Jahan, S., Mamud, Z.: A hybridized feature extraction approach to suicidal ideation detection from social media post. In: 2020 IEEE Region 10 Symposium (TENSYMP), pp. 985–988 (2020)
14. McHugh, C.M., Corderoy, A., Ryan, C.J., Hickie, I.B., Large, M.M.: Association between suicidal ideation and suicide: meta-analyses of odds ratios, sensitivity, specificity and positive predictive value. BJPsych Open **5**(2), e18 (2019)

15. Stone, D.M.: Changes in suicide rates—United States, 2018–2019. Morb. Mortal. Wkly Rep. **70**(8), 261–268 (2021)
16. Weiland, B.R.N., Peterson, F.: The Truth About Illness and Disease. Infobase Publishing, New York (2009)
17. De Choudhury, M.: Role of social media in tackling challenges in mental health. In: Proceedings of the 2nd International Workshop on Socially-Aware Multimedia, pp. 49–52 (2013)
18. Shing, H.-C., Resnik, P., Oard, D.W.: A prioritization model for suicidality risk assessment. In: Proceedings of the 58th Annual Meeting of the Association for Computational Linguistics, pp. 8124–8137 (2020)
19. Niederkrotenthaler, T.: Papageno effect: its progress in media research and contextualization with findings on harmful media effects. In: Media and Suicide: International Perspectives on Research, Theory, and Policy, pp. 133–158. Routledge, London (2017)
20. Chancellor, S., Birnbaum, M.L., Caine, E.D., Silenzio, V.M., De Choudhury, M.: A taxonomy of ethical tensions in inferring mental health states from social media. In: Proceedings of the Conference on Fairness, Accountability, and Transparency, pp. 79–88 (2019)
21. Coppersmith, G., Dredze, M., Harman, C., Hollingshead, K., Mitchell, M.: CLPsych 2015 shared task: depression and PTSD on Twitter. In: Proceedings of the 2nd Workshop on Computational Linguistics and Clinical Psychology: From Linguistic Signal to Clinical Reality, pp. 31–39 (2015)
22. Milne, D.N., Pink, G., Hachey, B., Calvo, R.A.: CLPsych 2016 shared task: triaging content in online peer-support forums. In: Proceedings of the Third Workshop on Computational Linguistics and Clinical Psychology, pp. 118–127 (2016)
23. Preotiuc-Pietro, D., Sap, M., Schwartz, H. A., Ungar, L.H.: Mental illness detection at the World Well-being Project for the CLPsych 2015 shared task. In: CLPsych@ HLT-NAACL, pp. 40–45 (2015)
24. Resnik, P., Armstrong, W., Claudino, L., Nguyen, T., Nguyen, V.-A., Boyd-Graber, J.: Beyond LDA: Exploring supervised topic modeling for depression-related language in Twitter. In: Proceedings of the 2nd Workshop on Computational Linguistics and Clinical Psychology: From Linguistic Signal to Clinical Reality, pp. 99–107 (2015)
25. Tsugawa, S., Kikuchi, Y., Kishino, F., Nakajima, K., Itoh, Y., Ohsaki, H.: Recognizing depression from Twitter activity. In: Proceedings of the 33rd Annual ACM Conference on Human Factors in Computing Systems, pp. 3187–3196 (2015)
26. De Choudhury, M., Kiciman, E., Dredze, M., Coppersmith, G., Kumar, M.: Discovering shifts to suicidal ideation from mental health content in social media. In: Proceedings of the 2016 CHI Conference on Human Factors in Computing Systems, pp. 2098–2110 (2016)
27. Sawhney, R., Joshi, H., Gandhi, S., Shah, R.R.: Towards ordinal suicide ideation detection on social media. In: Proceedings of the 14th ACM International Conference on Web Search and Data Mining, pp. 22–30 (2021)
28. Yang, Z., Yang, D., Dyer, C., He, X., Smola, A., Hovy, E.: Hierarchical attention networks for document classification. In: Proceedings of the 2016 Conference of the North American Chapter of the Association for Computational Linguistics: Human Language Technologies, pp. 1480–1489 (2016)
29. Demszky, D., Movshovitz-Attias, D., Ko, J., Cowen, A., Nemade, G., Ravi, S.: GoEmotions: a dataset of fine-grained emotions (2020). arXiv preprint arXiv:2005.00547
30. Kumar, A.: Real-time mental health analytics using IoMT and social media datasets: research and challenges (2021). Available at SSRN 3842818
31. Turcan, E., McKeown, K.: Dreaddit: a Reddit dataset for stress analysis in social media (2019). arXiv preprint arXiv:1911.00133

32. Matero, M., et al.: Suicide risk assessment with multi-level dual-context language and BERT. In: Proceedings of the Sixth Workshop on Computational Linguistics and Clinical Psychology, pp. 39–44 (2019)
33. Lin, C., et al.: SenseMood: depression detection on social media. In: Proceedings of the 2020 International Conference on Multimedia Retrieval, pp. 407–411 (2020)
34. Gui, T., et al.: Cooperative multimodal approach to depression detection in Twitter. In: Proceedings of the AAAI Conference on Artificial Intelligence, vol. 33, pp. 110–117 (2019)
35. Shen, G., et al.: Depression detection via harvesting social media: a multimodal dictionary learning solution. IJCAI **2017**, 3838–3844 (2017)
36. Tadesse, M.M., Lin, H., Xu, B., Yang, L.: Detection of depression-related posts in Reddit social media forum. IEEE Access **7**, 44883–44893 (2019)
37. Zogan, H., Razzak, I., Jameel, S., Xu, G.: DepressionNet: a novel summarization boosted deep framework for depression detection on social media (2021). arXiv preprint arXiv:2105.10878
38. Haque, A., Reddi, V., Giallanza, T.: Deep learning for suicide and depression identification with unsupervised label correction (2021). arXiv preprint arXiv:2102.09427
39. Zogan, H., Wang, X., Jameel, S., Xu, G.: Depression detection with multi-modalities using a hybrid deep learning model on social media (2020). arXiv preprint arXiv:2007.02847
40. Santos, C.M., Passos, A.M., Uitdewilligen, S.: When shared cognition leads to closed minds: temporal mental models, team learning, adaptation and performance. Eur. Manag. J. **34**(3), 258–268 (2016)
41. Naseem, U., Thapa, S., Zhang, Q., Rashid, J., Hu, L., Nasim, M.: Temporal tides of emotional resonance: a novel approach to identify mental health on social media. In: Proceedings of the 11th International Workshop on Natural Language Processing for Social Media, pp. 1–8 (2023)
42. Iorfino, F., et al.: The temporal dependencies between social, emotional and physical health factors in young people receiving mental healthcare: a dynamic Bayesian network analysis. Epidemiol. Psych. Sci. **32**(e56), 1–9 (2023). https://doi.org/10.1017/S2045796023000616
43. Jarman, H.K., McLean, S.A., Paxton, S.J., Sibley, C.G., Marques, M.D.: Examination of the temporal sequence between social media use and well-being in a representative sample of adults. Soc. Psych. Psych. Epidemiol. **58**, 1247–1258 (2023). https://doi.org/10.1007/s00127-022-02363-2

Author Index

A

Abhinash, S. II-231
Abraham, J. V. Thomas II-36
Adar, Uğur Güven I-19
Adla, Abdelkader II-117
Ahabchane, Chahid I-311
Aju, D. II-77
Al Rajab, Murad II-52
Albayrak Ünal, Özge I-86
Algburi, Raghad Mohammed Najm II-67
Alhamdany, Mohammed Noori II-137
Alhamdany, Saad Noori II-137
Alhamdany, Saba Noori II-137
Almado, Alaa Abdulkareem Ghaleb II-67
Awuor, Fredrick Mzee II-20
Ayaz, Halil Ibrahim I-32

B

Ballout, Khader II-52
Bayraktar, Irem II-199
Belhadi, Amine II-249
Ben Rejab, Fahmi I-189
Ben Romdhane, Syrine I-189
Benabdellah, Abla Chaouni II-249
Bizel, Gulhan I-336, II-36
Bouhalouan, Djamila II-117
Bouzid, Sarra I-67
Bui, Trung-Minh I-153

C

Çayli, Osman I-19
Chiddarwar, Shital I-133
Chitsaz, Ehsan II-98
Chuli, Anannya II-77

D

Dahbi, Houda II-249
Demiray, Abdullah II-214
Desai, Vaishnavi I-133
Do, Van-Dinh I-153, I-171

E

Ergün, Serap I-3
Erkayman, Burak I-86
Ervural, Bilal I-32

G

Ghatge, Anurupa II-182
Ghosh, Arpita I-274
Gorrab, Siwar I-292

I

Ijogun, Oluwaseyi I-212
Irani, Hamid Reza II-259

J

Jafari, Seyed Mohammad Ali II-98
Jagannathan, Sharath Kumar I-336, II-36
Jain, Aditi II-77
Janardhanan, Nethra II-231
Juneja, Sakshi II-182

K

Kasat, Kishori II-182
Khurana, Tanvi I-133
Kumar, Aditya II-231
Kumar, Shashwat II-77

L

Layeb, Safa Bhar I-311
Le, Van-Hung I-153, I-171
Limam, Hela I-257
Lupei, Maksym I-237

M

Mnasri, Khadija I-189
Mohta, Richa I-133
Mougari, Ouissem I-67
Mysiuk, Iryna I-121
Mysiuk, Roman I-121

N
Nachet, Bakhta II-117
Natsheh, Jannat II-52
Nhat, Do Hong II-154
Nouira, Kaouther I-292
Nozari, Hamed II-259

O
Obaid, Mahmoud II-52
Odeh, Suhail II-52
Oduor, Collins II-20
Orucho, Daniel Okari II-20
Oueslati, Wided I-257
Özmutlu, Seda I-104

P
Patra, Pradipta I-274
Pavlenchyk, Nataliia I-121

R
Rahmaty, Maryam II-259
Rahul, M. R. I-133
Raj, Shivam II-77
Ramdani, Nur Rahmah Syah II-3
Rao, Ijjada Sreenivasa I-46
Rebman Jr., Carl I-212
Rejab, Fahmi Ben I-292

S
Saadi, Sihem I-67
Shaikh, Naim II-182
Shliakhta, Myroslav I-237
Shuvar, Roman I-121
Sinha, Uday II-182
Sravani, Potula I-46
Sumalatha, M. R. II-231

T
Taslicali, Cahit II-214
Thorat, Nikhil II-182
Tsyuh, Svyatoslav I-121

V
Van Hop, Nguyen II-154
Voddi, Vijay Kumar II-36

W
Wimmer, Hayden I-212

Y
Yakoubi, Ghada I-311
Yesilkaya, Nimet Selen II-199
Yigit, Elif I-104
Yilmaz, Atınç I-19
Yuzevych, Volodymyr I-121

Z
Zahran, Isra' II-52
Zulkarnain, Zulkarnain II-3

Printed in the United States
by Baker & Taylor Publisher Services